Order within chaos
Towards a deterministic approach
to turbulence

Pierre Bergé
C.E.A. SACLAY, Service de physique du solide et de résonance magnétique

Yves Pomeau
C.E.A. SACLAY, Service de physique théorique

Christian Vidal
C.N.R.S., Centre Paul Pascal

Order within chaos
Towards a deterministic approach to turbulence

Preface by David Ruelle
Institut des Hautes Études Scientifiques

Translated from the French by **Laurette Tuckerman**

A Wiley-Interscience Publication

JOHN WILEY & SONS
New York · Toronto · Chichester · Brisbane · Singapore

HERMANN
publishers in arts and science   Paris

Editorial assistance: Chantal Pomeau

Copyright © 1984 by Hermann, Paris, France.

Published by Hermann and by John Wiley & Sons, Inc. Printed in France.

All rights reserved.

No part of this book may be reproduced by any means, nor transmitted, nor translated into a machine language without the written permission of Hermann.

L'ordre dans le chaos was originally published in French by Hermann in 1984.

> **Library of Congress Cataloging-in-Publication Data**
> Bergé, Pierre, 1934
>   Order within chaos.
>
>   Translation of: L'ordre dans le chaos
>   "A Wiley-Interscience publication"
>   Bibliography: p.
>   Includes index.
>   1. Differentiable dynamical systems. 2. Turbulence. 3. Chaotic behavior in systems. I. Pomeau, Yves. II. Vidal, Ch. (Christian) III. Title.

QA614.8.B4713 1986   530.1′5   86 9097
ISBN 0-471-84967-7
ISBN 2 7056 5980 3 (Hermann)
10 9 8 7 6 5 4 3 2 1

Table

Foreword XI
Preface by David Ruelle XIV

PART ONE : FROM ORDER...

Introduction — **Nonmonotonic evolution** 3

    I   Oscillation, vibration, nonmonotonic evolution 3
    II From the inanimate world... 3
    III ... To the animate world 7
    IV The key to the analysis : the oscillator 9

Chapter I — **Free oscillator — Damped oscillator** 11

  I.1 The free oscillator 11
    I.1.1 *Equation of the simple frictionless pendulum* 11
    I.1.2 *Description of motion in a phase space* 13
    I.1.3 *Canonical form of the equations of motion* 15

  I.2 Conservative systems 16
    I.2.1 *Conservation of energy* 16
    I.2.2 *Conservation of areas in phase space* 16
    I.2.3 *Invariance of the equations under time reversal* 18

  I.3. The damped oscillator 18
    I.3.1 *Dissipation of energy by friction* 18
    I.3.2 *Equation of the damped oscillator* 19
    I.3.3 *Phase portrait* 20

  I.4 Dissipative systems 22
    I.4.1 *Essential properties* 22
    I.4.2 *Contraction of areas in phase space* 22

Chapter II — **Forced oscillator — Parametric oscillator** 25

  II.1 The forced oscillator 25
    II.1.1 *The Van der Pol equation* 25
    II.1.2 *Phase portrait* 26
    II.1.3 *Amplitude of oscillation for $\varepsilon = 0_+$* 28

	II.2	The parametric oscillator	30
		II.2.1 *Equation of motion*	30
		II.2.2 *Stability of solutions*	32
		II.2.3 *The effect of damping*	35
		II.2.4 *Mechanism of parametric amplification*	36
	II.3	Introduction to bifurcations	38
		II.3.1 *Concept of bifurcation*	38
		II.3.2 *Hopf bifurcation*	38
		II.3.3 *Subcritical and supercritical bifurcation*	40
Chapter III	**The Fourier transform**		43
	III.1	Identification and characterization of a dynamical regime	43
	III.2	Discrete Fourier transform	44
		III.2.1 *Signal discretization*	44
		III.2.2 *Definition of the discrete Fourier transform*	44
		III.2.3 *The Wiener-Khintchin theorem*	46
		III.2.4 *The power spectrum*	48
	III.3	Different kinds of Fourier spectra	50
		III.3.1 *Periodic signal*	50
		III.3.2 *Quasiperiodic signal*	54
		III.3.3 *Aperiodic signal*	58
	III.4	Fast Fourier Transform (FFT)	61
Chapter IV	**Poincaré sections**		63
	IV.1	Definition of a flow	63
	IV.2	Poincaré sections	64
		IV.2.1 *Construction and properties*	64
		IV.2.2 *Practical interest*	65
	IV.3	Different kinds of Poincaré sections	67
		IV.3.1 *Periodic solution*	67
		IV.3.2 *Quasiperiodic solution*	68
		IV.3.3 *Aperiodic solution*	70
	IV.4	First return map	70
		IV.4.1 *Iteration of a one-dimensional map*	71
		IV.4.2 *Limit cycle of the Van der Pol oscillator*	72
		IV.4.3 *Reduction of a three-dimensional flow*	74
	IV.5	Practical implementation	76

Chapter V	**Three examples of dynamical systems**		79
	V.1 Introduction		79
	V.2 The compass		79
		V.2.1 *Description*	79
		V.2.2 *Evolution equation*	81
	V.3 Rayleigh-Bénard convection		83
		V.3.1 *The Rayleigh-Bénard instability*	83
		V.3.2 *Convective structures*	87
		V.3.3 *Evolution equations*	89
		V.3.4 *Experiment*	90
	V.4 The Belousov-Zhabotinsky reaction		91
		V.4.1 *Brief history of chemical oscillators*	91
		V.4.2 *Experimental set-up*	93
		V.4.3 *Evolution equations*	95

PART TWO : ... TO CHAOS

Introduction	**Temporal chaos in dissipative systems**		101
	I	Asymptotic behavior of a dissipative system	101
	II	The simplest attractors	102
	III	Pseudo-definition of chaos	103
	IV	Strange attractors	103
	V	Ways in which chaos appears	105
	VI	Floquet theory	106
	VII	Transitions resulting from the loss of linear stability	107
	VIII	Current limits of the theory	109
Chapter VI	**Strange attractors**		111
	VI.1 Dissipation and attractors		111
		VI.1.1 *The phenomenon of attraction*	111
		VI.1.2 *Two consequences of the contraction of areas*	112
		VI.1.3 *Nonintersection of phase trajectories*	114
	VI.2 Aperiodic attractors		117
		VI.2.1 *Characteristics of a chaotic regime*	117
		VI.2.2 *Properties of aperiodic attractors*	119

	VI.3	Examples of strange attractors	123
	VI.3.1	*The Lorenz attractor*	123
	VI.3.2	*The Hénon attractor*	130
	VI.3.3	*Experimental illustrations of strange attractors*	135
	VI.4	Measuring the dimension of strange attractors	144
	VI.4.1	*Problems of characterization*	144
	VI.4.2	*Fractal dimensions*	146
	VI.4.3	*Geometric characterization of an attractor*	150
	VI.4.4	*Implementation*	152
	VI.5	The horseshoe attractor	155

Chapter VII Quasiperiodicity ... 159

	VII.1	Hopf bifurcation from a limit cycle	159
	VII.2	The theory of Ruelle-Takens (R.T.)	161
	VII.2.1	*Description*	161
	VII.2.2	*Topological interpretation*	162
	VII.2.3	*Practical significance : a numerical simulation*	165
	VII.2.4	*Experimental illustration*	167
	VII.3	Transition to chaos from a torus T^2	168
	VII.3.1	*Curry-Yorke model*	168
	VII.3.2	*Behavior of the model as a function of* ε	171
	VII.3.3	*Some experimental illustrations*	175
	VII.4	Elements of a mathematical theory of transition to chaos from a torus T^2	182

Chapter VIII The subharmonic cascade ... 191

	VIII.1	Introduction	191
	VIII.1.1	*Subharmonic instability*	191
	VIII.1.2	*Period-doubling mechanism*	191
	VIII.1.3	*First return map*	194
	VIII.2	The subharmonic cascade	195
	VIII.2.1	*Quadratic mapping of the interval*	195
	VIII.2.2	*Period-doubling cascade*	196
	VIII.2.3	*Scaling laws*	200
	VIII.3	Characteristics of chaos	202
	VIII.3.1	*Beyond the subharmonic cascade*	202
	VIII.3.2	*Two fundamental properties of the mapping*	202
	VIII.3.3	*The inverse cascade*	205

| | | VIII.3.4 | *Windows of periodicity*........................ | 207 |
| | | VIII.3.5 | *The universal sequence* .... | 208 |

VIII.4 Experimental illustrations............................... 210
    VIII.4.1 *Nature of the observations*.................... 210
    VIII.4.2 *Subharmonic cascade : R.B. convection*........... 211
    VIII.4.3 *Inverse cascade : the compass* 213
    VIII.4.4 *Universal sequence : the B.Z. reaction* 217

Chapter IX Intermittency 223

IX.1 Introduction 223

IX.2 Type I intermittency....................................... 225
    IX.2.1 *General theorical considerations* 225
    IX.2.2 *Local analysis* 226
    IX.2.3 *Quantitative predictions of the model* 230
    IX.2.4 *Relaminarization* 231
    IX.2.5 *Intermittency in the Lorenz model*................ 241
    IX.2.6 *Intermittency in the B.Z. reaction* 244

IX.3 Type III intermittency 247
    IX.3.1 *Theory of type III intermittency*................. 248
    IX.3.2 *Type III intermittency in R.B. convection* 255

IX.4 Theory of type II intermittency 259

Conclusion Debate 265

Appendix A Local bifurcations of codimension one 271

    I   *Local bifurcations* 271
    II   *Codimension*.. 272
    III *Supercritical codimension-one bifurcations of fixed points* 272
    IV *Subcritical bifurcations* 276
    V   *Codimension-one bifurcations of periodic orbits*................ 278

Appendix B Lyapunov exponents 279

    I   *Description* 279
    II   *Analysis of simple cases* 281
    III *Methods of determination*..................................... 284
    IV *Characterization of an attractor* 286

Appendix C Synchronization of oscillators........................... 289

    I   *Generalities* 289

	II	*Analysis of the problem*	290
	III	*Winding number*	293
	IV	*Rational winding number. Frequency locking*	295
	V	*Irrational winding number. Quasiperiodic behavior*	297
	VI	*Structural stability and frequency locking. Devil's staircase*	298

Appendix D **The Lorenz model** 301

	I	*Rayleigh-Bénard convection*	301
	II	*Derivation of the Lorenz model*	302
	III	*Coherence of the model*	306
	IV	*Bifurcation diagram ($Pr = 10$; $b = 8/3$)*	308

Appendix E **Mathematical complements** 314

	I	*Matrices*	314
	II	*The Jacobian*	321
	III	*Homeomorphism, diffeomorphism*	324

References 325

Index 328

FOREWORD

This book is an introduction to the study of dissipative dynamical systems. It has been written, insofar as possible, at a relatively elementary level. It is addressed, first, to students, and secondly, to an audience which is scientifically cultivated, but not specialized in this discipline.

By *dynamical system* we mean any system, whatever its nature (physical, chemical, electromechanical, biological, economical, etc.) which can take various mathematical forms: ordinary differential equations (autonomous or not), partial differential equations, mappings (invertible or not) of the line or of the plane. The field of investigation thus appears quite large, since it encompasses the analysis of all time-dependent phenomena. Treating the major types of behavior or evolution without direct reference to the actual matter through which they are manifested, this body of doctrine exhibits a high degree of generality and of universality which, in this respect, resembles a physical theory as powerful and structured as, for example, thermodynamics. From this undoubtedly follows a large part of its interest.

At the outset, it is advisable to distinguish clearly between systems with friction and those without, for the two do not lead to the same class of problems. The presence of internal friction, in the most general sense of the term, has as a corollary the existence of an *attractor*, that is, of an asymptotic limit (as $t \to \infty$) of the solutions, a limit on which the initial condition — the point of departure — has no direct influence. In mechanics, where friction entails a continual decrease of the energy, the corresponding systems are for this reason called *dissipative*, an adjective which we will use from now on, no matter what the domain.

Frictionless systems, called *conservative* or *Hamiltonian* again in reference to mechanics, have their own interest and utility. In particular, certain questions of indisputable practical importance, like the evolution of the solar system or the behavior of a plasma in an accelerator, for example, come under this heading. This subject requires specific methods, notably because the absence of an attractor confers a determining influence on the initial condition.

Among the great unsolved problems of classical physics, turbulence is undoubtedly the oldest. It is striking to note that twenty years ago little more was known about this subject than at the beginning of the nineteenth century when Navier was setting down the equations governing the flow of a fluid. Here is something that seems extraordinary when we consider the striking progress accomplished since then in understanding the structure of matter as well as that of the universe. And yet fluid mechanics is a domain easily accessible to experiment: no laboratory machinery comes anywhere close — in complexity or in cost — to the accelerators used to study subnuclear particles! Despite its banality, this observation raises a question which historians of science will one day have to address: that of the underlying causes (circumstantial and epistemological) of the relative stagnation, in a discipline which has never lacked for practical and economic motivation.

In any case, the situation began to advance seriously starting in the 1960's. We are still far, even very far, from having solved the problem, that is, from being able to give a precise explanation of turbulence and the mechanisms of its genesis. Nevertheless, new methods have been developed to tackle the analysis of phenomena whose totally disordered appearance had heretofore posed a challenge to scientific description. The first but also the most impostant was the discovery of *strange attractors*. Found numerically in 1963 by E. Lorenz, the idea was elaborated mathematically in 1971 by D. Ruelle and F. Takens as a key element in understanding irregular behavior described by deterministic equations, notably turbulence. Thus was launched the study of what today is often called *deterministic chaos*, the first stage in the analysis of phenomena which are even more complex-depending on space as well as on time. This progress has implications for many fields besides fluid mechanics.

A decade or so after what must be called an essential renewal, the moment seems to have arrived to try to take stock of those ideas and methods which must henceforth be an integral part of the scientific educational program. This is the purpose that we have had in mind when preparing this book, which has, above all, a deliberate pedagogical bias. This has led us to make various choices that are arbitrary, and therefore subject to criticism. First, at the level of the method of presentation: we have constantly endeavored to be concrete, illustrating our remarks at every possible occasion with examples taken from experiment. In addition we have often substituted simplified "hand waving" calculations for rigorous mathematical proofs, whose necessity and sometimes beauty are incontestable. The absence of statements of theorems and lemmas must therefore not be attributed to any lack of rigor in the theory. Secondly, at the level of the questions treated: not attempting to be exhaustive, we have limited ourselves to the study of dissipative systems, by far the most widespread in nature. In addition, we have not tried to deal systematically with all of the subjects worthy of interest, preferring to present a panorama which is coherent and accessible, if not truly complete. Finally, at the level of the organization of the book itself into two parts: the basic vocabulary (Chapters I to V), the analytical methods (Chapters III and IV), the three experimental situations (a magnet placed in a magnetic field, thermoconvection in a fluid layer, a chemical reaction in an open medium) used for illustration (Chapter V), are presented in the first part. Entitled "From Order...", this first part deals essentially with regular, particularly periodic behavior. Having described the tools, we then use them in the second part, whose title "... To Chaos" reveals its content. The concept of the strange attractor is introduced in a very pragmatic way in Chapter VI. Afterwards the modes of transition leading from a periodic to a chaotic regime are discussed: quasiperiodicity (Chapter VII), the subharmonic cascade (Chapter VIII), intermittency (Chapter IX). A few topics, more specific or technical, have been relegated to appendices to avoid long interruptions to the train of thought. A short index and selected bibliography complete the whole.

In an educational book such as this one, there can be no question of making a

FOREWORD

complete review of the literature. In any case, it would not be easy, since the subject already counts hundreds of articles, and this flood, far from abating, increases with each passing day. We therefore simply give a few of the most important references in addition to the articles from which we have excerpted figures of numerical or experimental results. Similarly we have generally omitted authors' names in the text, preferring to emphasize the scientific content of their work. This book does not even begin to attempt to trace the sources of the different contributions. We merely indicate here that, taken as a whole,

— the study of dynamical systems is the fruit of the labor of a great many renowned scholars, particularly mathematicians and physicists, carried out all over the world,

— the contributions originating in France occupy one of the most honorable places in this domain, theoretically as well as experimentally.

Apart from a few books intended for specialists, there does not yet exist a book dedicated to a simple exposition of the theory of dynamical systems which can be used at the undergraduate level. As this book stands, which is to say, still too imperfect — and how could it be otherwise in such a new and prolific field? — this attempt has at least the merit, it seems to us, of being the first to try to fill this gap in the scientific literature. Our goal will have been reached if, in spite of its flaws, it contributes to spreading the ideas and methods recently developed to study deterministic chaos.

To Chantal Pomeau who undertook the typing, to Jacques Moineau, Bernard Ozenda, Jocelyne Dusautoir, and Madeleine Porneuf who, in diverse ways, helped in the preparation of this manuscript, we express our deep appreciation.

By their observations, criticisms, and suggestions, Anne-Marie Bergé, Monique Dubois, and Paul Manneville helped us to improve on several points of the original text; we thank them warmly for their valuable contributions.

David Ruelle agreed to write a preface, to place the book in its proper context; we wish to express our gratitude for this mark of esteem and of friendship.

Bordeaux, Paris, February 1984

Pierre Bergé, Yves Pomeau, Christian Vidal

PREFACE

In recent years, a new kind of physics has been emerging, which has been variously termed nonlinear, turbulent, or chaotic. Its extremely diverse subject matter includes hydrodynamic turbulence, chemical kinetics, and the study of electronic circuits. How can there be a common approach to questions, which appear to be so heterogeneous and complicated? In fact, attentive analysis has shown that the temporal evolution of these systems is similar, permitting unification of their study. But we are not referring here to a statistical or stochastic approach, nor to a superficial similarity at the descriptive level. The similarity of complicated behavior in hydrodynamics, in chemical kinetics, in mechanics and electronics concerns experimental details which have been accessible only recently. This similarity results from a profound mathematical theory: the modern theory of nonlinear systems or, more precisely, the qualitative theory of differentiable dynamical systems.

The new nonlinear physics concerns fascinating phenomena: turbulence and oscillations both physicochemical and biological. The chaotic and almost whimsical appearance of these phenomena is now giving way to an unexpected conceptual unification, using the ideas of bifurcations, strange attractors, Lyapunov exponents, and so on. Only a corner of the veil has been lifted and many phenomena remain to be explained and understood, but it seems that we have started along the path predicted by Feynman in his *Lectures on Physics*.

"The next great era of awakening of human intellect may well produce a method of understanding the *qualitative* content of equations. Today we cannot see that the water flow equations contain such things as the barber pole structure of turbulence that one sees between rotating cylinders. Today we cannot see whether Schrödinger's equation contains frogs, musical composers, or morality — or whether it does not. We cannot say whether something beyond it like God is needed, or not. And so we can all hold strong opinions either way."

A book could have been written on nonlinear dynamical systems with a mathematical emphasis. The present book — the first of its kind — is on the contrary centered on physics and chemistry. It has been written by a theorist, Y. Pomeau, and two experimentalists, P. Bergé and C. Vidal, who have all actively participated in elaborating the new ideas. They describe in a concrete way the important dynamical phenomena and the way in which they appear in experimental reality without getting lost in mathematical details. After reading this book, the physicist, the chemist, and the physiologist will themselves be able to apply the knowledge they have acquired to the interpretation of phenomena appearing in their specialities. They will also be able to further develop their mathematical knowledge on a solid foundation.

David Ruelle

PART ONE

From order...

PART ONE

INTRODUCTION

Nonmonotonic evolution

I Oscillation, vibration, nonmonotonic evolution

The ideas of oscillation and vibration are essential to the study of the dynamics of nonmonotonic evolution, characterized by to and fro motion, regular or irregular. There are two reasons for this. First of all, periodic phenomena are ubiquitous in the inanimate world as well as in living organisms, which makes the study of periodic phenomena intrinsically important. But, in addition, any time evolution can be analyzed as a sum of periodic contributions by means of the Fourier transform. These two elements combine to make periodic motion the cornerstone of any theory constructed to account for time evolution. We now understand immediately the importance of the role played by the oscillator, as an archetype of time-dependent behavior. Without going into detail, we want to emphasize in the introduction the scope of periodic phenomena in the world around us. Several examples taken from different domains will help to illustrate this point.

II From the inanimate world...

We recall that there exist several realizations of the oscillator in different branches of the physical sciences, in addition to the simple pendulum (to which we shall return in detail in Chapter I). Some familiar examples are: a weight suspended from an elastic spring, the traditional self-capacitance circuit, and the acoustic Helmholtz resonator are among the classical models whose evolution is governed by differential equations with periodic solutions. It would be a mistake to see these elementary oscillators as a simple exercise in physics without real utility or importance. On the contrary, they are involved in a large number of applications in mechanics, electricity, electronics, and acoustics — the list is long indeed. In the industrial era, the oscillator has become one of the basic components of a large number of man-made machines.

We mention also the oscillatory or vibrational phenomena that arise spontaneously. They in fact depend on the same analysis and their importance in daily life can hardly be exaggerated. Thus any device that rotates: a wheel, motor, helix, turbine, etc., gives rise to vibrations capable of perturbing its operation, even to the point of destroying itself: hence the necessity for perfect equilibration of the brace of a motor or of the wheels of a vehicle. The vibrations of shafts and beams commonly used in construction, (airplane wings, bridge structure, etc.) can have annoying, if not catastrophic consequences: after an accident that occurred near Manchester in 1831, no troop of soldiers would cross a bridge when marching in step, to avoid the occurrence of oscillations which may cause its collapse. Finally, we mention the vibrations arising in fluids, more particularly in water and in air, which can cause energy losses.

Before continuing, we briefly describe an oscillatory phenomenon in fluids. Figure 1 shows schematically the way in which streamlines surround a solid placed in a fluid flow, as a function of the relative velocity $\vec{v}$. We consider here simply a stationary cylinder, whose axis is perpendicular to the velocity $\vec{v}$ of the moving fluid. For velocities that are very small (fig. 1 a) or moderate (fig. 1 b) the streamlines are stationary — that is, they remain the same with the passage of time. On the other hand, above a certain velocity threshold, the recirculation vortices located in the wake of the cylinder (fig. 1 b) become unstable. The steady state then disappears, and is replaced by a set of vortices, turning alternately in one direction or the other, carried along by the moving fluid.

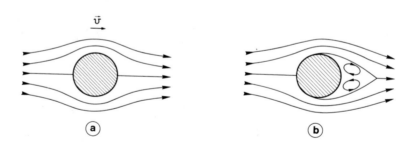

Figure 1 Schematic representation of the streamlines of the fluid surrounding a cylinder whose axis is perpendicular to the figure (the velocity v of the fluid is parallel to the plane of the figure). The regime of the flow is characterized by a nondimensional quantity, the Reynolds number, proportional to the velocity v:

$$\mathrm{Re} = \frac{v\phi}{\nu}$$

where ϕ is the diameter of the cylinder and ν the kinematic viscosity of the fluid.
a) very small velocity $\mathrm{Re} \simeq 10^{-2}$
b) moderate velocity $\mathrm{Re} \simeq 20$.

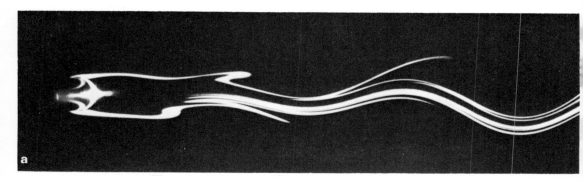

Photograph 1   Bénard-von Karmàn wake.
A cylinder is placed in an air flow. Smoke is emitted by two slits along the generators of the cylinder. The experimentalist shines a laser beam along the axis of the air flow perpendicular to the cylinder, allowing visualization of the motion of the smoke in the plane.
a) $Re - Re_c = 2.4$
b) $Re - Re_c = 13.6$
where Re is the Reynolds number relative to the diameter of the cylinder (here 10 mm) and Re_c is the critical Reynolds number for the appearance of the unsteady wake (here $Re_c \simeq 70$).
From L. Boyer, C. Mathis. See C. Mathis, Thesis, Université de Provence (1983).
Propriétés des composantes de vitesse transverses dans l'écoulement de Bénard-von Karmàn aux faibles nombres de Reynolds.

Photo 1 shows a snapshot of this particular wake configuration, known as the Bénard — von Karmàn vortex street. We emphasize the universality of these periodic phenomena: whether in the case of eddies behind the piers of a bridge, the structure of clouds produced by the wind on the isle of Madera, or even the vortices seen behind the red spot of Jupiter, one always finds the behavior described above. One of the consequences of this phenomenon is "Aeolian sound", a direct audible manifestation. When, due to wind, the vortices originating in the back of a solid are periodically detached, the solid is subjected to an alternating force, making it vibrate with the frequency of the vortex detachment. In consequence, the solid emits a characteristic sound, whose frequency increases with the wind velocity. Figure 2 shows a recorded segment of such an Aeolian sound. Depending on the circumstances, it can be pleasant (music) or, on the contrary, very annoying (noise)[1].

In addition to these physical processes, one also encounters less-known oscillators in chemistry. Both redox reactions in an acidic medium, and combustion in a gaseous phase, are capable of oscillating with time. The fact that what are called chemical oscillators were first discovered by chance does not diminish their conceptual importance[2]. This is especially so since, as we will see in the following section, there also

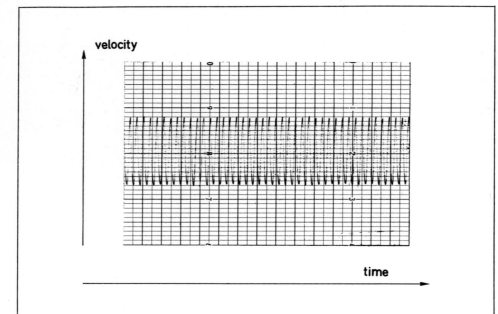

Figure 2 Aeolian noise: variation of the wind speed as a function of time. The experiment is that described in the caption of Photograph 1 and the Reynolds number is Re = 72.9 (here $\phi_{cyl} = 0.7$ cm and Re = 59.4).

1. Research for alternative energy sources has led to the consideration of high power Aeolian electrical plants, a kind of modern-day windmill. But since these Aeolians are by nature very noisy devices, that can be heard operating from very far away, the possibility of implementation is limited.

2. The behavior of these chemical systems is satisfactorily described by differential equations derived from thermodynamics and other branches of physics. They can then be studied mathematically, in particular from the dynamical systems point of view planned here. Moreover, these chemical systems are easily studied in laboratory experiments. The thermodynamics of irreversible processes shows that the second law does not forbid existence of nonmonotonic behavior, such as oscillation, as long as the system considered is sufficiently far from equilibrium.

exist biochemical and biological oscillators which can be better understood by the use of chemical oscillators as experimental models. All this explains why, since the end of the 1960's, oscillating reactions in aqueous media have been the subject of very active research. They are classified according to the oxidizing agent: iodate, bromate, or chlorite. The reducing agent (often an organic compound) and the possible catalyst can vary a great deal, allowing for many combinations[3].

III ... To the animate world

Chemistry leads us to the subject of biological systems, in which periodic behavior is as common as in inert matter. In biological systems too, the observation and analysis of periodic phenomena are the focus of much research. To this class belong biological rhythms, some of which have been known since antiquity: respiration, cardiac muscle contraction, alternation of watchfulness and sleep, reproduction cycles in plants, etc. Some of them, called exogenous, result from the regular variation of the environment in which the organism lives: seasons, lunar phases, night and day-time. We speak of circadian rhythms (a period near 24 hours) and of circannual rhythms (a period of about 365 days). Others, by contrast, have their source in the cellular or metabolic level: these are called endogenous. Their periods range from a few fractions of a second to twenty or thirty minutes. Thus, we observe in mammals the oscillation:
  — of certain smooth (stomach, intestine, uterus, etc.) or striated (hard) muscles and of muscle cells in culture as well;
  — of the central nervous systems (EEG), as well as of the action potential of neurons responsible for the transmission of nerve signals.

In animals or in plants, as well as in bacteria, protein synthesis or metabolic activities, such as glycolysis and photosynthesis, can occur in a periodic manner. Of all these oscillatory biochemical or biological reactions glycolysis is by far the most well known and understood. This interest can be understood in light of the role of this sequence of enzymatic reactions in cellular metabolism[4]. Discovered in 1957, glycolytic oscillation has since been observed *in vitro* and *in vivo*, as evidenced by Figure 3. Figure 3 shows a recording of the variation in time of certain properties, notably the fluorescence of NADH, one of the components of the glycolytic sequence:
  — in a population of mitochondria, the cellular organelles which form ATP (fig. 3 *a*),
  — in an isolated and intact cell of *Saccharomyces carl bergensis* (fig. 3 *b*).

---

3. Having chosen a chemical oscillator to illustrate certain aspects of this book, we refer the reader to Section V.4 for more details on the subject, which is still rather little-known outside specialized circles.
4. Recall that glycolysis is an important pathway in the breakdown of glucose, through which ATP molecules — the "fuel", one might say, of the cellular "motor" — are regenerated starting from ADP molecules. Glycolysis is thus an essential part of the cell's energy conversion process.

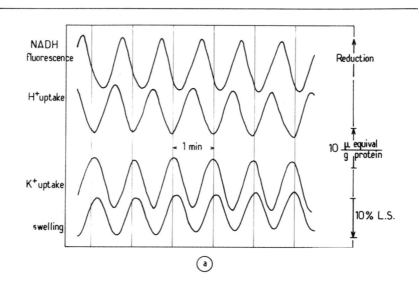

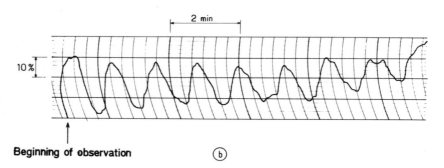

Figure 3 Oscillating glycolysis.
a) Simultaneous recording of the fluorescence of NADH (one of the components of the glycolytic chain), of the capture of H^+ and K^+ ions, and of the swelling of a population of mitochondria showing the oscillations caused by glycolysis.
From A. Boiteux and B. Hess (1975). "Oscillations in glycolysis, cellular respiration and communication", *Faraday Symposium of the Chemical Society* **9**, p. 202.
b) Recording of fluorescence of NADH in vivo from an intact cell of *Saccharomyces carlbengensis* two minutes after injection of glucose.
From B. Chance, G. Willianson, I. Y. Lee Mela, D. De Vault, A. Ghosh and E. K. Pye "Synchronization phenomena in oscillations of yeast cells and isolated mitochondria" in: *Biological and Biochemical Oscillators* (ed. B. Chance, E. K. Pye, A. Ghosh and B. Hess). Academic Press, New York (1973), p. 285.
In vitro (3 a) as well as *in vivo (3 b)* glycolysis proceeds in an oscillating manner with a period on the order of a minute.

We note that the period of oscillation is typically on the order of a minute in both cases.

Also periodic are the development and growth of numerous cells and unicellular species. On another organizational level, we have the periodic behavior of certain populations:

— the motion of slime moulds, such as *Dyctyostelium discoïdeum,* which aggregate in a sporangium when the medium in which they live becomes too poor nutritionally to allow them to survive dispersed,

— light emission by groups of fireflies,

— the development in phase opposition of numerous predator-prey pairs.

We see from this survey that periodic behavior is indeed very widespread, as much at the individual level — from the cell to the whole organism — as at the level of populations.

IV The key to the analysis: the oscillator

Since oscillatory phenomena are so frequent, it is natural for the oscillator to arouse intrinsic scientific interest. But interest in the oscillator goes much further than this, if we extend the preceding survey to all nonmonotonic evolution, of which oscillation is as the same time both an example and a component. We have said that the Fourier transform[5] leads to a description of an arbitrary evolution as an algebraic sum of periodic contributions. An understanding of oscillation is then necessary before we can construct any theory of time-dependent phenomena. This explains why indispensable analytic ideas naturally call for a treatment of the elementary oscillator such as the pendulum which we study in the first chapter. We see that the choice of this starting point is neither arbitrary nor purely pedagogical. It is in fact imposed by the nature of the problem: nonmonotonic temporal behavior or evolution.

5. See Chapter III.

CHAPTER I

Free oscillator — Damped oscillator

I.1 The free oscillator

I.1.1 EQUATION OF THE SIMPLE FRICTIONLESS PENDULUM

The simple pendulum is the oldest known example of the free oscillator. The pendulum consists of a mass m in a gravitational field of acceleration g suspended from a point O by a rigid wire of length ℓ (fig. I.1). This mass oscillates in the vertical plane. The two points along a vertical line from O are equilibrium points, of which R is stable and R' unstable. If $\theta(t)$ measures the angle at time t between the wire and the vertical OR, the fundamental law of mechanics leads to:

$$F = m\ell \frac{d^2\theta}{dt^2} = -mg \sin \theta$$

that is:

$$\frac{d^2\theta}{dt^2} + \frac{g}{\ell} \sin \theta = 0 \tag{1}$$

Note that equation (1) assumes an idealization of the device described. In particular, the point size of the mass is assumed implicitly, as well as the absence of all friction and the constancy of the various characteristics of the pendulum[6].

---

6. Still more important one substitutes for the real mechanical problem the analytic geometrical representation given by (1). Historically, there was an important difference, made for instance by Descartes, between geometrical curves described by the motion of a single point and mechanical ones resulting from the combination of the motion of many points. A circle drawn by a point attached to a fixed center belongs clearly to the geometrical curves, whereas cycloids (for instance) resulting from the combination of two independent circular motions are mechanical curves. The notion of continuum and consequently the advent of calculus stemmed partly from the necessity of fixing exactly the properties of any motion with respect to the others to get accurate "mechanical curves" although this is not needed for "geometrical curves".

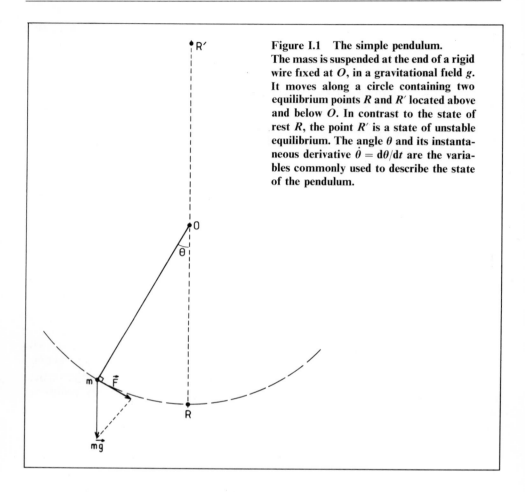

Figure I.1 The simple pendulum. The mass is suspended at the end of a rigid wire fixed at O, in a gravitational field g. It moves along a circle containing two equilibrium points R and R' located above and below O. In contrast to the state of rest R, the point R' is a state of unstable equilibrium. The angle θ and its instantaneous derivative $\dot{\theta} = d\theta/dt$ are the variables commonly used to describe the state of the pendulum.

Insofar as the angle θ remains small such that $\sin\theta \simeq \theta$ (linear approximation), equation (1) is easily integrated, with given initial conditions, leading to the classic result:

$$\theta = \theta_0 \cos(\omega t + \phi)$$

$$\omega = \sqrt{\frac{g}{\ell}} \quad ; \quad T = 2\pi\sqrt{\frac{\ell}{g}}$$

ω: angular frequency; T: period; ϕ: phase.

The motion is exactly periodic: these are the isochronous small oscillations useful in clock making.

Let us treat the problem in a more global manner, by seeking to specify what information is required to completely describe the instantaneous state of the

I.1 THE FREE OSCILLATOR

pendulum. It is clear that the knowledge of two quantities is both necessary and sufficient: the position $\theta(t)$ and the velocity $d\theta/dt = \dot{\theta}$. Consequently, instead of integrating equation (1), one can just as well represent the solution in the Cartesian plane $(\theta, \dot{\theta})$. In the idealized case considered, the solution is a closed curve, that is, an orbit in the plane. In practice, the kind of representation thus introduced plays an essential role in all studies of dynamics for the following simple reason: it provides an alternative when analytic integration of the differential equations proves impossible. Therefore, we will often use this kind of representation.

In general, we can define the *phase space*, the space whose axes are the coordinates of position and velocity[7] and a *phase trajectory* which is a curve in this space representing the evolution of the system. A set of phase trajectories constitutes a *phase portrait*. Another important idea is that of *degree of freedom*, which is not always defined in the same way. It is defined to be either :

— each pair of position-velocity coordinates associated with the displacement, or
— only one of the two elements, position or velocity.

Thus, a set of N bodies free to move in three spatial directions has either $3N$ or $6N$ degrees of freedom according to the definition chosen. In this book, we will use the second definition. It corresponds to the number of initial conditions that can be chosen independently, and so is in fact closer to the etymology of the phrase "degree of freedom". In the example of the oscillator developed in this chapter, the initial position and velocity are the two quantities that can be set arbitrarily; therefore we will say that the pendulum is a system with two degrees of freedom, whose phase space is two-dimensional.

I.1.2 DESCRIPTION OF MOTION IN A PHASE SPACE

Let us apply the preceding definitions and ideas to the study of the nonlinear equation (1). At each point in the phase plane $(\theta, \dot{\theta})$, the equation determines the orientation of the tangent to the trajectory. By successive extrapolation one is therefore able, at least in principle, to construct the phase trajectory[8]. In the example treated here, an important simplification appears, due to the fact that the quantity $E(\theta, \dot{\theta})$ defined by:

$$E(\theta, \dot{\theta}) = \frac{1}{2}\dot{\theta}^2 + \frac{g}{\ell}(1 - \cos\theta)$$

is a constant of the motion. Indeed:

$$\frac{dE}{dt} = \left(\frac{d^2\theta}{dt^2} + \frac{g}{\ell}\sin\theta\right)\dot{\theta} = 0.$$

---

7. More precisely the momentum (see Section I.2.1).
8. It is in this way that we proceed most of the time — in particular, each time that we do not know how to integrate the equation of motion.

Note that up to a factor of dimension $m\ell^2$, this quantity is none other than the energy of the pendulum, which by convention is zero at the state of rest (point R of fig. I.1):

$$E(0,0) = 0.$$

It follows that the trajectories that are solutions to (1) are the curves of constant energy — or level curves of E — in the phase plane. Because the energy E is a periodic function of θ, of period 2π, it suffices to represent the trajectories in the domain:

$$\theta \in [-\pi, +\pi] \quad ; \quad \dot{\theta} \in]-\infty, +\infty[.$$

To emphasize this property, Figure I.2 shows the graph for a more extended range of θ. Several remarks are worth making about Figure I.2. The circle centered about $\theta = 0$ (and more generally, about $\theta = 2\pi n$, n integer) describes the isochronous oscillations, whose period is independent of the amplitude and of the mass, as we have shown in the preceding section (linear approximation). These oscillations are associated with the smallest values of the energy. By contrast, in the nonlinear domain,

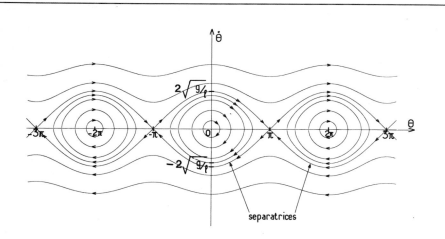

Figure I.2 Phase portrait for the simple pendulum.
The curves drawn in the $(\theta, \dot{\theta})$ plane are the lines of constant energy of the pendulum. The graph is periodic of period 2π along the θ axis. In the range of validity of the linear approximation of (1), these curves are circles centered about the points with coordinates $\theta = 0$, $\theta = \pm 2n\pi$, n integer. They correspond to oscillations whose period is independent of the amplitude (isochronism of small oscillations). For larger values of the energy the circles are replaced by ovals and the period is then a function of the amplitude. When the energy exceeds $2g/\ell$, the oscillations are replaced by continous rotation about the point O. Each of the two curves forming the boundary between the domains of oscillation and rotation is called a separatrix. The arrows indicate the direction of motion in time.

1.1 THE FREE OSCILLATOR

the oscillation period increases as the amplitude (and correspondingly the energy) increases. The relation between the period T and energy E is of the form:

$$T = 2\pi \sqrt{\frac{\ell}{g}} \cdot F\left(\frac{\ell E}{g}\right)$$

The function $F\left(\frac{\ell E}{g}\right)$ is an elliptic integral which diverges as we approach the *separatrix* — that is, the level curve passing through the unstable equilibrium points of the pendulum, with coordinates: $\dot{\theta} = 0; \theta = \pm(2n+1)\pi$ (the high position of the pendulum, point R' of Figure 1.1). On this separatrix the energy is equal to $\frac{2g}{\ell}$ and the period is infinite. *A priori* this second property may seem a little curious and even paradoxical, but it merely expresses the fact the pendulum takes an infinite time to attain the unstable equilibrium state when the available energy is exactly equal to that necessary to reach R'.

For a value of E greater than $\frac{2g}{\ell}$, there exist two possible trajectories, symmetric with respect to the θ axis. They correspond to rotation of the pendulum about 0, either in one direction or in the other.

1.1.3 CANONICAL FORM OF THE EQUATIONS OF MOTION

It is easy to transform equation (1) into two first-order differential equations:

$$\begin{aligned}\frac{d\theta}{dt} &= \dot{\theta} = \frac{\partial H}{\partial p_\theta} \\ \frac{dp_\theta}{dt} &= \dot{p}_\theta = -\frac{\partial H}{\partial \theta}\end{aligned} \qquad (2)$$

where we have introduced the angular momentum variable:

✗ $\quad p_\theta = \ell \dot{\theta}$

H is called the Hamiltonian function of the pendulum. One therefore sees that:

$$E(\dot{\theta}, \theta) = H(p_\theta, \theta) = \frac{1}{2}\frac{p_\theta^2}{\ell^2} + \frac{g}{\ell}(1 - \cos\theta).$$

The variables θ and p_θ are said to be canonically conjugate for the Hamiltonian H. Since:

$$\frac{d}{dt} H(p_\theta, \theta) = \frac{\partial H}{\partial p}\dot{p}_\theta + \frac{\partial H}{\partial \theta}\dot{\theta} = 0$$

we see that the energy remains constant during the motion, a result already established in the preceding section. This property places the pendulum in the class of conservative (or Hamiltonian) systems, characterized precisely by the invariance of energy.

I.2 Conservative systems

I.2.1 CONSERVATION OF ENERGY

The result at which we have arrived concerning the simple pendulum has far more general applications. Classical mechanics shows that it is in theory possible to describe the behavior of any dissipationless system by a Hamiltonian function H, which depends on the generalized space coordinates (denoted by q), the generalized canonically conjugate momentum coordinates (denoted by p), and, possibly, time[9]:

$$H(q_1, q_2, ..., q_n, p_1, p_2, ..., p_n, t).$$

We establish by the principle of least action that this function is a solution of the system of $2n$ differential equations:

$$\begin{aligned} \frac{dp_i}{dt} = \dot{p}_i &= -\frac{\partial H}{\partial q_i} \\ \frac{dq_i}{dt} = \dot{q}_i &= \frac{\partial H}{\partial p_i} \\ i &= 1, ..., n \end{aligned} \qquad (3)$$

of which (2) is a particular case. When H does not depend explicitly on time, one can verify that:

$$\frac{\partial H}{\partial t} = \frac{dH}{dt} = 0$$

from which $H(q, p) = E = $ constant.

Thus, all systems described by a time-independent Hamiltonian are conservative, the simple pendulum (Equation (1)) being one example.

I.2.2 CONSERVATION OF AREAS IN PHASE SPACE

One very important corollary to the preceding property is the conservation of areas in phase space. Starting from (3), it is possible to prove that the area of a surface element $\delta q_i \, \delta p_i$ of a phase plane is conserved by the motion as long as H does not depend explicitly on time[10]. What are the implications of such a result? Let us first

---

9. What follows is a simple reminder, without proofs, of the essentiel results of mechanics.
10. We refer the reader to books on mechanics for the proof of this property, which is simple enough, but relatively long to reproduce. The word "area" is generalized in the usual way to the measure of a two-dimensional surface. The two canonically conjugate variables q_i and p_i vary in a plane and the element of area $\delta p_i \cdot \delta q_i$ is conserved by equation (3). A set of initial conditions occupying a surface of measure σ in the (q_i, p_i) plane will continue to occupy a surface of the same measure during the course of the evolution. This property of Hamiltonian systems is much more restrictive than simple conservation of energy or reversibility of equations of motion. It entails the conservation of volume in phase space, since the volume element is the product of the area elements of each pair of conjugate variables, the phase space here being of even dimensionality.

examine the case where only one pair of canonically conjugate variagles (q, p) is involved. We consider a set of points occupying a surface $\sigma(0)$ in the neighborhood of an initial point. At a subsequent instant t, these points define a surface $\sigma(t)$ such that:

$$\sigma(t) \equiv \sigma(0)$$

according to the property stated above. For the preceding equality to be satisfied, the distance between neighboring trajectories must remain nearly constant or at least must not grow or decay too rapidly with time. Actually distances along the trajectories do not grow faster than linearly with time: two initial points along a trajectory follow that trajectory with velocities differing by some small constant. Thus the area is conserved if the distances perpendicular to these trajectories do not grow (or decay) faster than linearly with time (fig. I.3). However, this could be false for exceptional situations, such as nearby trajectories converging to fixed points. But all these properties are very specific to Hamiltonian systems with two degrees of freedom and are no longer necessarily true in higher dimensions. As soon as there are at least two pairs of canonically conjugate variables (q_1, p_1, q_2, p_2) the motion is on an energy surface[11]:

$$H(q_1, q_2, p_1, p_2) = E = \text{constant}$$

Since the area of a set of given initial conditions $\delta q_i \, \delta p_i$ is conserved in time, it is clear that the distance between initial conditions cannot decay, or otherwise the area would diminish with time. Nevertheless, the conservation of area car be guaranteed in one of two ways. Either:

— as before, the surface element considered follows the trajectory without being substantially deformed, or

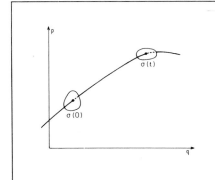

Figure I.3 Conservation of areas in the phase space (p, q).
The area element $\sigma(0)$, which can be seen as a set of initial conditions, is transformed by the motion into another element of area $\sigma(t)$. When the system is conservative, this element has the same area (but not necessarily the same shape) at all times t.

---

11. This surface is equivalent to the level curves of the simple pendulum. Since it results from imposing one condition in a four-dimensional space, the surface itself is of dimension three. It can therefore be mapped, at least locally, onto the ordinary space $\mathbb{R}^3$.

— it lengthens in one direction exponentially with time — that is like $e^{\lambda t}, \lambda > 0$ — while contracting exponentially like $e^{-\lambda t}$ in the perpendicular direction. The area is indeed conserved since $e^{\lambda t} \cdot e^{-\lambda t} \equiv 1$.

While under the first hypothesis two trajectories that are initially close remain so, under the second hypothesis they tend to separate exponentially. From the dynamical point of view, we emphasize that the difference is considerable: the trajectories are *stable* in the first situation, whereas they are *unstable* in the second, since a small initial difference is amplified very quickly in time.

I.2.3 INVARIANCE OF THE EQUATIONS UNDER TIME REVERSAL

Another important property of Equation (1) and of the more general form (3) is that changing the sign of the time (replacing $+t$ by $-t$) has no effect: the equations remain identical. Concretely this means that a film of the motion of the pendulum can be projected forwards or backwards without any discernable difference. The dynamics of conservative systems are said to be *reversible*.

To understand the singular nature of time reversal, we must bear in mind that it is not a property of most of the time evolution that we see. If we project any movie or television film backwards, the spectator realizes it in a few seconds: waves move away from the beach, smoke enters cigarettes, etc. Invariance under time reversal is, therefore, not at all common. It is characteristic of conservative systems and conveys the fact that the fundamental physical laws implied by (3) — inertia, gravitation — are of the same kind. We will see later that irreversibility appears as soon as the energy is dissipated instead of conserved.

From the phase space point of view, it is clear that conservation of areas is preserved by time reversal. Similarly, stable trajectories remain so: regardless of the direction of the flow of time they do not diverge from one another. In contrast when area is conserved at the cost of elongation in one direction and contraction in a perpendicular direction, time reversal implies the exchange of stability of these two directions. The direction which was originally stable (contraction) becomes unstable (elongation) and vice versa.

I.3 The damped oscillator

I.3.1 DISSIPATION OF ENERGY BY FRICTION

We have seen in Section I.1 that the description of the free oscillator was made at the cost of idealizing the simple pendulum. In particular all kinds of friction were neglected — that of the wire and of the mass in the air, as well as the friction at the

I.3 THE DAMPED OSCILLATOR

suspension point O. In practice, the motion of a real pendulum always[12] stops eventually due to friction; the amplitude of the oscillation decreases inexorably with time. This very general phenomenon of energy dissipation must be taken into account in the mathematical description. More specifically, Equation (1) must be modified. A simple remark helps to see in what way. We have emphasized that Equation (1) is invariant under time reversal. But any damping mechanism destroys this property since it eventually halts the oscillations. It is therefore necessary to break the time reversal invariance.

I.3.2 EQUATION FOR THE DAMPED OSCILLATOR

The simplest way to accomplish this is to add to Equation (1) a term proportional to an odd derivative in the angle θ, such as the first derivative, which in the linear approximation yields:

$$\frac{d^2\theta}{dt^2} + \gamma \frac{d\theta}{dt} + \omega^2\theta = 0 \tag{4}$$

γ = damping coefficient; $\omega^2 = g/\ell$: square of the angular frequency.

One obtains in this way the linearized equation of the damped pendulum. Experience shows this equation to be satisfactory when the friction is of "fluid" type[13]. Let us examine more closely the evolution of the energy which is now:

$$E(\theta, \dot{\theta}) = \frac{1}{2}(\dot{\theta}^2 + \omega^2\theta^2)$$

since (by the hypothesis necessary for linearization): $1 - \cos\theta \sim (\theta^2/2)$.

Using Equation (4) we see that:

$$\frac{dE}{dt} = -\gamma\dot{\theta}^2.$$

Thus energy is conserved if $\gamma = 0$ (no friction) and decreases if $\gamma > 0$. Since, by definition $\gamma \geq 0$, the energy tends to 0. The state of zero energy — the state of rest — is therefore stable, since the pendulum necessarily evolves towards that state.

---

12. Here we are excluding the erratic motion of the pendulum due to thermal fluctuations. This is entirely justified on our scale, due to the very small value of the Boltzmann constant.
13. That is, of intensity proportional to the speed. Examples:
 — a pendulum whose density is much greater than the fluid in which it is immersed (e.g. an ordinary pendulum in air),
 — an electric oscillator damped by the resistance of the circuit.
By contrast there also exists "solid friction" which is more difficult to include in equations, notably due to hysteresis effects.

Let us now disregard the manner in which we have arrived at Equation (4) to consider its intrinsic properties. Note that if γ is negative, the energy E increases *a priori* indefinitely, except of course starting from the state of rest, for which $\theta = \dot{\theta} = 0$ implies $dE/dt = 0$. In this case, however, the state of rest is unstable, since any slight displacement of the pendulum from the state of rest is amplified by the evolution.

I.3.3 PHASE PORTRAIT

The solutions to (4) are trajectories tangent at each point to the local velocity vector whose two components $(x = \theta, y = \dot{\theta})$ are, by definition:

$$\frac{dx}{dt} = \dot{\theta} = y$$

$$\frac{dy}{dt} = \ddot{\theta} = -\gamma\dot{\theta} - \omega^2\theta = -\gamma y - \omega^2 x.$$

In mathematical terms, the trajectories are the lines of force of this vector field. An essential point is that exactly one line of force passes through each point, except for singular points where the vector field is not defined. The origin ($\theta = \dot{\theta} = 0$) is the only singular point of the plane: only there are both velocity components zero.

We have already seen (fig. I.2) that for $\gamma = 0$, the trajectories form a continuous family of closed curves about the equilibrium point $\theta = \dot{\theta} = 0$. If γ is not zero, this family of curves is transformed into a set of trajectories that either converge to or diverge from the origin, depending on whether $\gamma > 0$ or $\gamma < 0$. For weak damping — that is, such that $|\gamma| \ll \omega$ — let us show that these trajectories are spirals. Indeed, for γ small, the time evolution of θ cannot be totally different from that when γ is zero. As a first approximation we may therefore set:

$$\theta \simeq \rho(t) \cos(\omega t + \phi)$$

where $\rho(t)$ is a function of time which varies slowly as compared to the period $T = 2\pi/\omega$. Using this approximate expression, let us evaluate the mean variation of the energy over one period in two different ways. First, we integrate the instantaneous variation:

$$\overline{\frac{dE}{dt}} = \frac{1}{T} \int_{t_0}^{t_0+T} \left(\frac{dE}{dt}\right) dt = -\gamma \overline{\dot{\theta}^2}.$$

Neglecting the weak variation of ρ over one period, we derive:

$$\overline{\dot{\theta}^2} = \frac{1}{T} \int_{t_0}^{t_0+T} \rho^2\omega^2 \sin^2(\omega t + \phi) = \frac{1}{2}\rho^2\omega^2$$

$$\overline{\frac{dE}{dt}} = -\frac{1}{2}\gamma\rho^2\omega^2.$$

1.3 THE DAMPED OSCILLATOR

Moreover, using the same approximations, the energy has the value:

$$E = \frac{1}{2}[\rho^2\omega^2 \sin^2(\omega t + \phi) + \rho^2\omega^2 \cos^2(\omega t + \phi)] = \frac{1}{2}\rho^2\omega^2$$

and its mean variation over a period is thus:

$$\overline{\frac{dE}{dt}} = \overline{\frac{d}{dt}\left(\frac{1}{2}\rho^2\omega^2\right)} = \frac{1}{2}\omega^2 \overline{\frac{d\rho^2}{dt}}.$$

Combining the two results we see that over a period:

$$\overline{\frac{d\rho^2}{dt}} = -\gamma\rho^2.$$

For γ very small compared to ω, we can integrate this relation over time, which yields:

$$\rho^2(t) = \rho^2(0) \exp(-\gamma t).$$

We have thus derived the parametric equation of the trajectories in the $(\theta, \dot\theta)$ plane:

$$\theta \simeq \rho(0) \exp\left(-\frac{\gamma t}{2}\right) \cos(\omega t + \phi)$$

$$\dot\theta \simeq -\rho(0)\omega \exp\left(-\frac{\gamma t}{2}\right) \sin(\omega t + \phi)$$

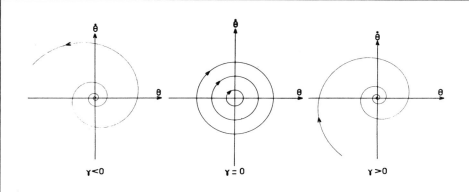

Figure I.4 Phase portrait of Equation (4) near the origin.
If $\gamma = 0$ the situation is the same as that of the simple pendulum in the linear approximation, so that we obtain a family of circles centered on the origin. By contrast, if γ is not zero, the trajectories are spirals leaving the origin ($\gamma < 0$: amplification) or ending at the origin ($\gamma > 0$: damping). For clarity of the figure, only one spiral is drawn for each of these two hypotheses.

This equation describes spirals, either convergent ($\gamma > 0$) or divergent ($\gamma < 0$), or else circles ($\gamma = 0$). Recall that here we are using the linear approximation (Equation (4)).

Figure I.4 shows schematically the type of phase portrait that we obtain in each of these three cases.

Whatever the value of γ, we note that the origin is a singular point corresponding to a stationary solution of the equation of motion. This stationary state is stable if $\gamma > 0$, unstable if $\gamma < 0$, since in the first case all trajectories lead to it, and, in the second, away from it. When γ is zero the origin is said to be *marginally stable*, since displacements are neither amplified nor diminished in time.

When γ is positive, all trajectories of the phase plane terminate at the origin, which is called, for this reason, an attracting point or *attractor*. This is a fundamental idea which can be immediately generalized to higher dimensional manifolds (curve, surface, etc.). We will use the concept of attractor a great deal in what follows.

I.4 Dissipative systems

I.4.1 ESSENTIAL PROPERTIES

The damped oscillator provides a typical example of a dissipative system. The dynamical properties of dissipative systems are in many ways the opposite of those of conservative systems. We can return to the considerations discussed in Section I.2. Time-independent Hamiltonians[14] do not generally exist for dissipative systems; therefore, energy is not conserved. On the other hand, in certain cases there exists a function of dynamical variables, called the *Lyapunov function*, which is positive and decreases monotonically with time (proving irreversibility). Such a Lyapunov function does not always exist. Dissipative systems can have a far more complicated evolutionary regime than simple decay, especially when the dynamics include both damping effects and also mechanisms maintaining the motion. In any case, whenever there is dissipation, the equations of motion change under time reversal; the evolution of dissipative systems is not reversible. Finally, another important point is that areas in phase space are no longer conserved.

I.4.2 CONTRACTION OF AREAS IN PHASE SPACE

In light of what we have already said this property is almost obvious. Let us return, for example, to Figure I.4 for the case of genuine friction ($\gamma > 0$). It is clear that a surface element $\delta\theta \cdot \delta\dot{\theta}$ cannot be conserved by the motion since all the trajectories which it contains end at the origin. We have here a property specific to dissipative systems: the

---

14. By increasing the number of degrees of freedom one can always embed a dissipative system in a larger Hamiltonian system. But this formal artifice is of no real utility in the study of dissipative systems.

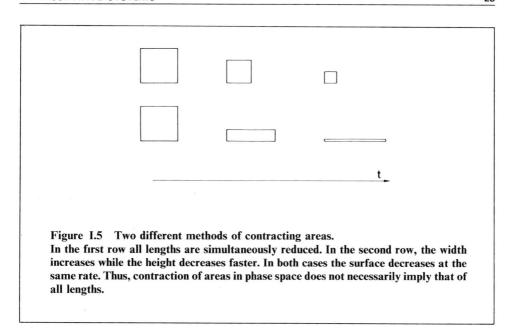

Figure I.5 Two different methods of contracting areas.
In the first row all lengths are simultaneously reduced. In the second row, the width increases while the height decreases faster. In both cases the surface decreases at the same rate. Thus, contraction of areas in phase space does not necessarily imply that of all lengths.

area of any set of initial conditions diminishes on the average in time, which is often expressed by saying that the flow contracts areas in phase space.

Several examples later on in the book will show us that this contraction can occur in different ways. Nevertheless, a mistake which must be avoided is to think that contraction of areas is necessarily synonymous with contraction of lengths. This is not at all the case. The result is entirely analogous to that already mentioned in Section I.2.2: conservation of areas can be satisfied when there is divergence in one direction if there is simultaneously convergence in another direction. Similarly here, contraction of areas can be obtained not only:
— by reduction of all lengths, but also
— by decrease of some of the lengths, accompanied by the less rapid increase of other lengths.

The simple diagram of Figure I.5 illustrates these two possibilities, using a square as the initial surface element.

This remark is very important for showing that divergence of phase trajectories in some direction remains possible, even in a dissipative system. Of course, knowledge of the way in which the contraction of areas takes place, and of its speed, are essential for a complete dynamical description. This is one reason for the utility and the importance of the Jacobian[15].

15. See Appendix E.

CHAPTER II

Forced oscillator — Parametric oscillator

II.1 The forced oscillator

II.1.1 THE VAN DER POL EQUATION

What is an appropriate description of the behavior of a forced oscillator? We have seen in the introduction that there exist numerous examples of forced oscillators in nature. Equation (4) is clearly not suited to this goal, and not only because of damping. The equation has two other shortcomings. One is that in the case $\gamma < 0$, the energy of the pendulum increases indefinitely, which has no physical meaning. The other flaw in Equation (4) is that, if $\theta(t)$ is a solution, then any product $\alpha\theta(t)$ is also a solution, for any real α, since (4) is linear in θ. Such invariance by dilation is naturally incompatible with the existence of oscillations of fixed amplitude. For the equation of a forced oscillator, we must modify (4) in such a way as to:
— destroy invariance under dilation,
— limit the increase in energy when $\gamma < 0$,
— introduce a continuous energy source[16] compensating for losses by viscous damping when γ is positive.

With this as a starting point, Van der Pol noticed that a mathematically simple change consisted of making the friction coefficient γ depend upon the amplitude θ of the oscillations. By arranging for the parameter γ to be negative for small amplitudes

---

16. The practical realization of the steady motion of a pendulum has had great historical importance, since it was necessary for the perfecting of mechanical clocks and consequently the precise measurement of time. We owe to the physicist Huygens the invention of the escapement, a device which regulates the motion of a pendulum. The eighteenth century saw the development of the torsion pendulum, a spiral which regulates mechanical watches, in which oscillations are maintained by a spring motor. Today the advent of electronics has led to the manufacture of auto-oscillating circuits of adjustable frequency, involving a piezoelectric quartz. These circuits are then tuned to vibrate with frequencies that are related to the differences between atomic energy levels by Planck's law ($\Delta E = h\nu$). This gives atomic clocks an extraordinary precision (10^{-13} s). As a concrete illustration, one would have to wait about 300 000 years for two atomic clocks to differ by a mere second.

and positive for large amplitudes, we will arrive at the desired goal. Since only the absolute value and not the sign of the amplitude is significant, it seems natural to adopt a dependence of $\gamma(\theta)$ on θ^2. The simplest possible expression satisfying these conditions is:

$$\gamma(\theta) = -\gamma_0 \left[1 - \frac{\theta^2}{\theta_0^2} \right]$$

$\gamma_0 > 0$; θ_0 : reference amplitude.

Therefore :

$\gamma < 0$ for $\theta^2 < \theta_0^2$
$\gamma > 0$ for $\theta^2 > \theta_0^2$.

By substituting this expression for γ in (4) we obtain what is called the Van der Pol equation[17]:

$$\frac{d^2\theta}{dt^2} - \gamma_0 \left[1 - \frac{\theta^2}{\theta_0^2} \right] \frac{d\theta}{dt} + \omega^2 \theta = 0.$$

It describes the behavior of a forced oscillator, in which small-amplitude oscillations grow and large-amplitude oscillations are damped. It is useful to put this equation in nondimensional form by taking $\theta_0 \sqrt{\omega/\gamma_0}$ as the unit of amplitude and $1/\omega$ as the unit of time. The equation becomes:

$$\frac{d^2\theta}{dt^2} - (\varepsilon - \theta^2) \frac{d\theta}{dt} + \theta = 0 \tag{5}$$

which contains only the nondimensional parameter $\varepsilon = \gamma_0/\omega$.

II.1.2 PHASE PORTRAIT

Given that ε is positive, the trajectories diverge as spirals in the neighborhood of the origin. In fact in this region θ is always small and the term in θ^2 remains negligible. We find ourselves exactly in the situation described by Figure I.4, with $\gamma < 0$. By contrast, far from the origin the trajectories tend to approach this singular point, since $\gamma(\theta)$ is then positive. Intuitively, we can hypothesize that there exists a closed trajectory between these two extremes encircling the origin. A more extensive study of the Van der Pol equation does establish the existence, uniqueness, and stability of such a trajectory which Poincaré called a *limit cycle*. All phase trajectories, located inside as well as outside the limit cycle, tend asymptotically towards it. Here, then, is a new example of an attractor (fig. II.1).

---

17. Van der Pol proposed this equation in 1922 to explain the triode oscillator.

II.1 THE FORCED OSCILLATOR

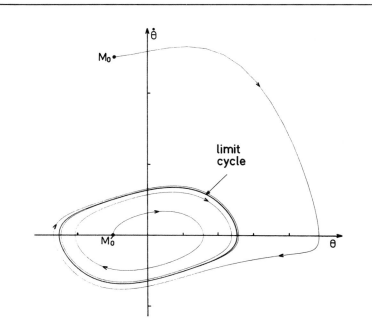

Figure II.1 Phase portrait of the Van der Pol equation.
One solution to this equation is represented in the $(\theta, \dot{\theta})$ phase plane by a closed curve, the limit cycle towards which all trajectories converge. We have chosen two initial conditions, one outside (M_0), the other inside (M'_0) the limit cycle for $\varepsilon = 0.4$.

The form of the limit cycle and of the oscillations depends strongly on the value of the positive parameter ε, as is shown by Figure II.2. There, we have depicted solutions to Equation (5) starting from a given initial condition, in the coordinate systems $(\theta, \dot{\theta})$ and (t, θ). Two values of ε are considered, one very small (fig. II.2 a), and the other much larger (fig. II.2 b). In the first case, the limit cycle is practically a circle and the oscillations are quasi-sinusoidal. In the second, the evolution takes place on two distinct time scales; a slow drift followed by a sudden variation in amplitude. This distinctive time dependence is given the name of *relaxation oscillations*.

The limit cycle regime is described by a periodic function $\theta(t)$ which can be expanded in a Fourier series. This is a very general property of all dynamical systems with this type of behavior. Therefore, as soon as we are in a limit cycle regime, the time dependence of any observable $X(t)$ is of the form:

$$X(t) = \sum_{n=0}^{\infty} x_n \sin(n\omega t + \phi_n)$$

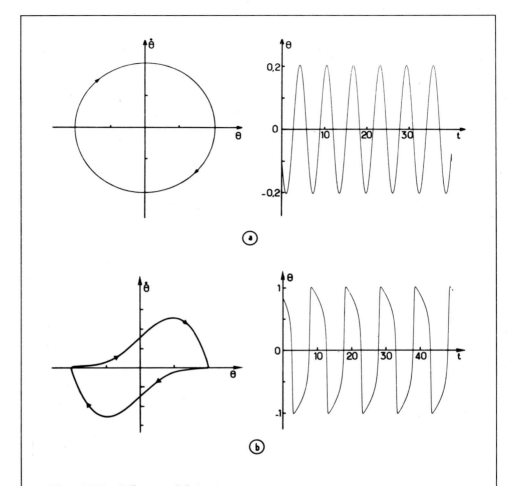

Figure II.2 Influence of the parameter ε.
a) For small parameter values ($\varepsilon = 0.01$), the temporal oscillations are quasi-sinusoidal, and the limit cycle has approximately the form of a circle.
b) When ε is much larger ($\varepsilon = 4.0$), the limit cycle is almost rectangular and its path causes two distinct time scales to enter in. The sawtooth variations of the amplitude θ are called relaxation oscillations.

II.1.3 AMPLITUDE OF OSCILLATIONS FOR $\varepsilon = 0_+$

In the units chosen, the energy is:

$$E(\theta, \dot{\theta}) = \frac{1}{2}(\dot{\theta}^2 + \theta^2).$$

II.1 THE FORCED OSCILLATOR

Its variation in time can be deduced from Equation (5):

$$\frac{dE}{dt} = \frac{1}{2}(2\dot\theta\ddot\theta + 2\theta\dot\theta) = (\varepsilon - \theta^2)\dot\theta^2.$$

The average energy is defined by:

$$\overline{\frac{dE}{dt}} = \lim_{\tau\to\infty}\frac{1}{\tau}\int_{t_0}^{t_0+\tau}\frac{dE}{dt}\,dt$$

$$\overline{\frac{dE}{dt}} = \lim_{\tau\to\infty}\frac{1}{\tau}[E(t_0+\tau) - E(t_0)]$$

According to the expression for dE/dt given above, we have:

$$\overline{\frac{dE}{dt}} = \varepsilon\overline{\dot\theta^2} - \overline{\theta^2\dot\theta^2} = 0.$$

Thus, the production of energy $\varepsilon\overline{\dot\theta^2}$ is, on the average, exactly compensated for by the dissipation $\overline{\theta^2\dot\theta^2}$ due to the nonlinearity. We now calculate the approximate amplitude of the limit cycle for $\varepsilon = 0_+$. In this case, we can assume that:

$$\theta(t) \simeq \rho \sin t$$

(see fig. II.2 a). It follows that:

$$\overline{\dot\theta^2} \simeq \frac{1}{2}\rho^2$$

$$\overline{\theta^2\dot\theta^2} \simeq \frac{1}{8}\rho^4.$$

The average energy balance worked out above shows that, in this range:

$$\rho \simeq 2\sqrt{\varepsilon}.$$

The size ρ of the limit cycle, nearly a circle, therefore varies as the square root of the parameter ε. We have established this result for the particular case of the Van der Pol equation, but its scope is far more general. The term in θ^2 can in fact be seen as representative of any nonlinearity as long as it remains small. It is the standard form of the first nonlinear term of the Taylor series expansion of an arbitrary function[18]

We point out that the dependence on $\sqrt{\varepsilon}$ of the size of the limit cycle, and of the

---

18. If we consider the Taylor expansion of the coefficient $\gamma(\theta)$ we might have thought that the expansion would begin with a linear term (whereas we have started with a quadratic term):

$$\gamma(\theta) = -\gamma_0[1 + \alpha\theta - (\theta^2/\theta_0^2)]$$

This linear term has no direct effect on the evolution of the energy, at least near $\varepsilon = 0$. In fact its contribution to the mean $\overline{dE/dt}$ is of the form $\alpha\overline{\theta\dot\theta^2}$, which is zero for $\theta = \rho \sin t$. This explains why we have not kept the linear term $\alpha\theta$ in the expression for $\gamma(\theta)$ and also why the relation $\rho \simeq \sqrt{\varepsilon}$ is indeed quite general.

amplitude of the corresponding oscillations, can be formally eliminated. By taking θ_0 as the unit of amplitude (instead of $\theta_0/\sqrt{\varepsilon}$), we obtain instead of (5):

$$\frac{d^2\theta}{dt^2} - \varepsilon(1 - \theta^2)\frac{d\theta}{dt} + \theta = 0.$$

Of course this form is exactly equivalent to (5) for a given value of ε. Nevertheless, repeating the preceding calculation, we conclude that the size of the limit cycle in the $(\theta, \dot{\theta})$ plane is now independent of ε. The variation in $\sqrt{\varepsilon}$ determined previously has been concealed by our adopting this unit of measure for θ.

II.2 The parametric oscillator

II.2.1 EQUATION OF MOTION

We return to the example of the simple frictionless pendulum placed in a gravitational field, and now imagine that we subject the pivot O (fig. I.1) to an alternating vertical motion. This is another means of maintaining the motion. In this way we define a parametric pendulum — that is, a pendulum one of whose parameters varies with time[19]. Equivalently, we could place a simple pendulum in a time-dependent gravitational field:

$$g(t) = g_0 + \beta(t)$$

In the case described above, $\beta(t)$ describes the vertical acceleration of the point O. In a linear approximation, the equation of motion deduced from (1) is:

$$\frac{d^2\theta}{dt^2} + \frac{g(t)}{\ell}\theta = 0 \tag{6}$$

Despite being formally very close to Equation (1), Equation (6) differs from it in an essential way: it is nonintegrable for an arbitrary function $g(t)$. This means that we do not know how to solve this ordinary differential equation by quadrature — that is, to express its solution $\theta(t)$ as an explicit function of time, in terms of integrals (or

---

19. Numerous other realizations are possible, for instance that of a self-capacitance circuit whose resistance is time-dependent. This situation, which could at first glance seem rather unusual, in fact corresponds to a very general problem: that of the behavior of any oscillator with a time-dependent parameter. A wonderful example of such a parametric pendulum is "O. Botafumeiro", a giant censer that hangs in the transept of the cathedral of Santiago de Compostela in the North West of Spain. (This cathedral is the goal of the most famous pilgrimage of Catholicism). Mechanically, O. Botafumeiro is a mass of 53 kg, hung in the vault of the cathedral by a rope of about 20 m. It is set into pendular motion by varying periodically the length of the rope with a device built during the 13th century. The amplitude of angular oscillation obtained through this parametric pumping can reach almost $\pi/2$ on both sides of the vertical.

quadratures) of the usual ordinary or transcendental functions such as rational fractions, exponentials, elliptic functions, etc. Equation (1) is, on the contrary, integrable, because the term in $g/\ell \sin\theta$ does not depend explicitly on time[20].

Despite this difference the state of rest $\theta = \dot\theta = 0$ is still a trivial solution to the above equation. Nevertheless, when g depends on time, the stability of this solution is not guaranteed and must be determined by detailed analysis. Although any form of the function $g(t)$ is *a priori* worth studying, we will consider only the case where $g(t)$ is a periodic function of time: the equation of motion is then called Hill's equation. Simplifying further and taking $\theta(t)$ to be a circular function, we arrive at a particular form called the *Mathieu equation*. Taking:

$$g(t) = g_0 + g_1 \cos(2\omega t)$$

we arrive at[21]:

$$\frac{d^2\theta}{dt^2} + \omega_0^2[1 + h\cos(2\omega t)]\theta = 0 \tag{7}$$

where:

$$h = g_1/g_0 \quad ; \quad \omega_0^2 = g_0/\ell.$$

The excitation represented by the term $h\cos(2\omega t)$ has period $T = \pi/\omega$. Without loss of generality we can assume $h \geqslant 0$. Note that ω_0 is the pendulum's angular frequency in the absence of external excitation ($h = 0$). Since Equation (7) is linear and of second order in θ, an initial condition $\theta(0)$, $\dot\theta(0)$ undergoes at each period (π/ω, $2(\pi/\omega)$, $3(\pi/\omega)$...) a linear transformation which is equivalent to multiplying the coordinates in the $(\theta, \dot\theta)$ plane by a matrix M (see Appendix E). Based on this observation, the Floquet theory of linear equations with periodic coefficients shows that the solutions to the Mathieu equation are of the form:

$$\theta(t) = e^{\mu t} \cdot P(t)$$

---

20. The concepts of integrability and nonintegrability can be immediately generalized to any set of ordinary differential equations. We mention that the solution by quadrature of an equation or system of equations presents mathematical problems addressed by the Galois theory of algebraic equations.

21. In addition to convenience in calculation, a deeper reason inspires this choice: the occasion to mention the general problem of coupled oscillators. Let us introduce the variable $\alpha = h\cos(2\omega t)$, and substitute for (6) the two equations:

$$\frac{d^2\theta}{dt^2} + \omega_0^2(1 + \alpha)\theta = 0$$

$$\frac{d^2\alpha}{dt^2} + 4\omega^2\alpha = 0$$

which describe the behavior of two oscillators of amplitude θ and α. They are coupled by the term $\omega_0^2\alpha\theta$. This formulation is almost intuitive since the periodic variation of $g(t)$ implies that it is controlled by an external oscillator with amplitude proportional to α. The external oscillator must be insensitive to the motion of the parametric pendulum, which explains the absence of a θ-dependent term in the second equation. This is a specific form of coupling, in which the first oscillator is in fact forced by the second.

Here μ is an eigenvalue of the matrix M, sometimes called the *characteristic exponent*, and $P(t)$ is a function of period $T = \pi/\omega$ — that is, such that:

$$P(t) = P\left(t + \frac{\pi}{\omega}\right).$$

We have separated the search for solutions to (7) into two parts. First, there is the determination of $P(t)$ which, as we have mentioned, cannot be expressed in terms of the usual functions: it is necessary to define a new class of transcendental functions, called *Mathieu functions*, which we do not plan to describe further. Secondly, there is the calculation of the characteristic exponent μ. The sign of the real part of μ determines whether the solution $\theta(t)$ grows or decays. When the real part of μ is positive, $\theta(t)$ grows indefinitely in time. In this case, the state of rest, and in fact every solution to (7) is unstable, because a small perturbation of the form $\theta(t)$ will grow without bound. It is this question of stability which we will now examine more closely.

II.2.2 STABILITY OF SOLUTIONS

For small values of the excitation (for h near zero), the exponent μ can be determined by a perturbative calculation[22]. We know that for $h = 0$, the general solution to (7) is:

$$\theta = \theta_0 \cos(\omega_0 t + \phi).$$

By continuity, for h close to, but not equal to 0, the solution $e^{\mu t} P(t)$ must necessarily be close to the expression above. We conclude that the approximation:

$$\theta(t) \simeq e^{\mu t} \cos(\omega t + \phi)$$

is satisfactory, provided that μ is small and ω close to ω_0. Substituting this approximate form into the Mathieu equation, we get:

$$\left[\frac{d^2}{dt^2} + \omega_0^2(1 + h \cos(2\omega t))\right] e^{\mu t} \cos(\omega t + \phi)$$

$$= \left[(\omega_0^2 - \omega^2 + \mu^2) \cos(\omega t + \phi) - 2\omega\mu \sin(\omega t + \phi)\right.$$

$$\left. + \frac{h}{2} \omega_0^2 \cos(\omega t - \phi) + \frac{h}{2} \omega_0^2 \cos(3\omega t + \phi)\right] e^{\mu t}.$$

---

22. In the particular case of the Mathieu equation, it is also known how to carry out the calculation in the limit $h \to \infty$, but using mathematical methods specific to this case. This is why we present only the method of calculation used for h close to 0, whose applicability is more general.

II.2 THE PARAMETRIC OSCILLATOR

Dropping the term in $\cos(3\omega t + \phi)$, which can be shown to be negligible by a more detailed calculation, let us look for the conditions under which the right hand side is zero for all t. For this to happen the multiplicative coefficients of the terms in $e^{\mu t}\cos(\omega t)$ and $e^{\mu t}\sin(\omega t)$ must cancel:

$$\left(\omega_0^2 - \omega^2 + \mu^2 + \frac{h}{2}\omega_0^2\right)\cos\phi - 2\omega\mu\sin\phi = 0$$

$$2\omega\mu\cos\phi + \left(\omega_0^2 - \omega^2 + \mu^2 - \frac{h}{2}\omega_0^2\right)\sin\phi = 0.$$

This system of two homogeneous linear equations in $\cos\phi$ and $\sin\phi$ has a solution only if the determinant of this system is zero, that is:

$$(\omega_0^2 - \omega^2 + \mu^2)^2 - \frac{h^2}{4}\omega_0^4 + 4\omega^2\mu^2 = 0$$

or:

$$\mu^4 + 2(\omega_0^2 + \omega^2)\mu^2 + (\omega_0^2 - \omega^2)^2 - \frac{h^2}{4}\omega_0^4 = 0.$$

This is a quadratic equation in μ^2 whose sum of roots is always negative and whose discriminant is always positive. There are two real roots, at least one of which is negative. Since a negative value of μ^2 is associated with a pure imaginary characteristic exponent μ, it cannot be the source of instability. On the contrary the solution is unstable when there exists a real positive value for μ^2 — and consequently for μ as well. The necessary condition for this is that the product of the roots be negative, that is:

$$h > 2\left|1 - \frac{\omega^2}{\omega_0^2}\right|.$$

In the Cartesian plane of the excitation parameters (h, ω) this defines a sector with vertex: $h = 0$, $\omega/\omega_0 = 1$ (fig. II.3).

Inside this sector (the hatched zone) μ^2 has a positive root and Equation (7) has no stable solution. What we mean by this is that there exist initial conditions very close to the state of rest from which originate trajectories that diverge exponentially from the initial condition. The state of rest is unstable with respect to an excitation of vanishing amplitude ($h \to 0$) if the condition $\omega = \omega_0$ is met, in which case, we say that there is *exact resonance*. This type of instability is often called *subharmonic*, since it leads the pendulum to oscillate with a frequency $\omega/2\pi$ equal to half the frequency $2\omega/2\pi$ of the external excitation.

Outside of the domain of instability, both of the possible values for μ^2 are negative. For ω close to ω_0 and h small, we can verify without difficulty that one of the imaginary

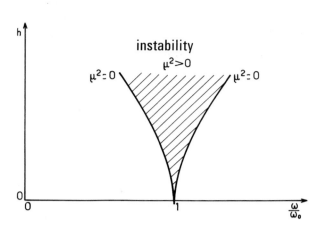

Figure II.3 Stability diagram of the Mathieu equation in the (h, ω) plane. Inside the hatched zones there exists a positive value of μ^2 and the equation has no stable periodic solution. In the neighborhood of an exact resonance ($\omega = \omega_0$), this is true even for a very weak excitation amplitude h, down to the limit 0. There exist other domains of subharmonic instability not represented in this figure. Their vertices have coordinates:

$$h = 0; \quad \omega/\omega_0 = 1/n, \quad n \text{ a positive integer.}$$

values of μ is close to 0, while the other is approximately equal to $2i\omega_0$. The resulting effect on the oscillations is as follows:

* the basic frequency is slightly modified by the presence of a characteristic exponent close to $2i\omega_0$,
* the small negative value of μ^2 leads to a low frequency modulation of the oscillations. In consequence, the behavior is, generally, quasiperiodic (see Section III.3.2 for a precise definition),
* the amplitude of this modulation is arbitrary since Equation (7) is linear in θ.

When the parameters of the excitation in the (h, ω) plane approach the domain of instability, the low frequency of modulation diminishes. It tends towards zero like the corresponding value μ^2 on the boundary of the domain. On the boundary, the exponential $e^{\mu t}$ of the Floquet theory is replaced by a term which is polynomial in time. Then, inside the unstable sector, the exponent μ becomes real. Summarizing, in moving through the parameter space we see that one frequency diminishes and then becomes zero, simultaneously with the appearance of an instability. The behavior which emerges in the domain of instability is governed by nonlinear effects not covered by the preceding analysis.

II.2.3 THE EFFECT OF DAMPING

Since all physical systems include friction we can ask what happens to the preceding results in the presence of damping. For weak fluid friction, that is, such that $0 < \gamma \ll \omega_0$ we can repeat all of the previous calculations starting from the equation:

$$\frac{d^2\theta}{dt^2} + \gamma \frac{d\theta}{dt} + \omega_0^2(1 + h \cos(2\omega t))\theta = 0.$$

If h is small, the existence of a solution of the form postulated by Floquet theory implies that:

$$(\omega_0^2 - \omega^2 + \mu^2 + \gamma\mu)^2 - \frac{h^2}{4}\omega_0^4 + (2\omega\mu + \omega\gamma)^2 = 0.$$

Marginal stability, corresponding to $\mu = 0$, is attained for:

$$\frac{h^2}{4} = \left(1 - \frac{\omega^2}{\omega_0^2}\right)^2 + \frac{\omega^2}{\omega_0^4}\gamma^2.$$

Thus, at exact resonance ($\omega = \omega_0$) the state of rest is called parametrically unstable if:

$$h > \frac{2\gamma}{\omega_0}.$$

A minimum level of excitation is necessary to destabilize this state. This is not surprising, and can be easily explained by the necessity to compensate for the damping of the oscillator. This time, the principal domain of subharmonic instability in the (h, ω) plane is bounded by a continuous curve inside the sector calculated in the absence of damping (fig. II.4).

We mention before concluding, that there exist other regions of instability in the (h, ω) plane. They correspond to sectors of width h^n not shown in Figure II.3 whose vertices are located at $h = 0$ and $\omega/\omega_0 = 1/n$ (n a positive integer)[23].

---

23. The knowledgeable reader may have noticed that up to differences in notation (t instead of x, h absorbed into the definition of g) the Mathieu-Hill equation is the same as the time-independent Schrödinger equation for one spatial dimension. There is a complete parallel between the theory of parametric resonance we have just developed and that of electrons — without interaction — in the periodic potential of a one-dimensional crystal. For historical reasons, the analogue to Floquet theory is called in the latter case Bloch-Wannier theory, despite the fact that Floquet's work largely predated the theory of electrons in metals. Analogous to the zones of parametric instability are, in the case of electrons, what are called forbidden energy bands.

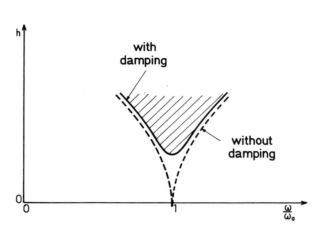

Figure II.4 Effect of damping.
In the presence of damping, the domain of subharmonic instability is reduced and is entirely contained in the instability zone that existed in the absence of damping. Note that a finite excitation threshold, no longer zero as before, is now needed to destabilize the parametric oscillator, even at the exact resonance.

II.2.4 MECHANISM OF PARAMETRIC AMPLIFICATION

Let us try to pin down the physical origin of subharmonic instability. Instead of taking a sinusoidal excitation, we will consider a function $g(t)$ of period $T = \pi/\omega$, made up of short impulses of length τ where τ, is much smaller than T (fig. II.5).

Let us then integrate Equation (6) over the interval τ of an impulse under the assumption that the amplitude of the impulse, while large, is not sufficient to cause a significant variation of θ during this time period. Under these conditions, for any function $f(\theta)$ we will have:

$$\int_{(\tau)} f(\theta)\, dt = \tau \cdot (\text{average value of } f(\theta)).$$

But we will also require in our approximation that:

$$\int_{(\tau)} \left(\frac{d^2\theta}{dt^2}\right) dt = \left[\frac{d\theta}{dt}\right]$$

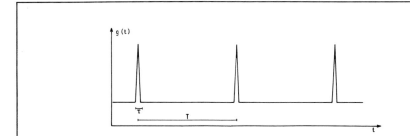

Figure II.5 Schematic representation of the function $g(t)$.
Each impulse is assumed to last a time τ which is negligible compared to the period T.

where $[d\theta/dt]$ is the change in velocity resulting from an impulse. From Equation (6) we then get:

$$\left[\frac{d\theta}{dt}\right] + G\theta(t) = 0$$

where:

$$G = \frac{1}{\ell} \int_{(\tau)} g(t)\, dt$$

$\theta(t)$ = average position of the pendulum during the impulse.

At the moment when the pendulum attains its maximum elongation, its velocity and kinetic energy are zero, whereas its potential energy is proportional to $(1/2)\theta_{max}^2$. Suppose that an impulse is applied just at this moment. The potential energy does not change since the position remains constant. However, the velocity is modified by a quantity $[d\theta/dt] = -G\theta_{max}$ and the kinetic energy after the impulse is therefore $(1/2)G^2\theta_{max}^2$. The net result is that, the impulse increases the energy of the oscillator by multiplying it by a factor $(1 + G^2)$ necessarily greater than 1. If the next impulse occurs just at the moment where the elongation is again maximal, the pendulum will be destabilized. In order to have loss of stability it is necessary that successive impulses be separated by a time interval equal to an integer multiple of half-periods of the pendulum undisturbed motion. *A priori* this multiple can be arbitrary; nevertheless it is clear that this parametric amplification process is most efficient when an impulse takes place at each half-period, as we have already seen. This shows us why the stability diagram of the Mathieu equation has in the (h, ω) plane a succession of unstable domains issuing from the points with abcissas $1/n$. In the preceding interpretation, this corresponds to an impulse each half-period, each period, each period and a half, and so on.

II.3 Introduction to bifurcations

II.3.1 CONCEPT OF BIFURCATION[24]

The description of the forced and parametric oscillators allows us to tackle another idea which is basic to the study of dynamical systems: that of bifurcation. We have already indicated above, for example, that in moving through the (h, ω) plane of the parametric pendulum, we observed a complete change in behavior upon crossing the boundary separating the stable and unstable solutions. This observation must be generalized: whenever the solution to an equation or system of equations changes qualitatively at a fixed value — called a *critical value* — of a parameter, this will be called a *bifurcation*. A point in parameter space where such an event occurs is defined to be a *bifurcation point*. From a bifurcation point emerge several (two or more) solution branches, either stable or unstable. The representation of any characteristic property of the solutions as a function of the *bifurcation parameter* constitutes a *bifurcation diagram*.

In order to clarify these ideas, let us demonstrate them for the forced oscillator, more precisely for Equation (5), which we will now consider independently of the physical significance of the parameter ε. For all values, positive or negative, of the bifurcation parameter ε, the origin $\theta = \dot{\theta} = 0$ is always a singular point. The origin is stable — and therefore an attractor — for $\varepsilon < 0$, but becomes unstable for $\varepsilon > 0$ and is replaced by another attractor, the limit cycle. A change in the nature of the stable solution thus occurs for the critical parameter value $\varepsilon_c = 0$. Graphing a typical property of the solution, for example the maximum elongation, as a function of ε, we construct the diagram of Figure II.6 *a*. To emphasize the meaning of this bifurcation diagram, we have also drawn the form of the trajectories in the $(\theta, \dot{\theta})$ phase plane.

For $\varepsilon = \varepsilon_c$, the oscillator is in a state of marginal stability. Linear stability analysis says nothing about the fate of small perturbations around $\theta = \dot{\theta} = 0$. However, by a continuity argument, it is not difficult to understand that, if for $\varepsilon > 0$ a small limit cycle of order $\varepsilon^{1/2}$ in size grows out from the $\theta = \dot{\theta} = 0$ state, then for $\varepsilon = 0$, this small limit cycle collapses to $\theta = \dot{\theta} = 0$ which is still an attracting fixed point. Although small perturbations near this fixed point decay as $e^{\varepsilon t}$, when ε is small and negative, one can show that they decrease for large times as $t^{-1/2}$ if ε is strictly zero. This transition from exponential to algebraic behavior just at a transition point is also rather common in many branches of physics.

II.3.2 HOPF BIFURCATION

When, as in the preceding example, a point gives rise to a limit cycle when a critical parameter value is crossed, we are dealing with a distinctive phenomenon called a *Hopf*

---

24. A more complete presentation of local bifurcations of codimension one is given in Appendix A.

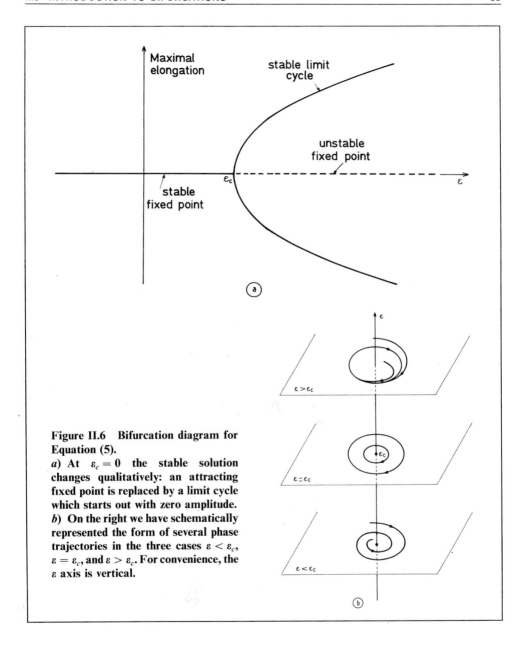

Figure II.6 Bifurcation diagram for Equation (5).
a) At $\varepsilon_c = 0$ the stable solution changes qualitatively: an attracting fixed point is replaced by a limit cycle which starts out with zero amplitude.
b) On the right we have schematically represented the form of several phase trajectories in the three cases $\varepsilon < \varepsilon_c$, $\varepsilon = \varepsilon_c$, and $\varepsilon > \varepsilon_c$. For convenience, the ε axis is vertical.

bifurcation. This important idea has a much larger range of applications than the limited presentation made here. Hopf bifurcations occur in vector fields as well as in diffeomorphisms; the mechanism is independent of both the number of degrees of freedom, and of the detailed form of the equations or applications involved. For our purpose, it suffices to mention the two principal properties of the limit cycle in the neighborhood of the Hopf bifurcation point. One of these properties was demonstrated in Section II.1.3, starting from Equation (5): the amplitude of the limit cycle is proportional to $|\varepsilon - \varepsilon_c|^{1/2}$. That is, the amplitude grows like the square root of the distance from the bifurcation point[25]. On the other hand, it can be shown that the period of oscillation is independent of $\varepsilon - \varepsilon_c$, at least in a first approximation (in fact, to second order). These are two typical characteristics of great practical interest since they can be used to identify a Hopf bifurcation.

II.3.3 SUBCRITICAL AND SUPERCRITICAL BIFURCATION

The bifurcation to which Equation (5) gives rise at $\varepsilon = \varepsilon_c = 0$ is called *supercritical* or *normal*. The limit cycle is created with a zero amplitude and at the bifurcation point, the system is in a state of marginal stability. This results from the presence of the dissipative term in $\theta^2 \dot{\theta}$ whose effect, as we have seen, counteracts that of the instability. Consequently, the solution does not undergo a sudden transformation at the bifurcation but changes gradually. This is generally the case whenever the nonlinear terms of lowest order are such as to counteract the instability (see Appendix A).

The diametrically opposed situation, where the nonlinear terms tend to amplify the instability is easily imagined (it suffices, for example, to change the sign of the term $\theta^2 \dot{\theta}$ in (5) and to add higher order terms like $\theta^4 \dot{\theta}$ to damp the large amplitude oscillations). This happens in other equations as well and leads to a subcritical bifurcation which is pictured schematically in Figure II.7. In this case, because of the amplification of the instability by the nonlinearity the stable solution is always at a finite distance from the solution which becomes unstable at the bifurcation. In traversing the ε axis from right to left, we see that ε_c, an unstable fixed point gives rise to a stable fixed point and to an unstable limit cycle. This is the inverse of the normal bifurcation of Figure II.6, and is in fact also called an *inverse* or subcritical bifurcation. In the interval $[\varepsilon'_c, \varepsilon_c]$ two stable solutions coexist: one is stationary (an attracting fixed point), the other oscillatory (a limit cycle).

From a practical point of view we must emphasize that:
— these two solutions are not simultaneously observable in the same system and it is the initial condition which determines the solution attained;
— the stable limit cycle appears at onset with a finite amplitude and also disappears with a finite amplitude.

Let us now imagine varying the parameter ε successively in one direction and then the other. We will observe the hysteresis phenomenon shown schematically in

---

25. In the particular case of Equation (5), $\varepsilon_c = 0$, which explains the dependence on $\varepsilon^{1/2}$ that we derived.

II.3 INTRODUCTION TO BIFURCATIONS

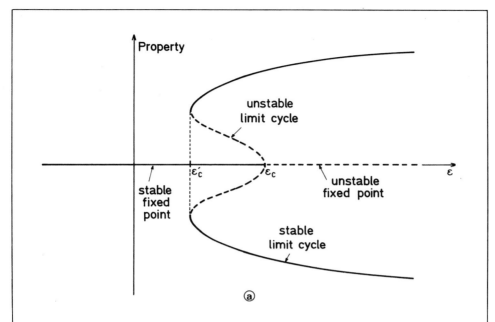

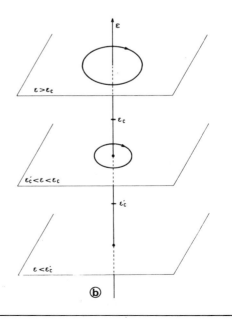

Figure II.7 Example of inverse or subcritical bifurcation.
a) The graph is of the same type as Figure II.6. In the interval $[\varepsilon'_c, \varepsilon_c]$, two solutions are stable; here, one is stationary and the other periodic.
b) To simplify the figure, only the attractors in the phase plane (fixed point and limit cycle) are shown, as a function of ε.

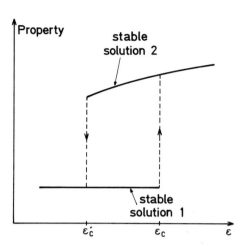

Figure II.8 Hysteresis phenomenon associated with a subcritical bifurcation. Because two stable solutions coexist on a finite interval of the bifurcation parameter, the transitions between them — when the parameter crosses the boundaries of this interval — take place at different values, depending on the direction of crossing: ε_c as ε is increased, ε'_c as ε is decreased (see arrows).

Figure II.8. When we increase the bifurcation parameter, the transition from the fixed point to the limit cycle takes places at $\varepsilon = \varepsilon_c$. The inverse transition, obtained by diminishing ε, occurs instead at $\varepsilon = \varepsilon'_c$. Thus, we see that the absence of marginal stability, a nonzero oscillation amplitude at transition, and hysteresis are properties of a subcritical bifurcation that are qualitatively different from those of a normal (supercritical) Hopf bifurcation.

CHAPTER III

The Fourier transform

III.1 Identification and characterization of a dynamical regime

When, as a result of an experiment or numerical simulation, we have a time-dependent signal $x(t)$ — called a time series — one of the essential tasks is to determine the kind of evolution that produced it. The purpose is to compress information in a way that emphasizes the most significant dynamical characteristics. Are we dealing with an oscillation, more or less complicated in shape, but with a perfectly well-defined period? Are we dealing with a more or less linear superposition of several different oscillations? Or is it something else entirely?

The answer to these questions is not at all trivial, except in very simple situations such as those discussed in the previous chapter. There we were able to ascertain the existence of periodic solutions to the equation of motion of a forced oscillator. A description of these solutions includes a period (or, equivalently, the frequency) and an amplitude: essentially, the amplitude of the limit cycle[26], and the time taken to traverse it. These two characteristics emerge almost automatically as soon as the equations of motion are known. It is important to emphasize that these characteristics remain, and play just as essential a role, for any periodic phenomenon whose underlying mechanism is unknown, or described by equations that are not soluble analytically, by far the most frequent situation in practice.

Certain dynamical regimes are a superposition of oscillations which differ in amplitude, period, ratio of harmonics, etc. We have already seen an example of this in the Mathieu equation and will see others later on. The associated attractor is no longer a limit cycle, but rather a *torus*. This type of regime is called quasiperiodic.

Other regimes are of a nature more difficult to grasp. Given their completely disordered appearance, we call them "chaotic"[27]. When the dynamics are

---

26. The existence of a limit cycle — or of any kind of attractor — necessarily implies that the system is dissipative, which we will henceforth assume. A conservative system can never have an attractor, since its evolution does not involve contraction of areas in phase space.
27. These regimes are also called turbulent, or aperiodic, or nonperiodic. These adjectives are more or less synonymous, as the terminology has not yet been precisely decided upon. In any case, it is impossible to detect any long term regularity in the corresponding time series $x(t)$.

deterministic (that is, representable by a finite number of nonlinear coupled differential equations or the equivalent), the trajectories in phase space converge onto a *strange attractor*, whose topological properties are radically different from those of a torus.

To answer the questions asked in the beginning of the chapter, we must use "objective" methods of analysis, not merely the observer's judgement of the regularity of the time series. There are several ways to identify and to characterize a dynamical regime. We will explain two frequently used methods: first, the Fourier transform in this chapter[28]; then the Poincaré section, the subject of Chapter IV. In Chapter VI we will see a method developed recently for studying an attractor directly.

III.2 Discrete Fourier transform

III.2.1 SIGNAL DISCRETIZATION

The rapid development of computational methods has meant that a signal $x(t)$ — a continuous function of time — is very often measured by sampling and discretizing. Therefore, an experiment generally provides a discrete sequence[29] of real numbers $x_j (j \in Z)$ regularly spaced at time intervals of Δt (see fig. III.1).

In practice this sequence of numbers is necessarily finite, containing n values for a total length of time $t_{max} = n \cdot \Delta t$. The choice of the two quantities n and Δt is determined by practical considerations, such as the acceptable duration of the experiment, and the capacity for storage and processing of the measurements. Fourier transforms can, of course, be applied to continuous functions, as well as to discrete sequences, with integrals replacing summations. We will, however, limit our presentation to discrete sequences, with the understanding that the signal $x(t)$ is, mathematically, an integrable and square integrable function.

III.2.2 DEFINITION OF THE DISCRETE FOURIER TRANSFORM

We define the Fourier transform of a discrete time series x_j to be the operation creating a corresponding discrete series $\hat{x}_k$ such that:

$$\hat{x}_k = \frac{1}{\sqrt{n}} \sum_{j=1}^{n} x_j \exp\left(-i \frac{2\pi jk}{n}\right) \qquad (8)$$

$k = 1, ..., n$

$i = \sqrt{-1}$ pure imaginary.

---

28. The Fourier transform was developed in the nineteenth century by the mathematician Jean Baptiste Fourier during the course of his work on the heat equation. Other kinds of transforms — Rademacher and Hadamard transforms — with similar properties, can also be utilized. Rather than describing them, or the numerous mathematical results on the Fourier transform, we will present here only the practical results most relevant to our subject.

29. The disadvantages of discretization must be compared to the considerable advantage of being able to process the signal entirely numerically. With modern electronic technology, it is possible to sample with a time interval Δt of less than 10^{-6} second and to digitize the signal in real time.

III.2 DISCRETE FOURIER TRANSFORM

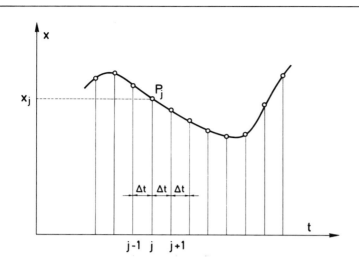

Figure III.1 Discretization of a continuous function.
The continuous function — for example, an experimental signal — is represented by the solid line. The discrete values x_j are taken at the instants $...j-1, j, j+1...$ equally spaced at time intervals of Δt.
The continuous function is hereafter replaced by the series of points P_j.

For convenience we have taken Δt as the unit of time, so that incrementing j corresponds to evolution over a time interval Δt. We can consider the transformation defined by (8) as a kind of rotation, mapping the vector $(x_1, x_2, ..., x_j, ..., x_n)$ into the vector $(\hat{x}_1, \hat{x}_2, ..., \hat{x}_k, ..., \hat{x}_n)$. It is not a rotation in the usual sense, since the Euclidean length $\Sigma_j x_j^2$ is not conserved. What is conserved by (8) is the Hermitian length[30], as expressed by the Parseval-Plancherel equation:

$$\sum_{j=1}^{n} |x_j|^2 = \sum_{k=1}^{n} |\hat{x}_k|^2.$$

We point out that $\hat{x}$ is a function of the variable conjugate to the time, which is the frequency f. The frequency also varies by the discrete interval Δf:

$$\hat{x}_k = \hat{x}(k \cdot \Delta f) \quad ; \quad \Delta f = \frac{1}{t_{max}}.$$

If the series x_j is real-valued, as we have assumed, knowledge of the complex components $\hat{x}_k$ implies that of $2n$ real quantities (real and imaginary parts of each $\hat{x}_k$).

---

30. The hermitian length is the quantity $\Sigma x x^* = \Sigma |x|^2$. The notation x^* designates the complex conjugate of x.

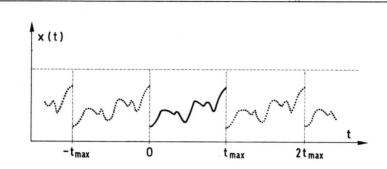

Figure III.2 Sampling of a function $x(t)$ over a finite time. The measured interval $[0, t_{max}]$ is equivalent to one period of a function defined for all time $t \in]-\infty, \infty[$ and with period t_{max}.

Clearly, there is redundancy in this data, which is expressed by the easily established relation:

$$\hat{x}_k = \hat{x}^*_{n-k}.$$

By the inverse transformation we can go from the vector $\hat{x}$ back to the initial vector:

$$x_j = \frac{1}{\sqrt{n}} \sum_{k=1}^{n} \hat{x}_k \exp\left(i\frac{2\pi k j}{n}\right). \tag{8'}$$

Note that this relation can be used to define components x_j not only for $j \in [1, n]$ but for all integer values of j. This function x is periodic in n (in fact $n \cdot \Delta t$) since:

$$x_{j+n} = x_j.$$

This explains why any signal $x(t)$ can be expressed as a sum of periodic functions as in (8'). The finite interval $[0, t_{max}]$ can simply be considered as representing one period of a function, which, while defined for all t, is periodic in t_{max}, as shown in Figure III.2.

III.2.3 THE WIENER-KHINTCHIN THEOREM

We define the *autocorrelation function* of the signal x_j by:

$$\psi_m = \frac{1}{n} \sum_{j=1}^{n} x_j x_{j+m}.$$

The unit of time is still Δt, so that:

$$\psi_m = \psi(m \cdot \Delta t).$$

III.2 DISCRETE FOURIER TRANSFORM

Physically, this function represents the average of the product of the signal values at a given time and at a time $m \cdot \Delta t$ later. We can therefore deduce from ψ_m whether, and for how long, the instantaneous value of the signal depends on its previous values: hence its name. Or else, we can say that it is a measure of the degree of resemblance of the signal with itself as time passes [31].

Since the series x_j is periodic in n, the autocorrelation function necessarily has the same property:

$$\psi_m = \psi_{m+n}.$$

By applying the inverse Fourier transform, we get:

$$\psi_m = \frac{1}{n^2} \sum_{j=1}^{n} \sum_{k,k'=1}^{n} \hat{x}_k \hat{x}_{k'} \exp\left[i \frac{2\pi}{n} (jk + (j+m)k')\right].$$

Using the property $\hat{x}_{k'}^* = \hat{x}_{n-k'}$ and summing over the indices j and k', we establish that:

$$\psi_m = \frac{1}{n} \sum_{k=1}^{n} |\hat{x}_k|^2 \cos\left(\frac{2\pi mk}{n}\right).$$

This signifies that, up to a factor of proportionality, the auto-correlation function is merely the Fourier transform of $|\hat{x}_k|^2$.

Let us now find the inverse relation between $|x_k|^2$ and ψ_m. Let S_k be the function defined by:

$$S_k = \sum_{m=1}^{n} \psi_m \cos\left(\frac{2\pi mk}{n}\right).$$

Substituting into this definition the expression for ψ_m calculated above, we get:

$$S_k = \sum_{\ell=1}^{n} |\hat{x}_\ell|^2 \frac{1}{n} \sum_{m=1}^{n} \cos\left(\frac{2\pi mk}{n}\right) \cos\left(\frac{2\pi m\ell}{n}\right).$$

Using the equality:

$$\cos\left(\frac{2\pi mp}{n}\right) = \frac{1}{2}\left[\exp\left(i\frac{2\pi mp}{n}\right) + \exp\left(-i\frac{2\pi mp}{n}\right)\right]$$

the sum over m becomes a geometric series with n terms. From which:

$$\frac{1}{n} \sum_{m=1}^{n} \cos\left(\frac{2\pi mk}{n}\right) \cos\left(\frac{2\pi m\ell}{n}\right) = \frac{1}{4}[\delta^{(n)}_{k+\ell} + \delta^{(n)}_{k-\ell} + \delta^{(n)}_{-k-\ell} + \delta^{(n)}_{-k+\ell}]$$

---

31. As long as the autocorrelation function is appreciable, the signal remains relatively predictable; knowledge of the signal for a sufficiently long time allows us to calculate with sufficient confidence its value at a later time by extrapolation. On the other hand, when ψ_m approaches zero, the temporal similarity of the signal with itself disappears, and prediction of its evolution becomes impossible.

where $\delta_j^{(n)}$ is 1 if j equals 0 modulo n, and 0 otherwise. Joining this result to the symmetry relation $|\hat{x}_{n-\ell}|^2 = |\hat{x}_\ell|^2$, we find:

$$S_k = |\hat{x}_k|^2 = \sum_{m=1}^{n} \psi_m \cos\left(\frac{2\pi mk}{n}\right)$$

which is the sought-after inversion relation. This result constitutes one form of the Wiener-Khintchin [32] theorem, which says that the function $|\hat{x}_k|^2$ is proportional to the Fourier transform of the autocorrelation function ψ_m of the signal. The graph representing $|\hat{x}_k|^2$ as a function of the frequency f ($f = k \cdot \Delta f$) is called the *power spectrum*.

The power spectrum of a real function has the property:

$$|\hat{x}_k|^2 = |\hat{x}_{n-k}|^2$$

which comes from the equality $\hat{x}_k = \hat{x}_{n-k}^*$. This expresses the obvious fact that information about the phase of a component $\hat{x}_k$ is lost [33] when we consider $|\hat{x}_k|^2$.

III.2.4 THE POWER SPECTRUM

Let us examine in closer detail the characteristics of the power spectrum and the kind of information it conveys about the signal $x(t)$. We begin with a time series of n equidistant points, separated by an interval of Δt (fig.III.3 a). By applying relation (8) we calculate the $\hat{x}_k$, and then the $|\hat{x}_k|^2$, giving rise to a new function on n discrete points of the abscissa $k/n \Delta t$ ($k = 1, ..., n$). The abscissa has the dimension of inverse time — that is, of frequency, from which comes the name of spectrum given to the graph of $|\hat{x}_k|^2$. The interpretation of the ordinate axis depends on the nature of the signal measured.

It is conventional to call this a power spectrum, by analogy to the case when the ordinate represents power, i.e. energy per unit time. Consider a signal $x(t)$ originating in the reception of waves (electromagnetic, sound, etc.) by an antenna. If the detector is linear and does not introduce distortion in the frequency band considered, then the quantity it measures is proportional to the variation of the electric field or of the pressure in the vicinity of the antenna. The theory of waves shows that the power carried by a wave is proportional to the square of its amplitude (just as the energy of a

---

32. This theorem was proved by Norbert Wiener after work by G. Taylor, who analogically measured a quantity similar to $|\hat{x}_k|^2$ from a turbulent signal (variations in the resistance of a hot-wire anemometer).
33. We note incidentally that loss of phase information is commonly observed in physics in, for example, electron diffusion or in collisions between elementary particles at high energy. The information contained in the phase is associated with the mean reversibility of the signal. In functions such as $|\hat{x}_k|^2$ or ψ_m, which are unchanged when the order of the indices of x_j is reversed, this information cannot be preserved. On the other hand, this would not be true of a correlation function ϕ_m defined by:

$$\phi_m = \frac{1}{n} \sum_{j=1}^{n} (x_j^2 x_{j+m} - x_{j+m}^2 x_j)$$

which changes sign when the time series is treated in the opposite order.

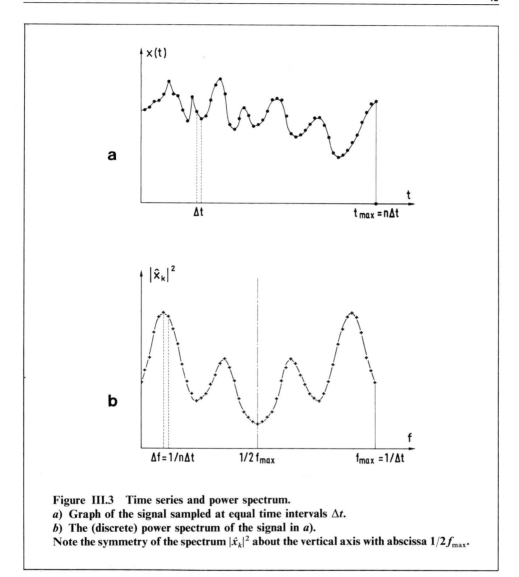

Figure III.3 Time series and power spectrum.
a) **Graph of the signal sampled at equal time intervals Δt.**
b) **The (discrete) power spectrum of the signal in *a*).**
Note the symmetry of the spectrum $|\hat{x}_k|^2$ about the vertical axis with abscissa $1/2 f_{max}$.

harmonic oscillator is proportional to the square of its amplitude) averaged over time. By the Parseval-Plancherel formula, we can replace the average over time by an average over frequency. This justifies the appellation "power spectrum", which we keep, for convenience, in all cases [34].

[34]. In fact, the relationship between power — in the sense of energy received per unit time — and the amplitude of the spectrum of the signal output by amplifiers or, more generally, by a sequence of detectors, can be rather complicated. This is all the more so if the detectors perform a nonlinear transformation of the signal.

The power spectrum in Figure III.3 b is that of the signal in Figure III.3 a. The step size $f = 1/(n \cdot \Delta t)$ along the abscissa corresponds to the spectral resolution. To improve the resolution, the product $n \cdot \Delta t$ must be increased. The highest frequency of the spectrum is $f_{max} = 1/\Delta t$. To enlarge the frequency domain explored, Δt must be reduced. In any case, because of the relation:

$$|\hat{x}_k|^2 = |\hat{x}_{n-k}|^2$$

the spectrum is symmetric[35] with respect to the vertical line $f = 1/2 f_{max}$. Consequently, the effective useful frequency domain — that containing non-redundant information — extends only from 0 to $1/(2\Delta t)$.

To be completely rigorous, the spectrum is composed of a sequence of "steps", each of width Δf (fig. III.4). In practice, one often merely draws a vertical line segment of height $|\hat{x}_k|^2$ at $f = k \cdot \Delta f$. Most of the time, one in fact joins the successive points $(k \cdot \Delta f, |\hat{x}_k|^2)$ by line segments (fig. III.3 b). It is not uncommon for the amplitude of $|\hat{x}_k|^2$ to vary over several orders of magnitude, in which case a logarithmic scale is used for the ordinate. The unit of measure along this axis is then the decibel, which is equal to 10 times the decimal logarithm. A difference of α decibels corresponds to an amplitude ratio of $10^{0.1\alpha}$.

III.3 Different kinds of Fourier spectra

III.3.1 PERIODIC SIGNAL

The appearance of the power spectrum clearly depends on the way in which the signal $x(t)$ evolves over time. The interest of the Fourier spectrum is, in fact, that it reveals properties of the evolution which would otherwise remain undetected.

Let us first consider the simple case in which $x(t)$ is a periodic signal of period T, that is:

$$x(t) = x(t + T) = x\left(t + \frac{2\pi}{\omega}\right)$$

such as the signal engendered by a Van der Pol oscillator in the limit cycle regime.

An extreme situation is that in which the period is exactly equal to the duration of measurement:

$$T = t_{max} = n \cdot \Delta t.$$

Then, according to (8), the Fourier components are located exactly at the frequencies:

$$\frac{1}{T}, \frac{2}{T}, \frac{3}{T}, ..., \frac{n}{T}.$$

---

35. Recall that this property follows from the fact that the series x_j is assumed to be real-valued.

III.3 DIFFERENT KINDS OF FOURIER SPECTRA

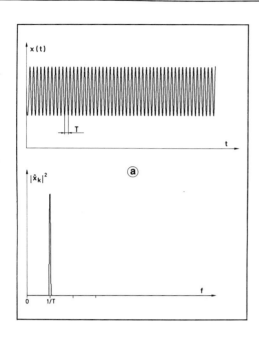

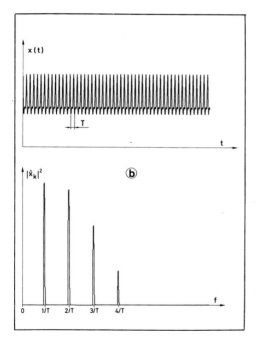

Figure III.4 Periodic functions and their Fourier spectra.
a) Pure sinusoidal function.
b) Periodic function containing harmonics. Only the first four harmonics are shown. (Unless otherwise stated, a logarithmic scale is used for the power in this and subsequent Fourier spectra).

In the simplest case, if $x(t)$ is a circular function — sine or cosine — the spectrum consists of only one nonzero component, with abscissa $1/T$ (or $k = 1$, fig. III.4 a). For a signal of a different form, such as a relaxation oscillation, the amplitude of the harmonics of the frequency $(2/T, 3/T, ...)$ is no longer zero. This is why the presence of harmonics in the spectrum is indicative of the non-sinusoïdal nature of the evolution (fig. III.4 b).

The preceding results can be easily generalized to the case where the duration of measurement t_{max} is an integer multiple of the signal period T:

$$t_{max} = pT; \; p \text{ a positive integer} > 1.$$

In this case, the nonzero components of the spectrum are still at $1/T, 2/T, ...$, but the frequency resolution is p times better. It follows that all the components $|\hat{x}_k|^2$ for which k is not an integer multiple of p are zero.

In fact, the situation we have just described is rather academic, since, in practice, the period of the signal is either unknown or known imprecisely. The ratio t_{max}/T is therefore generally not an integer. What are the consequences on the shape of the spectrum? To answer this question, let us calculate the Fourier transform for the simple case of a circular function:

$$x(t) = \exp\left(i\frac{2\pi t}{T}\right)$$

$$x_j = \exp\left(i\frac{2\pi j \, \Delta t}{T}\right)$$

$$\hat{x}_k = \frac{1}{\sqrt{n}} \sum_{j=1}^{n} \exp\left(i\frac{2\pi j \, \Delta t}{T}\right) \exp\left(-i\frac{2\pi j k}{n}\right).$$

Setting: $\phi_k = \Delta t/T - k/n$

$$\hat{x}_k = \frac{1}{\sqrt{n}} \sum_{j=1}^{n} \exp(i2\pi\phi_k j) = \frac{1}{\sqrt{n}} \exp(i2\pi\phi_k) \frac{\exp(i2n\pi\phi_k) - 1}{\exp(i2\pi\phi_k) - 1}$$

$$|\hat{x}_k \hat{x}_k^*| = |\hat{x}_k|^2 = \frac{1}{n} \frac{\sin^2(n\pi\phi_k)}{\sin^2(\pi\phi_k)}.$$

Note that $n\phi_k = (t_{max}/T) - k$ is the difference between the integer k and the (noninteger) ratio of measurement duration to signal period. Therefore $n\phi_k$ is finite for all k.

Under the additional hypothesis of large n — thus small $\pi\phi_k$ — we have, asymptotically:

$$|\hat{x}_k|^2 \simeq n \frac{\sin^2(n\pi\phi_k)}{(n\pi\phi_k)^2}.$$

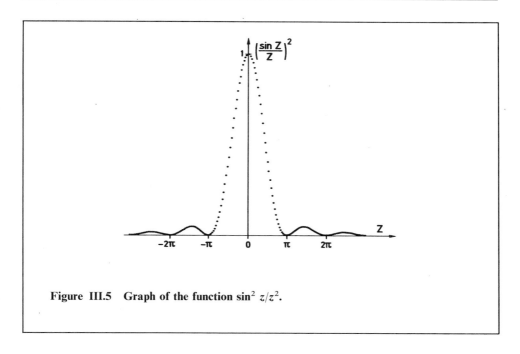

Figure III.5 Graph of the function $\sin^2 z/z^2$.

Therefore the behavior of $|\hat{x}_k|^2$ is like that of the function $\sin^2 z/z^2$, whose graph appears in Figure III.5.

This function has a maximum amplitude of one at $z = 0$, and a series of secondary maxima at $\pm(\ell + 1/2)\pi$ (ℓ a positive integer), whose amplitude decreases like $1/z^2$. It follows that $|\hat{x}_k|^2$ is maximum when k is equal to k_0, the integer closest to $n\,\Delta t/T$; it is there that the discrete variable ϕ_k is close to zero. Moreover, since the integer values of k close to k_0 correspond to the lateral arches of the function $\sin^2 z/z^2$, the amplitude of $|\hat{x}_k|^2$ for $k = k_0 \pm \ell$ is not zero, even though it decreases rapidly with ℓ. The end result is that in addition to the single peak obtained when there is resonance between the signal period and the duration of measurement, there are also secondary peaks, called sidelobes, centered around the frequency $1/T$. The linewidth is of the order of several frequency units Δf, and is widest when the ratio $n\,\Delta t/T$ is a half-integer. The diagram in Figure III.6 illustrates this possibility, contrasting it with the case of resonance.

In summary, the spectrum of a periodic signal of period T is made up of a peak at the frequency $1/T$, its sidelobes, and possibly a certain number of other peaks (and their sidelobes) that are harmonics of the fundamental frequency.

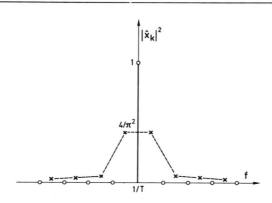

Figure III.6 Form of one spectral component under two different sampling conditions.
— solid line (circles) satisfies the resonance condition $n\,\Delta t/T$ integer,
— dashed line (crosses) satisfies $n\,\Delta t/T$ half-integer.

III.3.2 QUASIPERIODIC SIGNAL

A function y of r independent variables $t_1, t_2, ..., t_r$ is said to be periodic, of period 2π in each of its arguments, when increasing one of these variables by 2π does not change its value:

$$y(t_1, t_2, ..., t_j, ..., t_r) = y(t_1, t_2, ..., t_j + 2\pi, ..., t_r) \qquad j = 1,$$

Such a function is called quasiperiodic in time if its r variables are all proportional to the time t:

$$t_j = \omega_j t \qquad j = 1, ..., r.$$

A quasiperiodic function has r fundamental frequencies:

$$f_j = \frac{\omega_j}{2\pi} \qquad j = 1, ..., r$$

and therefore r periods: $T_j = 1/f_j = 2\pi/\omega_j$.

Before arriving at the form of the Fourier spectrum of such a function, recall that the formal definition above must be related to phenomena with multiple periodicity. We cite one example among many: the astronomical position of a point on the surface of the earth is described by a quasiperiodic law, since it results from the rotation of the earth about its axis ($T_1 = 24$ h), the rotation of the earth around the sun ($T_2 = 365.242$ days), and finally the precession of the earth's axis of rotation (T_3

III.3 DIFFERENT KINDS OF FOURIER SPECTRA

= 25 800 years, sometimes called a platonic year)[36]. We mention also that the phase trajectory associated with a quasiperiodic function is defined on a *torus*[37] of dimension r, written T^r.

As one might guess, the Fourier spectrum of a function which is quasiperiodic in time generally has a relatively complex appearance. One exception exists, nevertheless: when the quasiperiodic signal $x(\omega_1 t, ..., \omega_r t)$ is the sum of periodic functions:

$$x(\omega_1 t, ..., \omega_r t) = \sum_{i=1}^{r} x_i(\omega_i t).$$

Then, using the linearity of Equation (8), the power spectrum is the sum of the r spectra of each of the functions $x_i(\omega_i t)$. It is therefore composed of a set of peaks, located at the fundamental frequencies $f_1, f_2, ..., f_r$, and of their harmonics:

$$m_1 f_1, m_2 f_2, ..., m_r f_r$$

($m_1, m_2, ..., m_r$ positive integers). But there is, in general, no reason why the signal should be the sum of r independent periodic terms. If the quasiperiodic function includes a term like the product of circular functions (for example: $\sin(\omega_i t) \sin(\omega_j t)$) then the Fourier spectrum contains fundamental frequencies $|f_i - f_j|$ and $|f_i + f_j|$ and their harmonics, since:

$$\sin(\omega_i t) \sin(\omega_j t) = \frac{1}{2} \cos(|f_i - f_j| 2\pi t) - \frac{1}{2} \cos(|f_i + f_j| 2\pi t).$$

Generalizing this result, we see that the Fourier spectrum of a quasiperiodic function $x(t)$, which depends nonlinearly on periodic functions of the variables $\omega_i t$, contains components at all frequencies of the form:

$$|m_1 f_1 + m_2 f_2 + ... + m_r f_r|$$

where m_i are arbitrary integers.

To further the discussion, we limit ourselves for the moment to the biperiodic case ($r = 2$) and set aside the fact that the resolution Δt of the power spectrum is finite. Each nonzero component of the spectrum of the signal $x(\omega_1 t, \omega_2 t)$ is a peak with abscissa $|m_1 f_1 + m_2 f_2|$ which we may abbreviate by (m_1, m_2). The ratio f_1/f_2 can be either a rational or an irrational number. In the latter case it is shown in number theory that sums of the kind $|m_1 f_1 + m_2 f_2|$ constitute a dense set over the positive reals. In other words, any real positive number is as close as one wishes to such a sum. As a consequence, the spectrum of the signal is itself dense. But this does not mean that it is represented by a continuous function. Indeed, two peaks that are close together on the

---

36. We neglect the very small perturbations to these fundamental motions. Given the order of magnitude difference in the time scales of the motions, their interaction remains weak.

37. In the simplest case $r = 2$, one can show under fairly general conditions that the trajectory described parametrically by the three functions of time $x(\omega_1 t, \omega_2 t)$, $(d/dt) x(\omega_1 t, \omega_2 t)$, and $(d^2/dt^2) x(\omega_1 t, \omega_2 t)$ is indeed located on a torus in $\mathbb{R}^3$, that is, a torus in the usual sense of the word.

frequency axis have no *a priori* reason to have amplitudes that are close. Most of the time, one therefore observes a very limited number of frequencies for which the lines have a significant amplitude, because the high-order lines — that is, those corresponding to values of m_1 and m_2 greater than ten or so — have amplitudes too low to be detected[38]. In practice, one identifies a quasiperiodic spectrum by looking for the two fundamental frequencies f_1 and f_2 which allow the frequencies of the high amplitude lines to be represented by simple combinations $|m_1 f_1 + m_2 f_2|$ with m_1 and m_2 small: $0, \pm 1, \pm 2 ...$ (see fig. III.7).

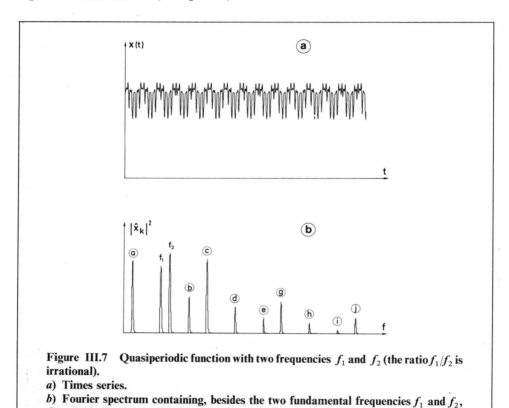

Figure III.7 Quasiperiodic function with two frequencies f_1 and f_2 (the ratio f_1/f_2 is irrational).
a) Times series.
b) Fourier spectrum containing, besides the two fundamental frequencies f_1 and f_2, the principal peaks of frequency $f = m_1 f_1 + m_2 f_2$.

$a \longrightarrow f_2 - f_1$ $e \longrightarrow 5f_1 - f_2$ $j \longrightarrow 5f_1 + f_2$
$b \longrightarrow 3f_1 - f_2$ $g \longrightarrow 3f_1 + f_2$
$c \longrightarrow f_1 + f_2$ $h \longrightarrow 5f_1$
$d \longrightarrow 3f_1$ $i \longrightarrow 7f_1 - f_2$

38. Very generally (but not always, from a strict mathematical point of view), the amplitudes of high-order peaks decrease like $\exp(-a_1|m_1| - a_2|m_2|)$ where a_1 and a_2 are positive. For large $|m_1|$ and $|m_2|$, these amplitudes therefore become negligible. They are practically undetectable as soon as the indices m_1 and m_2 significantly exceed a_1^{-1} and a_2^{-1} in absolute value.

III.3 DIFFERENT KINDS OF FOURIER SPECTRA

Under the hypothesis of f_1/f_2 rational, the Fourier spectrum is not dense. It follows that the spectrum is definitely not represented by a continuous function. Since:

$$f_1/f_2 = n_1/n_2 \ (n_1, n_2 \text{ integers}).$$

the quasiperiodic signal is in fact periodic with period $T = n_1 T_1 = n_2 T_2$. Indeed, according to the definition, one has:

$$x(\omega_1 t, \omega_2 t) = x(\omega_1 t + 2\pi n_1, \omega_2 t + 2\pi n_2)$$

$$x(\omega_1 t, \omega_2 t) = x\left(\omega_1\left(t + \frac{n_1}{f_1}\right), \omega_2\left(t + \frac{n_2}{f_2}\right)\right).$$

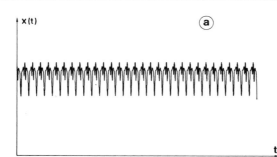

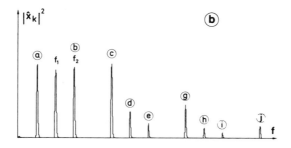

Figure III.8 Quasiperiodic function (f_1/f_2 rational).
a) Time series.
b) Fourier spectrum. The function is almost identical to that shown in Figure III.7, but we have changed f_1 so that $f_1/f_2 = 2/3$. Note that under these conditions, all the peaks are harmonics of the frequency $f = f_2 - f_1 = 1/3 f_2$

$a \longrightarrow f_2 - f_1$ $d \longrightarrow 6f$ $h \longrightarrow 10f$
$b \longrightarrow 3f(= f_2)$ $e \longrightarrow 7f$ $i \longrightarrow 11f$
$c = 5f$ $g \longrightarrow 9f$ $j \longrightarrow 12f$

One says that there is *frequency locking* of f_1 with f_2. All the lines of the Fourier spectrum are harmonics of the lowest frequency:

$$f_0 = \frac{1}{T} = \frac{f_1}{n_1} = \frac{f_2}{n_2}.$$

Figure III.8 illustrates this possibility. Consecutive lines of the spectrum are always separated by the same distance of $1/T$.

III.3.3 APERIODIC SIGNAL

When the signal $x(t)$ is neither periodic nor quasiperiodic, it is called *aperiodic* (or sometimes nonperiodic). We will meet with several examples of this situation later in the book, and we will see that the Fourier spectrum is then continuous[39]. The real difficulty is that a Fourier spectrum which looks continuous cannot be automatically attributed to an aperiodic signal, because this is also the appearance of the spectrum of a quasiperiodic signal with a very high number of frequencies (in the infinite limit).

If we suppose that the signal is truly aperiodic, we must still decide, for methodological reasons, whether the number of degrees of freedom is small (say, less than ten) or, on the contrary, very large. In the first case, we have the capability of developing a completely deterministic description of the system. By contrast, in the second case, only a probabilistic approach can be achieved with the current state of knowledge. This leads us to introduce the notion of chance, but without challenging the (hidden) determinism of the phenomena. Although we are now venturing beyond the subject of this book, it is appropriate to give an idea of what we mean by the distinction between "deterministic" and "random".

The extreme case of a random signal is what is called "white noise". This name is evocative, alluding to noise — in the sense of sound without any harmonic structure — and to the absence of color in light. One can thus "see" white noise — a white light — or "hear" it — it would be, typically, the sound of a waterfall. White noise is produced by a multitude of independent agents: the atoms of a heated filament emitting white light, or droplets in a waterfall falling on a boulder.

One can consider the signal $x(t)$ resulting from white noise as being "new"[40] at each instant, at least in a first approximation. Let us show that the corresponding

---

39. The spectrum of an aperiodic signal approaches a continuous function only in the mean. In fact, if we calculate the discrete spectrum of a signal of duration t_{max} by the sampling method described, we find that the amplitudes of neighboring $|\hat{x}_k|^2$ vary a great deal, no matter how large we take t_{max}. The difference between $|\hat{x}_k|^2$ and $|\hat{x}_{k\pm1}|^2$ is typically comparable to these quantities themselves. To obtain a continuous spectrum (or at least a reasonable approximation to it) starting from the sequence $|\hat{x}_k|^2$, one must therefore either average over many spectra, each taken over a time window of length t_{max}/N (N large), or else, equivalently, average the $|\hat{x}_k|^2$ locally over a certain number of consecutive values of k.

40. To be completely rigorous, the noise is white if, in addition to being independent, the individual emitters each produce a signal which is an impulse of infinitesimal duration. If each of these individual signals has a spectral structure, the light or noise becomes *colored*.

III.3 DIFFERENT KINDS OF FOURIER SPECTRA

power spectrum is "flat", — in other words, that the amplitude is independent of the frequency — and so possesses the essential characteristic of being devoid of all harmonic structure.

Let $x(t)$ be a signal of zero mean and of variance $\overline{x^2}$. The absence of correlation between signal values at different instants is expressed by the condition[41]:

$$\langle x_j x_{j+m} \rangle = \overline{x^2} \delta_m = \langle \psi_m \rangle$$

where δ_m is the Kronecker delta function ($\delta_0 = 1$ and $\delta_m = 0$ for $m \neq 0$). Using the Wiener-Khintchin theorem, we find:

$$\langle |\hat{x}_k|^2 \rangle = \sum_{m=1}^{n} \langle \psi_m \rangle \cos\left(\frac{2\pi mk}{n}\right) = \overline{x^2}$$

showing that the average amplitude of the signal's power spectrum is independent of k. In other words, the average amplitude is independent of the frequency for white noise.

The situation we have just described is quite common in practice; the high frequency structure of numerous natural phenomena is that of white noise. Indeed, noise due to molecular agitation involves the action of an almost infinite number of independent agents: molecules, conductance electrons, elastic vibrations of the atoms in a network, etc. The theory of thermodynamic noise is one of the great success stories of physics in the first half of twentieth century. As a concrete illustration, we consider the classical example of Nyquist noise, named after the person who explained it. Let an RC circuit (R = resistance, C = capacitance) be at thermal equilibrium (in the absence of an externally applied voltage). A sensitive voltmeter measures the voltage fluctuation $V(t)$ across the capacitor or the resistor. The energy stored in the capacitor of capacitance C, as a function of the voltage across it, is equal to $CV^2/2$. This energy is due to thermal fluctuations, and by Ehrenfest's equipartition theorem, its average value is equal to $kT/2$, where k is the Boltzmann constant ($k = R_J/N$, R_J = Joule's constant, N = Avogadro's number) and T is the absolute temperature of the system. Then:

$$\langle V^2 \rangle = \frac{kT}{C}$$

the average being taken either over time, or over the thermodynamic ensemble. But this is an average of instantaneous values, and we would like to know the frequency structure of this noise. We see this as follows. If $V(t)$ is the voltage across the capacitor at an instant t, the voltage measured at a later time $t + \tau$ will come from two factors.

---

41. In the expressions which follow, the averages are *ensemble averages*. We consider implicitly autocorrelation functions ψ_m and Fourier transforms $|\hat{x}_k|^2$ obtained from series of n values of x, giving an individual realization of these quantities. We then take their arithmetic average, denoted by $\langle \ \rangle$. Note that for a given realization, the mean square distance between $|\hat{x}_k|^2$ and its average value is of the same order of magnitude as the average. This justifies, as we have already said, the fact that spectra of turbulent signals must be either smoothed, or else averaged over an ensemble of realizations, to appear as continuous functions of the frequency.

On the one hand, the capacitor will discharge to the resistor R, giving a voltage $V(t)\, e^{-\tau/RC}$, where (RC) is the time constant of the circuit. On the other hand, new electric charge is deposited on the capacitor plates by thermal fluctuations. These fluctuations are uncorrelated with the initial charge of the capacitor, and thus also uncorrelated with $V(t)$. Since the RC circuit is linear, we have:

$$\langle V(t)V(t+\tau) \rangle = \langle V(t)^2 \rangle \exp^{(-|\tau|/RC)}$$

where we have used the absolute value $|\tau|$ in case τ is negative. The spectrum of the noise is given by the Wiener-Khintchin theorem for continuous variables:

$$S(\omega) = \frac{1}{\pi} \int_0^\infty d\tau \, \langle V(t)V(t+\tau) \rangle \cos(\omega\tau)$$

$$S(\omega) = \frac{2RkT}{\pi(1+\omega^2 R^2 C^2)}.$$

With this formalism, the Parseval-Plancherel relation is written:

$$\int_0^{+\infty} S(\omega)\, d\omega = \langle V^2(t) \rangle = \frac{kT}{C}$$

This formula gives the spectral distribution of the voltage fluctuations across the capacitor of the RC circuit. In the limit of zero capacitance, it suffices to set $C = 0$ in the expression for $S(\omega)$ which gives the classic expression for Nyquist noise:

$$S(\omega) = \frac{2RkT}{\pi}.$$

The thermal noise across a resistor is therefore a white noise, since $S(\omega)$ is independent of ω. The noise amplitude is proportional to the absolute temperature and the resistance. Note that in this limit $(C = 0)$, the average V^2 diverges, since $\langle V^2 \rangle = kT/C$. This divergence does not occur in practice, since at very high frequencies, one always finds parasitic capacitance in a circuit, so that C is never exactly zero.

The noise produced by a simple resistor is white noise, whose spectral power is independent of the frequency. By contrast, the thermal fluctuations of the voltage of an RC circuit constitute a *colored noise:* the spectrum is still a continuous function of frequency, but no longer constant.

Having completed this digression on random signals, we return to our main goal, which is to study the signals produced by deterministic dynamics. As we have already indicated, an aperiodic signal has a power spectrum of continuous appearance, like the example of Figure III.9. Therefore, this method of analysis does not distinguish between an aperiodic and a random signal. This limits the application of the Fourier transform, leading us to turn to other methods, notably that of Poincaré sections.

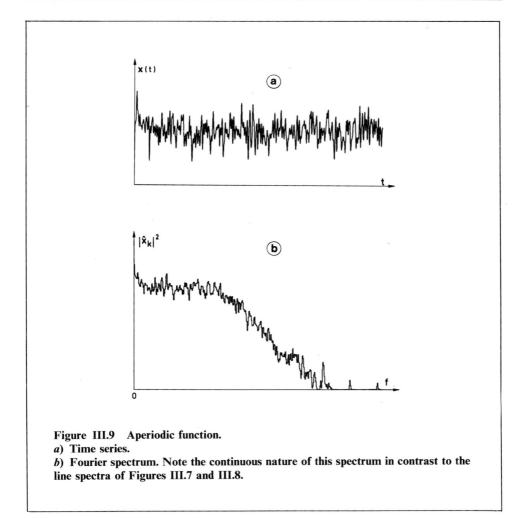

Figure III.9 Aperiodic function.
a) Time series.
b) Fourier spectrum. Note the continuous nature of this spectrum in contrast to the line spectra of Figures III.7 and III.8.

III.4 Fast Fourier Transform (FFT)

We have seen that there exists a close relationship between the periodic, quasiperiodic, or aperiodic nature of a signal $x(t)$, and the form of its power spectrum. To identify the nature of a dynamical regime, it remains to calculate $|\hat{x}_k|^2$ explicitly. In theory, there is no difficulty in using the formula given by Equation (8). But in practice, the extent of the task leaps to the eye as soon as n takes on appreciable values. Yet it is naturally advantageous to choose n large and Δt small to represent the signal $x(t)$ as faithfully as possible. With $n = 10^3$, which is still relatively small, we must already calculate 1 000 sums, each of which contains 1 000 terms! More generally, we see that calculation of the n components of a spectrum requires a number of operations (additions and multiplications) which is on the order of n^2. This rapid increase is a severe limiting factor which was, for a long time, a considerable deterrent to the use of Fourier transforms. Even the advent of the first computers did not greatly improve the situation.

However, when n is a power of two, an algorithm, called the FFT[42], permits calculation of the spectrum with many fewer operations, on the order of $n \log_2 n$. Then, for $n = 2^{10} = 1\,024$, the gain is already a factor of 100; it attains 7 000 for $n = 2^{18}$. The difference in cost which results is considerable. It is in fact thanks to this fantastic economy of scale that analysis by Fourier transform is a routine procedure in research laboratories, enabling great progress to be made in many domains[43], in addition to the study of dynamical systems. We mention, among others, different kinds of spectroscopy: IR, Raman, X, magnetic resonance, photon beating, etc. Digital Fourier analyzers are commercially available, which, combining the FFT algorithm with other tricks, provide the power spectrum of a signal containing several thousand points in a few seconds.

---

42. FFT abbreviates *Fast Fourier Transform*. This algorithm was developed in 1965 by Cooley and Tukey. They later discovered that the same result had been obtained as early as 1942 by other researchers whose article had received no attention at the time.

43. Signal processing methods (notably filtering and increase of the signal-to-noise ratio) have undergone considerable technical improvement to which implementation of the FFT has greatly contributed.

CHAPTER IV

Poincaré sections

IV.1 Definition of a flow

We now adopt a rather global point of view beginning with the remark that, as seen in the three previous chapters, the evolution of numerous systems is described by a set of n first order ordinary differential equations:

$$\frac{d}{dt}\vec{X}(t) = F(\vec{X}, t) \qquad (9)$$

$\vec{X}$ is a vector in $\mathbb{R}^n$ (the phase space) and F is a vector field over this space. The laws controlling the behavior of different oscillators are, we have seen, precisely of this form[44].

A system of differential equations such as (9) is called a *flow* in $\mathbb{R}^n$. When F does not depend explicitly on time, but only on $\vec{X}$:

$$F = F(\vec{X}(t))$$

the flow is said to be *autonomous*. Equation (5) is a typical example of an autonomous flow. On the other hand, when F does depend explicitly on time, the flow is *nonautonomous*. This is the case of the Mathieu equation (7).

Solutions to (9) have an analytic expression only in particular well-defined situations where the flow is integrable. Most of the time the flow is not integrable and we must study each solution by considering its trajectory in phase space. As this often proves difficult to carry out, we simplify the task by using a method developed by Henri Poincaré.

---

44. This is with the understanding that there always exists an appropriate change of variables transforming an nth order differential equation into a system of n first order differential equations. We have already used this property in constructing the phase portrait for the Van der Pol oscillator.

IV.2 Poincaré sections

IV.2.1 CONSTRUCTION AND PROPERTIES

In theory, there is no restriction on the dimension n of the phase space. However we will limit ourselves in this chapter to the three-dimensional case[45] for didactic and practical reasons. In addition, only the asymptotic behavior, as t tends to infinity, will be considered; the study of transient states, which is less generally applicable, will be set aside.

Rather than directly studying the solution to (9) in $\mathbb{R}^3$, it can be fruitful to observe the points of intersection of the trajectory with a plane. This construction is indicated schematically in Figure IV.1. There we have taken, for illustrative purposes, the plane S defined by $x_3 = $ constant, and we have placed the points of intersection corresponding to a given direction of the evolution ($\dot{x}_3 < 0$). The height h of the plane is chosen so that the trajectory Γ intersects S at P_0, P_1, P_2, ..., where the dynamics are assumed to be such that x_3 continually crosses from one side of S to the other.

S can be any plane, but an appropriate choice yields sections that are more easily analyzed. Starting with an initial condition, we thus obtain a set of points comprising the *Poincaré section* — that is, a graph in two dimensions.

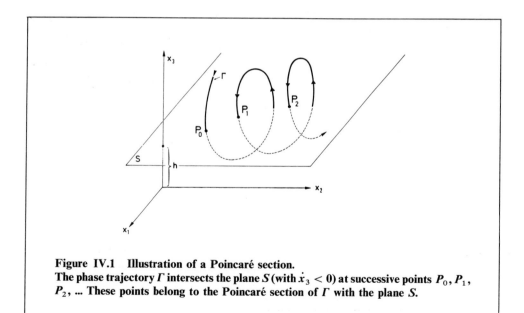

Figure IV.1 Illustration of a Poincaré section.
The phase trajectory Γ intersects the plane S (with $\dot{x}_3 < 0$) at successive points P_0, P_1, P_2, ... These points belong to the Poincaré section of Γ with the plane S.

---

45. Either n is equal to three, or else we consider a projection of the phase trajectories onto a three-dimensional subspace.

The transformation leading from one point to the next is a continous mapping T of S into itself called the *Poincaré map*:

$$P_{k+1} = T(P_k) = T(T(P_{k-1})) = T^2(P_{k-1}) = \ldots \qquad (10)$$

Since the solution to (9) is unique, the point P_0 completely determines P_1, which in turn determines P_2, and so on. If, inversely, P_1 uniquely determines P_0 by reversing the sign of t in (9), then T is an *invertible* mapping of S into itself.

We then note that the Poincaré section replaces the continuous-time evolution of (9) with a discrete-time mapping (10). Aside from exceptional (but nevertheless interesting) cases, the time interval between two successive points is not constant.

Finally, we emphasize that, by construction, the Poincaré section and map have the same kind of topological properties as the flow from which they arise. For example, if the flow (9) is dissipative, so that volumes in phase space are contracted, then T contracts areas in the plane S. Conversely, T conserves areas if the flow (9) is conservative or Hamiltonian. Similarly, if the flow has an attractor, its structural characteristics are also found in the Poincaré section.

IV.2.2 PRACTICAL INTEREST

The method of Poincaré sections simplifies the study of continuous flows for three reasons. First, we pass from a flow in $\mathbb{R}^3$ to a mapping on the plane, reducing the number of coordinates by one[46]. Secondly, the time is discretized, and the differential equations are replaced by difference equations defining the Poincaré map $P \longmapsto T(P)$. These algebraic equations are substantially easier to solve. Finally, the quantity of data to be manipulated is greatly reduced, since almost all the points on the trajectory can be ignored. We will now indicate advantages arising from these simplifications.

The first advantage is at the level of mathematical modelling. Iterating a mapping of the plane:

$$x_i(k+1) = T(x_i(k)) \quad i = 1,2$$

is incomparably easier than integrating a flow such as (9), in terms of the necessary computational time and power[47]. Consequently, iterated maps allow the testing, at little cost, of the relevance of a model.

In Chapter VI we will see in an example how fundamental properties can be discovered using this method. Although *a priori* the domain of the Poincaré section of a three-dimensional flow is a surface, one can often reduce the problem further to a (non-invertible) mapping of the line $\mathbb{R}^1$ onto itself.

---

46. More generally, a continuous flow in $\mathbb{R}^n$ is reduced to a discrete mapping of $\mathbb{R}^{n-1}$ onto itself.
47. In the case of a mapping of the plane onto itself, the calculations can practically be carried out on a pocket calculator. This is how Hénon discovered the Hénon attractor, defined by iteration of the mapping $(x, y) \longmapsto (\beta y, 1 - \alpha x^2 + y)$ with $\alpha = 1.4$, $\beta = 0.3$ (see Chapter VI).

The method of Poincaré has additional advantages beyond that of facilitating mathematical modelling. In a three-dimensional space the trajectories can only be represented in perspective or by a projection, from which it is difficult to get a feeling for their organization. On the other hand, the Poincaré section of the trajectories rapidly gives a good picture. Figure IV.2 illustrates this. Here we have represented the projection onto a plane of a three-dimensional trajectory. The intertwining of the curve does not reveal much about the dynamics (except that they are complicated). But the Poincaré section on the figure shows immediately that this trajectory can probably be inscribed on a torus T^2, which we know to be the attractor of a biperiodic regime.

In the same spirit, we mention another way in which the Poincaré section can characterize a three-dimensional flow. The points of the Poincaré section can be either located along a single curve, more or less complicated, or else distributed on a surface as shown in Figure IV.3. In the second case we can conclude with certainty that the flow is aperiodic. However, if the Poincaré section looks like a simple curve, we are dealing either with a dynamical regime which is truly quasiperiodic, or else which is aperiodic but strongly contracting. In fact, if the contraction of areas in phase space is too rapid, the "lateral" extent (i.e. the direction in which the contraction occurs) of the Poincaré section of the attractor is concealed[48]. A more detailed analysis is then necessary to determine the nature of the flow.

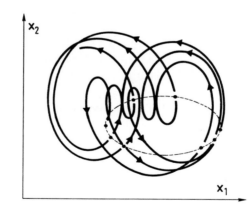

Figure IV.2   Poincaré section for a quasiperiodic regime.
A portion of a trajectory in three-dimensional space projected onto the (x_1, x_2) plane is shown, along with the corresponding Poincaré section (dashed curve).

---

48.   See the example of the Lorenz flow in Appendix D.

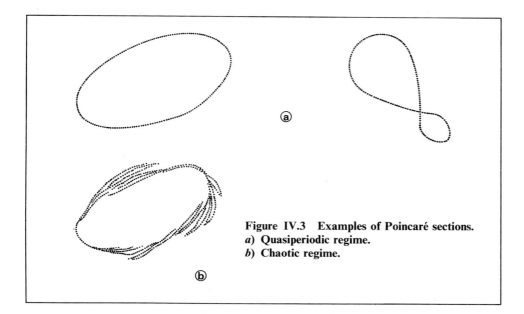

Figure IV.3 Examples of Poincaré sections.
a) Quasiperiodic regime.
b) Chaotic regime.

IV.3 Different kinds of Poincaré sections

IV.3.1 PERIODIC SOLUTION

Following the procedure of Section III.3 we will examine the appearance of the Poincaré section of an attractor according to the dynamical properties of the corresponding solution. When the solution is periodic, we have already indicated that the phase trajectory is a closed orbit, the limit cycle. The corresponding Poincaré section is very simple, reducing to a single point P_0 (fig. IV.4), or possibly several points when the limit cycle has a tortuous form. This point is a fixed point of the Poincaré map T since:

$$P_0 = T(P_0) = T^2(P_0) = \ldots$$

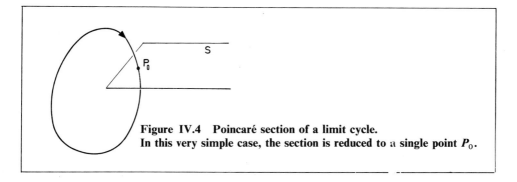

Figure IV.4 Poincaré section of a limit cycle.
In this very simple case, the section is reduced to a single point P_0.

Because of this property, it is possible to go one step further and to investigate the periodic solution's stability. Due to its importance later on, we will explain the problem here (see also Appendix E).

Let us then discuss the stability of the closed trajectory with respect to infinitesimal perturbations. A linear stability analysis, which is limited to terms of order one in the perturbation, is then sufficient. The Poincaré map T is described (to first order) by a matrix M defined in the neighborhood of P_0 by:

$$M = \left[\frac{\partial T}{\partial x_i}\right]_{x_i^0} \quad i = 1, 2.$$

This matrix, called a Floquet matrix, is such that after one circuit (or one period) the image of a point $P_0 + \delta$, very close to P_0, is now at a distance[49]:

$$T(P_0 + \delta) - P_0 \simeq M\delta \quad \|\delta\| \longrightarrow 0.$$

The eigenvalues of M determine the stability of the trajectory. After m periods:

$$T^m(P_0 + \delta) - P_0 \simeq M^m \delta$$

so that the initial displacement δ is multiplied by M^m. As a result, the displacement decreases exponentially in time if the eigenvalues of M are all of modulus less than one — that is, if they are all contained within the unit circle of the complex plane. Then the periodic trajectory is linearly stable, since any displacement from the fixed point P_0 tends to decrease. In the opposite case, when at least one eigenvalue of M has a modulus greater than one, the displacement can grow exponentially in time: the limit cycle is unstable. Of course the exponential growth does not continue indefinitely, but is checked by nonlinear effects neglected here. The loss of stability of the limit cycle thus corresponds to the crossing of the circle by one or more eigenvalues of the Floquet matrix. We will return to this important point in the second part of the book.

IV.3.2 QUASIPERIODIC SOLUTION

For a biperiodic solution with two fundamental frequencies f_1 and f_2, we know that the attractor is a torus T^2 that can be drawn in $\mathbb{R}^3$. Any trajectory on the surface of the torus can be seen as the superposition of two motions[50]: one, rotation along the

---

49. For simplicity, we do not present the calculation in terms of the coordinates of the intersection plane. Although what follows describes essentially what happens along an eigenvector of the Floquet matrix, it applies equally well to general displacements.

50. If (x, y, z) are coordinates in the phase space $\mathbb{R}^3$, one possible parametric representation of the trajectory might be:

$$x = a_1 \sin(2\pi f_1 t + \alpha_1) + a_2 \sin(2\pi f_2 t + \alpha_2)$$
$$y = b_1 \sin(2\pi f_1 t + \beta_1) + b_2 \sin(2\pi f_2 t + \beta_2)$$
$$z = c_1 \sin(2\pi f_1 t + \gamma_1) + c_2 \sin(2\pi f_2 t + \gamma_2)$$

where the a_i, b_i, c_i, α_i, β_i, and γ_i are real constants, and f_1 and f_2 the two frequencies of the motion. In this special case, the motion is simply a sum of two periodic motions. In the general case, the superposition of the two frequencies is more complicated.

IV.3 DIFFERENT KINDS OF POINCARÉ SECTIONS

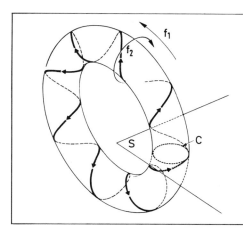

Figure IV.5 Torus T^2 and its Poincaré section with the plane S.
The two frequencies with which the torus is traversed are designated by f_1 and f_2. Under the assumption that f_1/f_2 is irrational, the Poincaré section is the closed curve C.

larger dimension, and the other, rotation about the axis of the "cylinder" forming the torus (fig. IV.5). Each fundamental frequency f_1 and f_2 is associated with one of these rotational motions. The points of intersection of a trajectory with a plane of section S appear at regular time intervals, equal to the period of the first motion (here, $T_1 = 1/f_1$). The points are located on a closed curve C whose form can be either:
— simple: that is, with no point of self-intersection (circle, ellipse, etc.),
— more complicated (figure eight, cycloid, etc.) in the presence of harmonics of f_1 and of f_2.

The exact form of the Poincaré section depends on the ratio f_1/f_2. If it is irrational, the trajectory never closes on itself and densely covers the surface of the torus. One also says in this case that the two frequencies are *incommensurate*. The closed curve C is then continuous. Since each of its points is the image under T of another point of C, the curve is invariant under the mapping T:

$$T(C) = C.$$

The curve C, though itself continuous, is not traversed continuously by successive intersection points of the trajectory with the plane S. On the contrary, the mapping T corresponds to a finite shift along C.

When the ratio f_1/f_2 is rational, the Poincaré section is composed of a finite set of points distributed along C. However, C is no longer a continuous curve, for the trajectory is not dense on the torus. There is frequency locking between f_1 and f_2: the ratio f_1/f_2 is equal to that of two integers n_1 and n_2. After having accomplished n_1 "circuits" and n_2 "rotations" the trajectory closes upon itself; we are in fact dealing with a periodic solution of period $T = (n_1/f_1) = (n_2/f_2)$. The Poincaré section contains only n_1 points, such that:

$$P_i = T^{n_1}(P_i).$$

Figure IV.6 summarizes these two possibilities.

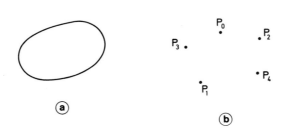

Figure IV.6 Poincaré section for a quasiperiodic regime with two frequencies f_1 and f_2.
a) f_1/f_2 is irrational.
b) f_1/f_2 is rational, 3/5 in this example. The index specifies the order in which the points P_0, P_1, P_2, P_3, P_4 are traversed.

IV.3.3 APERIODIC SOLUTION

To discuss the appearance of the Poincaré section of an aperiodic solution to a three-dimensional flow, an understanding of the layered structure of strange attractors is indispensable. The treatment of this question is therefore postponed until Chapter VI which is dedicated to the description of these complicated attractors. For the moment, we recall that, when the flow is very dissipative and results in a rapid contraction of areas, its Poincaré section can practically be considered to be a set of points distributed along a curve (a line segment, an arc of a curve, etc.). In this case, one defines a coordinate x for each point on the curve, and studies how x varies with time. The Poincaré map on this one-dimensional graph is called a *first return map*. A natural extension of the study of the Poincaré section is therefore the analysis of the first return map that is, of the iteration:

$$x_{k+1} = f(x_k)$$

expressing the relationship between the coordinates of a point and its antecedent.

IV.4 First return map

In the second part of the book, we will often study one-dimensional nonlinear maps of the form:

$$x_{k+1} = f(x_k)$$

and we will show, in particular, that such a map can account for chaotic behavior.

IV.4 FIRST RETURN MAP

Chapter IX will demonstrate the relevance of first return maps to the description of intermittency phenomena. Before arriving at this point, we can begin by discussing the use of first return maps in simpler situations.

IV.4.1 ITERATION OF A ONE-DIMENSIONAL MAP

Classically, the tools used to study a one-dimensional map are the graphs in the (x_k, x_{k+1}) plane of the function $f: x_{k+1} = f(x_k)$ and of the identity map: $x_{k+1} = x_k$. To illustrate the method, we will take as an example the nonlinear mapping[51]:

$$x_{k+1} = 4\mu x_k(1 - x_k), \qquad x \in [0, 1].$$

The graph of the function $f(x) = 4\mu x(1 - x)$ for a given value of μ between 0.25 and 0.75 ($\mu = 0.7$ on fig. IV.7) is zero at $x = 0$ and $x = 1$, and has a maximum equal to μ at $x = 0.5$. With the help of this graph, let us now study the iteration defined above, starting from an initial condition x_0. The first iterate x_1 is at the intersection of the graph f with the vertical line with abscissa x_0 (see fig. IV.7). Similarly, the second iterate $x_2 = f(x_1)$ is located at the intersection of f with the vertical line with abscissa x_1, and so on. A simple and efficient method of constructing the successive iterates consists of using the identity map, or diagonal, $x_{k+1} = x_k$. Indeed the horizontal line $x_{k+1} = x_1$ intersects the diagonal $x_{k+1} = x_k$ at a point with $x_k = x_1$. It then suffices to draw a vertical from this point without referring to the abscissa axis.

By repeating the sequence of operations:
– draw a vertical from the diagonal till its intersection with the graph of f,
– from the point obtained, draw a horizontal until its intersection with the diagonal,

we obtain the successive iterates of the mapping. We ascertain from Figure IV.7 that the iteration converges to the point with abscissa x^*, the intersection of the diagonal with the graph of f. The reader can easily verify that any initial condition x_0 chosen converges to x^* under iteration of f with the exception of the endpoints 0 and 1 of the interval. It is clear that any point of intersection of $f(x)$ with the identity map is its own iterate; it is a *fixed point* of f. This is the case for the origin: taking $x_0 = 0$, we find $x_1 = 0$, $x_2 = 0$, etc. However, for a value of x_0 that is arbitrarily close but not equal to 0, the iteration converges to x^*. This means that any displacement from the origin, no matter how small, is amplified by the iteration. A point at a small distance from the origin moves further away from this fixed point, which is therefore called unstable. In contrast, the fixed point x^* towards which the iteration converges for any initial condition in $]0, 1[$ is a stable fixed point.

There is a criterion for determining whether a fixed point is stable or unstable which does not require tedious calculation. The graphical construction shows that if

---

51. This mapping – the quadratic iteration on the unit interval – has been shown to have extremely rich and varied dynamical behavior, as we will see in Chapter VIII.

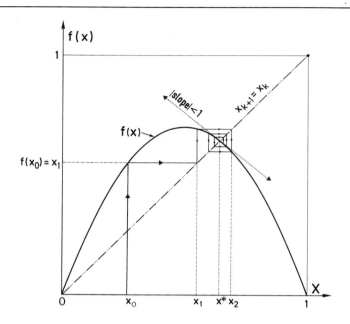

Figure IV.7 Graph of $f(x)$ for $\mu = 0.7$.
The fixed points of the map are the intersections of the graph with the diagonal, or identity map (see text for the geometric method of constructing successive iterates of f). There is a stable fixed point x^* towards which the iterates of all initial conditions in the interval $]0, 1[$ converge. The origin is an unstable fixed point. In this representation, the stability of a fixed point can be determined from the slope of the tangent of the curve at that point: if the absolute value of the slope is less than one, the fixed point is stable (x^*), and otherwise it is unstable (0).

the slope of $f(x)$ at the fixed point is of absolute value greater than one, then the fixed point is unstable: this is the case for the origin in Figure IV.7. On the other hand, if the slope of f is less than one in absolute value, then the iteration converges towards the fixed point: this is the case for x^*.

These results on the existence and stability of fixed point of a one-dimensional mapping concern not only the quadratic function chosen to present them, but more general maps a well.

IV.4.2 LIMIT CYCLE OF THE VAN DER POL OSCILLATOR

We are now in a position to easily establish a result stated in Section II.1.2, but without proof: the existence of a stable periodic solution to Equation (5) for $\varepsilon > 0$. This equation defines a flow in the Cartesian coordinate plane $(\theta, \dot{\theta})$. We have already said

IV.4 FIRST RETURN MAP

that the phase trajectories are spirals, divergent near the origin and convergent far from it.

Consider, in the phase plane, a ray at the origin, and its intersections with trajectories starting at points P_0 and P'_0 (see fig. IV.8). In doing this, we are actually constructing a Poincaré section in $\mathbb{R}^2$; here the "plane" of section is reduced to a half-line, or ray.

We can represent each point on the ray by the coordinate x measuring its distance from the origin (i.e. its radius in the $(\theta, \dot{\theta})$ plane). We define the function $x \longmapsto f(x)$ which assigns to each point on the ray another point which is the first intersection with the ray of the trajectory emerging from x. This function f has certain easily verifiable, but important properties:

* f is well-defined, i.e. f maps each point of the ray into one and only one other point of the ray,
* f is continuous,
* near the origin, the slope of f exceeds one. Otherwise, we would have in this region $f(x) < x$, so that the successive iterates:

$$x_1 = f(x_0) < x_0$$
$$x_2 = f(x_1) < x_1 < x_0$$
$$\dots \quad \dots$$

would converge towards the origin, contradicting the fact that the trajectories are divergent,

* for large x, the slope of f is less than one (convergent spirals),
* f is an invertible function whose slope is always of the same sign (positive). Indeed, we can "go back in time" along a trajectory in a unique way (Equation (5)). Therefore, to each value of $f(x)$ corresponds one and only one value of x.

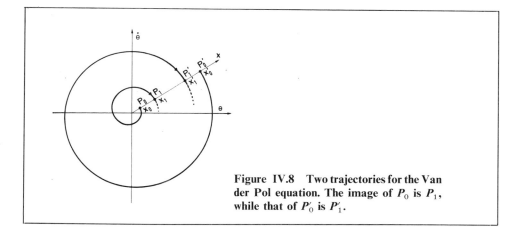

Figure IV.8 Two trajectories for the Van der Pol equation. The image of P_0 is P_1, while that of P'_0 is P'_1.

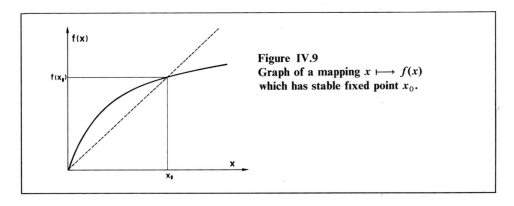

Figure IV.9
Graph of a mapping $x \longmapsto f(x)$ which has stable fixed point x_0.

Taking into account all of these properties, the simplest form for the graph of f is that shown in Figure IV.9. By the continuity of f, there necessarily exists at least one [52] positive value x_0 where f intersects the identity:

$$f(x_0) = x_0.$$

Therefore the first return map on the Poincaré section has a fixed point. The flow from which it arises must therefore have a limit cycle.

By virtue of the condition $|df/dx|_{x_0} < 1$, the fixed point x_0 is an attractor, which implies that the periodic solution is stable.

IV.4.3 REDUCTION OF A THREE-DIMENSIONAL FLOW

Some, but by no means all, three-dimensional flows can be reduced to lower-dimensional mappings. As an illustrative example, we consider the following flow, put forth by Rössler:

$$\begin{aligned} \dot{x} &= -y - z \\ \dot{y} &= x + ay \\ \dot{z} &= b + xz - cz \end{aligned} \qquad (11)$$

where x, y and z are variables, a, b, and c parameters to be chosen. For certain parameter values (example: $a = b = 0.2$, $c = 5.7$) solutions to Equations (11) exhibit aperiodic or chaotic behavior. If we look at Figure IV.11, the evolution of the variables $x(t)$, $y(t)$ and $z(t)$ seems quite incoherent, whereas equations themselves are perfectly straightforward. Figure IV.12 a shows, for this case, the projection of a phase trajectory on the (x, y) plane. Among the many possible Poincaré sections, there is one which is very simple to realize in practice — that corresponding to the plane defined by:

$$y + z = 0.$$

---

52. In fact, further calculation shows that there is only one fixed point for the Van der Pol equation.

IV.4 FIRST RETURN MAP

The first equation of (11) indicates that x is then at an extremum since $\dot{x} = 0$. By numerical integration of (11) we can obtain the value of x each time it goes through an extremum of given type — a maximum, for example. Let x_k be this sequence of values. *A priori* there is no reason for there to be any relation between two consecutive values. And yet, if we graph x_{k+1} as a function of its antecedent x_k, we obtain the striking result shown in Figure IV.10. Instead of being scattered throughout the plane, all the points lie along a curve with a "hump". Two points can be made.

First of all, there is no doubt of the close connection between the flow (11) and the curve in Figure IV.10. Therefore, it must be possible to draw conclusions about the flow from studying the first return map, without having to completely analyze the flow. The simplification brought about by reducing the three-dimensional flow to a mapping on the interval is considerable. Second, and more important, we reiterate the point made in the previous paragraph: that a relatively simple deterministic description (Equation (11)) can give rise to behavior without any apparent regularity (fig. IV.11). We allude here to the essential question — that of the existence of deterministic chaos — to which the second part of this book is devoted.

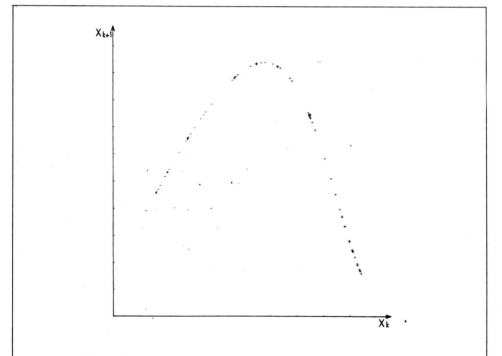

Figure IV.10 For an aperiodic solution of the flow (11), successive maxima of x are plotted against the value of the immediately preceding maximum.
Note the regularity of the curve obtained and the existence of a maximum.

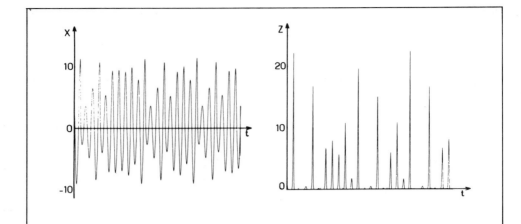

Figure IV.11 Aperiodic solution of the flow (11) obtained by numerical integration. Despite the simplicity of the differential equations, the evolution is far from being regular, as shown here by the functions $x(t)$ and $z(t)$. This conclusion is confirmed by the Poincaré section as well as by Fourier analysis.

IV.5 Practical implementation

The ideas that we have just explained about the method of Poincaré presuppose knowledge of the phase space of the dynamical system under consideration. In practice however, this is information to which we never have access, at least for the experimental study of real systems. Most often one measures the variation of only one property, related in a simple or complicated way to one of the independent variables of the problem (if not to several of them). This might seem to imply that it is hopeless to expect the practical implementation of the preceding analysis.

Must we then conclude that the method of Poincaré is applicable only to numerical simulations of flows such as (11), and cannot be used to study the behavior of real systems? Certainly not, at least while the number of degrees of freedom remains small, which is the case for the systems discussed in this book. We then expect each of the variables to reflect sufficiently well the overall behavior to permit a complete analysis. Hence the following suggestion: starting from observations of only the signal $X(t)$, it should be possible to reconstruct the topology of the attractor, by taking as the phase space either $X(t), X(t+\tau), X(t+2\tau), \ldots$ or $X(t), \dot{X}(t), \ddot{X}(t)\ldots$ In other words, we can consider the signal $X(t)$ to be independent of the same signal at a later time $X(t+\tau)$ where τ is an arbitrary constant called the delay. Or else we can treat $X(t)$ and its time derivative $\dot{X}(t)$ as independent signals. This does not mean that the attractor obtained in the new space is identical to that in the original phase space, but merely that the new representation of the attractor retains the same topological properties, which may suffice for studying its essential characteristics.

When the attractor is simple — a fixed point or a limit cycle — the topological equivalence of the two representations is nearly obvious. It is no longer obvious for a more complicated attractor — a torus or a strange attractor — and, of course, it is these we wish to study. Lacking precise bounds on the domain of validity for the equivalence of the two representations, it is useful to look at an example. For the flow of (11), numerical integration yields all of the variables:

$$x(t), \quad y(t), \quad z(t), \quad \dot{x}(t), \quad \dot{y}(t), \quad \dot{z}(t), \quad \ddot{x}(t), \quad \ddot{y}(t), \quad \ddot{z}(t).$$

Starting from here, we can compare the characteristics of the attractor in the phase space (x, y, z) with those of any other representation. Figure IV.12 compares the projection of the phase trajectories onto the (x, y) plane with the projection onto the $(x, \dot{x})$ plane of the trajectories in the $(x, \dot{x}, \ddot{x})$ space[53]. The resemblance between the topologies of these two graphs is clear. In addition, the largest Lyapunov exponent[54] also shows excellent quantitative agreement: the values calculated from the two representations differ by less than 1 %!

It remains to put the procedure into practice. One method consists of numerically processing the signal $X(t)$ to determine $X(t + \tau)$ or $\dot{X}(t)$, which is easy when the signal has been stored on a computer. Other methods, more physical than computational, are also possible. For example, one can directly obtain both $X(t)$ and its derivative $\dot{X}(t)$ at each instant by an analog electronic device. Or it is sometimes possible to arrange that the measurement of the signal at two different well-chosen points of the system correspond to a time-delay between two signals of the same kind. All of these difficulties are diminished when two independent quantities can be simultaneously measured. In any case, we see that the experimentalist has a certain latitude in his choice of methods.

Finally, we note that for a signal with periodicity T — a periodic or especially a quasiperiodic signal — a simple way of constructing a Poincaré section is to record the signal values at times $T, 2T, ..., nT$. Since T is exactly the time interval necessary to complete one cycle on the attractor (limit cycle or torus), we are effectively

---

53. For the Rössler equations we can write explicitly the relations transforming from $(x, \dot{x}, \ddot{x})$ coordinates to the original phase space coordinates (x, y, z):

$$x = x$$
$$y = -\frac{\ddot{x} + (c - 2a - x)\dot{x} + x + b}{c - a - x}$$
$$z = \frac{\ddot{x} - a\dot{x} + x + b}{c - a - x}.$$

Formally, the change of variables $(x, y, z) \longrightarrow (x, \dot{x}, \ddot{x})$ is a curvilinear coordinate transformation in phase space. Under natural hypotheses of continuity, such a transformation preserves the topology of objects such as the attractor of flows.

54. This quantity is a measure of the mean speed with which two trajectories diverge from one another (see Appendix B).

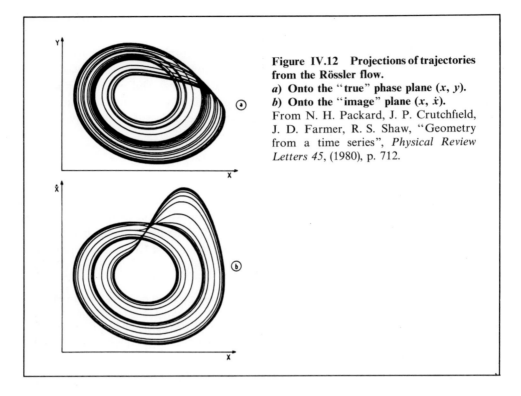

Figure IV.12 Projections of trajectories from the Rössler flow.
a) Onto the "true" phase plane (x, y).
b) Onto the "image" plane $(x, \dot{x})$.
From N. H. Packard, J. P. Crutchfield, J. D. Farmer, R. S. Shaw, "Geometry from a time series", *Physical Review Letters 45*, (1980), p. 712.

stroboscoping the signal. As this task is easy to carry out, the stroboscopic[55] method is often used to identify and study quasiperiodic solutions.

Having reviewed the different means of characterizing the time-dependent behavior of dynamical systems, we shall now illustrate the methods in concrete cases. We have chosen several archetypes to shed light on both the methods of analysis and on their results. The three experimental situations chosen are described in detail in the next chapter.

---

55. This method is equally appropriate for analyzing systems that are periodically forced. In this case, the forcing frequency provides a natural interval for stroboscopic sampling.

CHAPTER V

Three examples of dynamical systems

V.1 Introduction

To illustrate concretely the theoretical ideas to be developed in the rest of the book, we have decided to call upon three different experimental situations. The richness of their dynamical behavior and the abundance of results have guided our choice.

The first example is an electromechanical apparatus with three degrees of freedom, consisting of a magnet placed in a magnetic field, which we will designate as a compass. The motion of a horizontal layer of fluid, heated from below, will provide a second example, this one from hydrodynamics. The key phenomenon is called the Rayleigh-Bénard instability, which causes spontaneous motion of the fluid. We will use the abbreviation R.B. to refer to this phenomenon. Finally, our third example comes from chemistry. It is the oscillating chemical reaction which is the best known and understood: the Belousov-Zhabotinsky reaction, designated by B.Z.

In this chapter, we propose to state what is known of the differential equations describing each of these oscillators and to outline the methods and the equipment used to study them.

V.2 The compass

V.2.1 DESCRIPTION

The compass is simply a skillful realization of the parametric pendulum, allowing convenient and precise observation of the behavior of a system with three degrees of freedom. Like other similar devices, it was developed partly for purposes of demonstration. One of its most attractive aspects is that it offers the possibility of greatly varying the importance of friction. This permits the observation and analysis of the influence of energy dissipation as we go from a conservative or quasi-conservative situation (zero or negligible friction) to one which is more and more dissipative.

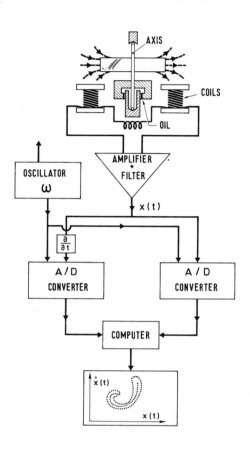

Figure V.1 Schematic diagram of the experimental set-up used to study the behavior of the compass.

The magnet rotates about its axis under the combined influence of two magnetic fields, one stationary and one rotating (see also fig. V.2). The magnitude of the dissipation is determined by the viscosity of the oil in the bearing, whose temperature can be varied for fine adjustment. The motion of the magnetized bar is measured by the voltage induced in the pick-up coils. This voltage $x(t)$ is a function of the magnet's instantaneous angular velocity. An analog device yields the time derivative $\dot{x}(t)$. Two analog-to-digital converters are turned on and off with the same frequency as that of the rotating magnetic field. In this way, $x(t)$ and $\dot{x}(t)$ are recorded in the memory of a microcomputer at each period of the field's rotation. The Poincaré section of the trajectories is then obtained by graphing the points $(x(t), \dot{x}(t))$ recorded during an experiment of sufficient length.

V.2 THE COMPASS

A diagram of the compass is presented in Figure V.1. A magnet, free to rotate about a vertical axis, is placed between two pairs of Helmholtz coils with horizontal perpendicular axes. An alternating sinusoidal current of angular frequency ω and variable intensity is supplied to each pair of coils. Because the currents are in phase quadrature, these pairs of coils generate both a fixed magnetic field and a field which rotates about the axis of the magnet with the angular frequency ω of the current. The rotating magnetic field tends to set the magnet into motion. Pick-up coils, perpendicular to the Helmholtz coils, are located near the magnet and detect its motion via the induced electromotive force. The signal collected $X(t)$ is a function of the instantaneous angular velocity of the magnet. After being amplified and filtered, the signal is sent to a Fourier analyzer which calculates the power spectrum.

Obtaining a Poincaré section of the phase trajectories is straightforward, since the compass is a forced oscillator. Indeed, the rotation period of the imposed magnetic field provides a natural time interval for stroboscopy. In other words, the period of the alternating current through the Helmholtz coils is used as a sampling interval. So as to have a two-dimensional representation, two quantities are digitized and recorded simultaneously: the signal X itself, and its time derivative $\dot{X}$, obtained analogically.

The pivots of the axis of the magnet are made of ruby, to minimize the friction between the solids. By immersing the lower support of the axis in oils with different viscosities, one can vary the importance of fluid friction and the associated energy dissipation. The same apparatus can thus be made very dissipative or almost conservative.

V.2.2 EVOLUTION EQUATION

For the moment, we consider the case in which friction is negligible. The motion of the compass results from the forces applied. The magnet, with a moment of inertia J and a magnetic moment $\mathcal{M}$, is placed in a superposition of two magnetic fields: one is stationary with induction B_1, and the other, with induction B_0, rotates with angular velocity ω (see fig. V.2). Then the variation of the angle θ between the magnet and a fixed direction in the horizontal plane is governed by the differential equation:

$$J \frac{d^2\theta}{dt^2} = - \mathcal{M} B_1 \sin \theta - \mathcal{M} B_0 \sin (\theta - \omega t).$$

We nondimensionalize the equation by setting:

$$M = \mathcal{M} B_1 / J\omega^2; \quad P = \mathcal{M} B_0 / J\omega^2$$

and taking $1/\omega$ as the unit of time. We get:

$$\frac{d^2\theta}{dt^2} = \ddot{\theta} = - M \sin \theta - P \sin (\theta - t).$$

To arrive at this expression it is necessary to make the assumption — amply justified, in practice — that the magnetic fields are independent of the position of the magnet.

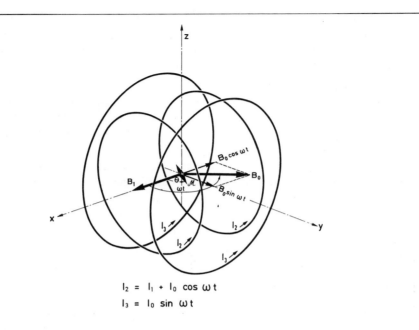

$I_2 = I_1 + I_0 \cos \omega t$
$I_3 = I_0 \sin \omega t$

Figure V.2 Diagram of the arrangement of the magnet M with respect to the field-producing coils.
The rotating magnetic field B_0 is produced by the two pairs of Helmholtz coils with perpendicular axes, supplied with sinusoidal currents $I_0 \sin \omega t$ and $I_0 \cos \omega t$, respectively. The stationary magnetic field B_1 is produced by a constant electric current sent through one of the coil pairs.
From V. Croquette.

The effect of fluid friction is, as we know, to slow the motion with a force proportional to the speed. Consequently, when friction is no longer negligible, we must add to the preceding equation a term proportional to $\dot\theta$:

$$\ddot\theta + \alpha\dot\theta = - M \sin \theta - P \sin (\theta - t) \qquad (12)$$

This nonautonomous second order differential equation is equivalent to a flow in $\mathbb{R}^3$, which we write with the three variables y, θ, ϕ:

$$\begin{aligned}\dot y &= -\alpha y - M \sin \theta - P \sin (\theta - \phi)\\ \dot\theta &= y\\ \dot\phi &= 1.\end{aligned}$$

We see that we indeed have a nonlinear system with three degrees of freedom[56].

---

56. More generally, we see that any nonautonomous flow can be put into the form of an autonomous flow in a higher dimensional phase space by an appropriate change of variables.

Physically, the variable ϕ (related to ωt) represents the angle between the rotating magnetic field and a fixed direction of the plane. This is why the solution to the flow is always periodic — of period 2π in units of $1/\omega$ — in the ϕ component. We note that with no rotating field, i.e. $P = 0$, (12) is reduced to the equation for the simple damped pendulum[57], whose linear approximation was studied in I.3.2.

Once the magnet is chosen, and thus the values of J and of $\mathcal{M}$, and the angular frequency ω of the sinusoïdal current is fixed, the control parameters remaining are the induction B_1 and B_0 of the fixed and rotating magnetic fields. These are controlled by the intensity of the electric current through the Helmholtz coils. The experimenter can therefore vary the parameters M and P at will, to observe the different possible dynamical regimes and the transitions and bifurcations between them.

V.3 Rayleigh-Bénard convection

V.3.1 THE RAYLEIGH-BÉNARD INSTABILITY

In a thermally expansive fluid, any temperature difference creates a density difference. If the fluid is placed in a gravitational field, the field in turn gives rise to forces that can lead to fluid motion called thermal convection. To study this phenomenon quantitatively, one must carefully control the temperature difference applied to the fluid, since it is the temperature difference which drives the motion. Figure V.3 shows a schematic diagram of a typical device used for this purpose. A layer of fluid of thickness d is confined between two rigid horizontal plates of high heat conductivity. The upper plate is maintained at the fixed temperature T_0, and the lower plate at temperature $T_0 + \delta T$ ($\delta T > 0$). The layer is bounded by vertical walls (not shown in Figure V.3) whose thermal conductivity is, in general, close to that of the fluid. From the experimental point of view, it is essential that the thermal exchange between the fluid and the plates be of high quality, so that a linear vertical temperature gradient is established in the layer, without any discontinuity at the boundaries. Compared with other, more or less natural, types of convection, this geometry is characterized by the absence of a horizontal temperature gradient before the onset of convection.

Any thermally expansive fluid placed between these two plates is subject to two conflicting tendencies. The cold liquid, which is dense and located in the upper part of the layer, tends to fall, whereas the lower part, warmer and less dense, tends to rise.

---

57. Note that we arrive at an analogous result when the stationary field vanishes ($M = 0$) if the motion is observed in a reference frame which rotates with angular velocity ω. The action of the two fields is symmetric: the rotating field becomes stationary in a reference frame rotating with its angular velocity, and vice versa. Note also that in the absence of friction and in the presence of only one magnetic field, the magnetized bar oscillates regularly about the field. At small amplitudes the oscillation has an angular frequency $\sqrt{M}$ if the stationary field is present, and $\sqrt{P}$ if it is the rotating field.

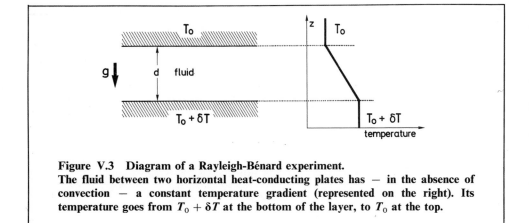

Figure V.3 Diagram of a Rayleigh-Bénard experiment.
The fluid between two horizontal heat-conducting plates has — in the absence of convection — a constant temperature gradient (represented on the right). Its temperature goes from $T_0 + \delta T$ at the bottom of the layer, to T_0 at the top.

Still, as long as the temperature difference δT remains small, no convective motion appears, due to the stabilizing effects of friction [58]. The origin of the instability can be seen as follows. If fluid elements are displaced along the paths HH' and BB' as in Figure V.4, a torque which would amplify the displacement is created. After the displacements HH' and BB' have occurred, the temperature difference between the two fluid elements diminishes, due to the thermal diffusivity D_T, with a characteristic time on the order of:

$$\tau_{th} \sim \frac{d^2}{D_T}.$$

The characteristic time of the displacement $HH' - BB'$ depends on the forces acting on the fluid — that is, the buoyant force due to the density difference and the viscous frictional force. This characteristic time τ_m behaves like:

$$\tau_m \sim \frac{\eta}{\rho_0 \, g \alpha d \, \delta T}$$

η : dynamic viscosity
α : expansion coefficient } of the fluid
ρ_0: mean density
g : gravitational acceleration.

---

58. This is not so if the temperature gradient is horizontal rather than vertical. In this case, the liquid tends to rise along the warmer boundary and fall near the colder boundary. The forces are no longer opposed, but instead create a nonzero torque. No matter how small the temperature difference δT, even *infinitesimal*, this torque induces motion in the fluid. Hence we see the necessity of avoiding horizontal temperature gradients in a R.B. experiment if we wish to have a well-marked convective threshold.

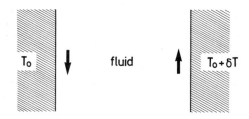

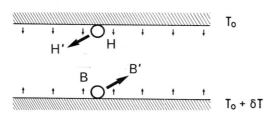

Figure V.4 Diagram of the generation of motion in the fluid.
We consider two fluid elements H and B displaced to H' and B' respectively. This motion is damped or maintained, according to the value of δT.

The condition for the onset of sustained motion is that the time τ_{th} (lifetime of the cause) be greater than τ_m (time for appearance of the effect). Hence the condition for sustained convection is:

$$\frac{\rho_0 g \alpha d^3}{\eta D_T} \delta T \geqslant \text{constant.}$$

The left hand side, called the Rayleigh number Ra, is a nondimensional measure of the temperature difference δT. The inequality above states that there exists a critical Rayleigh number Ra_c (or equivalently, a critical temperature difference δT_c) above which the state of rest ceases to be stable and convection begins. Convective instabilities were first clearly observed experimentally by Bénard in 1900 and first interpreted by Lord Rayleigh in 1916; it is for this reason that the two names are associated to the phenomenon.

Above the convection threshold, the fluid motion is not at all disordered. On the contrary, a regular structure of rolls with parallel horizontal axes is formed[59] (see fig. V.5). At mid-height we see a succession of alternately rising and descending currents. The currents are equidistant from one another, the distance separating two neighboring currents being on the order of d. Two adjacent rolls necessarily rotate in opposite directions. However, if we consider the entire set of rolls, we see that it is possible to reverse the direction of rotation of each roll without globally changing the geometric and dynamical properties of the moving fluid. This is why, in a perfect R.B. experiment, one has strict equiprobability of observing one or the other sense of rotation. Equivalently, at a point in the fluid, the velocity can point in one direction just

---

59. The original experiments by Bénard used a thin layer of whale (spermaceti) oil, with a free upper surface. Under these conditions, hexagonal cells with vertical axes appear, due to surface tension effects at the oil/air interface, rather than rolls. This type of instability, now called the Marangoni effect, is due to the temperature dependence of the surface tension.

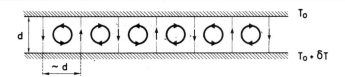

Figure V.5 Diagram of the organization of fluid motion into convective rolls, seen in vertical cross-section.
The rolls turn in alternate directions, much like the meshing of gears. Note that the distance between two adjacent vertical currents (which can be considered to be the diameter of a roll) is on the order of d, the distance between the two rigid horizontal plates.

as well as in the opposite direction. The bifurcation diagram of Figure V.6 shows this property. On this graph, we have plotted the value of the velocity v at one point in the layer, as a function of the Rayleigh number Ra. For Ra $<$ Ra$_c$, the only possible state is the state of rest ($v = 0$). At the instability threshold Ra$_c$, a bifurcation occurs, above which, for a given value of Ra, two convective states, of equal but opposite velocities, are accessible[60]. For Ra slightly above Ra$_c$, the convection rolls have a stationary configuration: the velocity and temperature are functions that are independent of time. From the dynamical systems point of view, the transition at Ra$_c$ is a supercritical

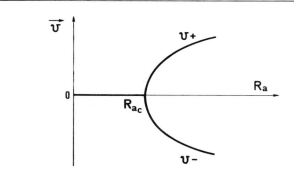

Figure V.6 Bifurcation diagram near the threshold Ra$_c$ of the thermoconvective instability.
In an experiment free of imperfections, the rolls engendered at Ra$_c$ have equal probability of rotating in either direction. This is expressed by the existence of two branches denoted by v_+ and v_-.

---

60. For higher values of Ra, other instabilities appear, which depend a great deal on experimental conditions. Their diversity is so great that an entire volume would barely suffice to describe them all. Several examples which serve to illustrate the theory of dynamical systems are presented in the second part.

V.3 RAYLEIGH-BÉNARD CONVECTION

bifurcation between two stationary states: the state of rest and the convective state. The spatial structure of these states is, of course, quite different.

We mention here one of the major advantages of R.B. convection over other hydrodynamical situations: the absence of any mechanical motion maintained externally. This remarkable property makes R.B. convection a good system for studying the onset of turbulence.

V.3.2 CONVECTIVE STRUCTURES

In a sufficiently wide layer, the convection rolls can adopt multiple configurations or "convective patterns" in the horizontal plane. The process of pattern selection is a

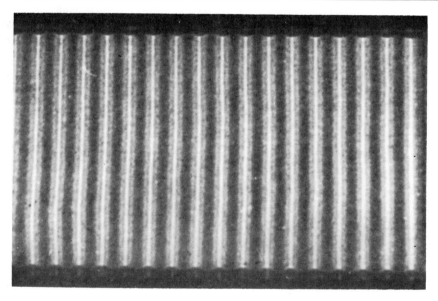

Photograph V.1 Convective pattern of ordered rolls.
Convection is induced in a layer of silicone oil of depth 2.5 mm and horizontal dimensions 98 mm by 60 mm. By using a transparent upper plate and a lower plate of optically polished copper, we can visualize the rolls by the refraction of a light beam passing through the layer. Around the colder descending convective currents, horizontal temperature gradients create gradients in the refraction index which cause light to converge. Thus a wide beam of light shone vertically through the layer of oil, undergoes periodic focusing, revealing the descending currents as light stripes on a dark background.
A small temperature gradient along the two smaller lateral boundaries has been introduced so as to stabilize the phase of the lateral rolls and make the structure perfectly periodic.

complex problem which is the subject of current research. In the case of rectangular containers[61] the pattern can be composed primarily of straight equidistant rolls, whose axes are parallel to the shorter sides of the rectangle. We can see in such a structure an analogy to a one-dimensional crystal (see Photograph V.1).

When the temperature difference δT is increased, the pattern first becomes more complicated while retaining a certain regularity. As δT is further increased, the pattern ends up being completely destroyed, "melted" one might say. The ordered stationary pattern has been replaced by a disordered configuration in perpetual motion: the fluid motion has become turbulent. This complex unpredictable behavior is clearly caused by the multiplicity of possible spatial patterns, each with its own evolution. From their interaction results turbulent behavior which is said to be chaotic, with a large number of degrees of freedom[62].

How can the order of the first convective pattern be preserved, and the manifestation of this type of chaos prevented? It suffices to drastically reduce the number of possible configurations which are compatible with the imposed contraints. A natural idea is to place the fluid in a cell whose horizontal dimensions L_x, L_y are on the same scale as the height. The *aspect ratio*, defined as $\Gamma = L/d$, is small in such a cell and the number of rolls necessarily limited. Thus, in a cell of aspect ratio $\Gamma = 2$, only two rolls can be present (fig. V.7). Experiments show that with this type of cell, spatial order is indeed preserved over a very large range of Ra. This does not mean that the dynamical behavior is independent of Ra. Indeed, what are called secondary

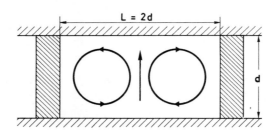

Figure V.7 Diagram of cell with small aspect ratio ($\Gamma = L/d = 2$).
The presence of lateral plexiglass boundaries separated by a distance of only twice the depth stabilizes the convective structure and reduces the number of rolls to two, at least for moderate values of Ra. Note that in the direction perpendicular to the figure, the distance L' between the plexiglass boundaries is even smaller; for the silicone oil experiments described in the text, $L' = 1.2d$.

---

61. In fact the convective structure must be at first induced to take on the pattern shown.
62. The definition of the degree of freedom used here is different from that in Chapter I.

instabilities are observed as Ra is increased, which affect the temporal behavior of the convective pattern. For example, one bifurcation leads to exactly periodic oscillation of the velocity at one point. Such a system is said to have a small number of degrees of freedom (like the number of spatial modes involved in the convective pattern), in contrast to the situation encountered in large aspect ratio cells.

V.3.3 EVOLUTION EQUATIONS

The equations describing convective phenomena couple the velocity $\vec{v}$ of a fluid element to its temperature perturbation[63] θ. Established in the nineteenth century, the equations are derived by considering the local balance of momentum (Navier-Stokes equation), of mass (continuity equation), and of heat. A few approximations, quite acceptable for treating thermal convection under the usual experimental conditions, lead to the following nondimensional equations[64]:

— Navier-Stokes equation:

$$\text{Pr}^{-1}\left(\frac{\partial \vec{v}}{\partial t} + \vec{v} \cdot \nabla \vec{v}\right) = -\nabla p + \theta \vec{\lambda} + \nabla^2 \vec{v} \tag{13 a}$$

— incompressibility of the fluid (if all the velocities are small compared to the speed of sound):

$$\nabla \cdot \vec{v} = 0 \tag{13 b}$$

— heat propagation:

$$\frac{\partial \theta}{\partial t} + \vec{v} \cdot \nabla \theta = \text{Ra}\, \vec{\lambda} \cdot \vec{v} + \nabla^2 \theta \tag{13 c}$$

where p is the hydrostatic pressure, $\vec{\lambda}$ the unit vector along the vertical axis (the direction of gravity), and Pr is the ratio of the kinetic viscosity v of the fluid to its thermal diffusivity:

$$\text{Pr} = \frac{v}{D_T}.$$

This nondimensional quantity, called the Prandtl number, depends on the nature of the fluid, and, to a lesser extent, on its temperature.

---

63. The perturbation θ is the difference between the temperature in the presence of convection and that which the fluid would have in its absence. In the state of rest, θ is thus identically equal to zero everywhere.
64. ∇ designates the spatial derivative operator in Cartesian coordinates:

$$\nabla = \frac{\partial}{\partial x}\vec{i} + \frac{\partial}{\partial y}\vec{j} + \frac{\partial}{\partial z}\vec{k}$$

• is the inner or dot product.

There are two nonlinear terms in the evolution equations: $\vec{v} \cdot \nabla \vec{v}$ (in 13 a) and $\vec{v} \cdot \nabla \theta$ (in 13 c). Their relative importance depends on the value of Pr. If Pr is small (such as for liquid helium, where Pr < 1), it is the term $\vec{v} \cdot \nabla \vec{v}$ which predominates. We must then expect the secondary instabilities occurring in a steadily convecting fluid to be essentially of hydrodynamic origin. They are caused by the nonuniformity of the velocity field and result from the natural tendency of a hydrodynamic velocity field to become uniform. On the other hand, in a fluid with high Prandtl number such as certain silicone oils (Pr > 100), $\vec{v} \cdot \nabla \theta$ is the dominant nonlinear contribution, and secondary instabilities are primarily of thermal order. Between these two extremes, there are liquids like water (Pr = 5-10 depending on the temperature), where the hydrodynamic and thermal effects compete, leading to more complicated behavior[65].

V.3.4 EXPERIMENT

For a given fluid and aspect ratio, the control parameter remaining is the Rayleigh number Ra. It is by varying the applied temperature difference that the experimentalist can either induce a transition from one dynamical regime to another, or else modify the relative importance of nonlinear effects within a given regime. The choice of apparatus is dictated by the fluid, which also determines the time scale for convective phenomena. There is quite a difference between studying the behavior of liquid helium at temperatures on the order of a few degrees Kelvin, and that of water, mercury or silicone oils near room temperature, which is 300 °K. We will not ge into detail concerning the nature of the cells and the means of temperature control. Table V.1 summarizes the experimental conditions under which were obtained the results to be cited in the second part of the book.

Table V.1.

Fluid	Γ	Conditions	Pr	d (cm)	T_{mean} (°K)	δT_c (°)	$(Ra/Ra_c)_{max}$	δT_{max} (°)
Mercury	2		0.031	0.8	273	0.91	6	~ 6
Mercury	4	Magnetic field	0.031	0.7	273	~ 1.3	4	~ 5.5
Helium	3	$P = 3$ bars	0.5	0.125	3.5	$2.5 \, 10^{-3}$	43	~ 0.1
Helium	2 to 3	$P = 1$ to 5 bars	0.4 to 0.8	0.125	[2.5; 4]	q.q. 10^{-3}	80	~ 0.2
Silicone oil	2	$v = 0.03$ sto	38	1.25	~ 300	~ $3 \, 10^{-2}$	450	~ 13
Silicone oil	2	$v = 0.1$ sto	130	2	~ 300	~ $2 \, 10^{-2}$	600	~ 12

---

65. For pedagogical reasons, the observations to be cited in this book refer to small aspect ratio cells, at either very low or very high Prandtl number.

The variables to be measured are $\vec{v}$ and θ, or at least properties closely related to them[66]. In an optically transparent medium, laser Doppler anemometry is the best technique for determining the local velocity $\vec{v}$ at a point in the fluid. Since this technique yields directly one of the dynamical variables, it is used whenever possible. However, this type of measurement is contaminated by experimental noise which may be too high for certain purposes. Another optical method, with a higher signal-to-noise ratio, is often used. The method uses the fact that a temperature gradient in the fluid causes a gradient in the index of refraction. One then measures the angle of deflection of a light ray due to the nonuniformity of the refraction index. This easily implemented technique provides a signal corresponding to the mean temperature gradient traversed by the light ray over its entire path through the fluid. It is therefore only a semi-local measure, not related in a simple way to the variable θ, but is nonetheless sometimes used successfully. Use of the light deflection technique is limited to transparent fluids of sufficiently high viscosity, since it is only in such fluids that appreciable gradients of temperature, and thus of the index of refraction, can exist.

In opaque fluids (mercury, liquid potassium, etc.) the two preceding techniques are not feasible. The idea of optically measuring the local velocity $\vec{v}$ must be abandoned. On the other hand, it is impossible to introduce a temperature probe (thermometer, thermistor, thermocouple, etc.) without disturbing the fluid motion. The only possible location for a non-perturbative temperature probe is the boundary of the fluid layer, at the plates. But in theory, the temperature perturbation θ is always zero there. One solution to this dilemma has been to place bolometers at the boundaries[67]. This type of detector measures the heat flux rather than the temperature perturbation θ itself. This information is nevertheless valuable and can provide the basis for analyzing the dynamical behavior of the fluid as a function of Rayleigh number.

Whatever the method of measurement, the signal is then analyzed by the methods described in Chapters III and IV: Fourier transform, investigation of the topology of the attractor. These procedures are carried out either by digital or analog means (see figs. V.8 a et b).

V.4 The Belousov-Zhabotinsky reaction

V.4.1 BRIEF HISTORY OF CHEMICAL OSCILLATORS

It is in studying the oxidation of citric acid by potassium bromate, catalyzed by the redox pair Ce^{3+}/Ce^{4+}, that the Russian biochemist Belousov, by chance, observed

---

66. Naturally, the measurement must be non-perturbative. That is, it must not influence the phenomena which we seek to observe. This obvious requirement is difficult to satisfy, especially near the threshold of an instability.

67. A bolometer is a component whose electrical resistance depends strongly on temperature. By inserting a bolometer in a Wheatstone bridge, it is possible to measure temperature variations with great accuracy.

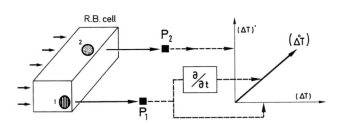

Figure V.8 a Poincaré sections for R.B. convection. Experimental set-up.
The convective regime here is biperiodic, and the thermoconvective oscillators are designated by 1 and 2. Two light beams are shone through the fluid, one near each of the oscillators. Two photodiodes P_1 and P_2 are located along the axes of the light beams. The time-dependent thermal gradient due to the thermoconvective oscillators causes the light beams to be deflected, modulating the photocurrent received by P_1 and P_2. Let (ΔT) and $(\Delta T)'$ be the signals measured by P_1 and by P_2 respectively, and let f_1 and f_2 designate the characteristic frequencies of the oscillators 1 and 2. Then (ΔT) will be more modulated at f_1 than at f_2 and $(\Delta T)'$ more modulated at f_2 than at f_1.
From M. Dubois.

oscillations manifested by the regular alternation of color of the solution, between yellow and clear. When he published in 1958 — eight years, it would seem, after the observation — this kind of reaction was nevertheless not completely unknown. In 1921, Bray had indicated that the decomposition of hydrogen peroxide became oscillatory in the presence of IO_3^- ions. Here too, the observation was fortuitous and dated from 1917. Despite the astonishing singularity of this behavior — most chemical reactions evolve monotonically[68] — these two discoveries of oscillation in homogeneous reactions in liquids aroused no attention. The young Soviet chemist Zhabotinsky devoted his thesis in the early 1960's to the study of the mechanism and properties of the reaction discovered by Belousov. But it was not until later developments in the thermodynamics of irreversible processes, and the discovery of oscillations on the cellular or metabolic level in biology, that the importance of this neglected work was fully appreciated. Since then, research on this subject has flourished, in chemistry as well as in biochemistry.

The B.Z. reactive medium is incontestably the most widely studied of oscillating chemical reactions. The behavior of this reaction is quite remarkable, both in time and

---

68. The linear thermodynamics of irreversible processes shows that oscillation cannot take place about an equilibrium state. However, in the nonlinear domain, the second law of thermodynamics does not necessarily imply monotonic evolution. Therefore, far from equilibrium nothing forbids a reaction from oscillating due to the dissipation of available chemical energy.

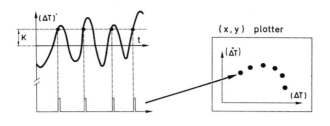

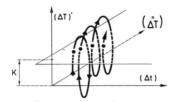

Figure V.8 b Poincaré sections for R.B. convection. Theory underlying the method. A three-dimensional phase space can be defined by the coordinates (ΔT), $(\dot{\Delta T})$ and $(\Delta T)'$ (where $(\dot{\Delta T})$, the time derivative of (ΔT), is obtained by analog means). We obtain projections of trajectories in this space by using (ΔT) and $(\dot{\Delta T})$ as inputs to the two channels of an x-y plotter.
To draw Poincaré sections, the $(\Delta T)'$ signal is used as a reference: the plotter pen is lowered when $(\Delta T)'$ attains a given value K. In particularly favorable cases, we can measure a signal $(\Delta T)'$ which essentially depends only on the second oscillator. Then the technique described is the same stroboscopic method as for the compass: one point (ΔT), $(\dot{\Delta T})$ is plotted at each period of $(\Delta T)'$.
From M. Dubois.

in space. Starting from a uniform mixture of the B.Z. reactants, a spatial structure appears in several minutes, in the form of cylindrical rings in a tube, or of concentric circular waves in a thin layer. We can easily see the interest that this extraordinary reaction would hold for the scientific world.

V.4.2 EXPERIMENTAL SET-UP

The B.Z. reactants — sulfuric acid H_2SO_4, malonic acid $CH_2(COOH)_2$ or other organic reductants, sodium bromate $NaBrO_3$, cerium sulfate $Ce_2(SO_4)_3$ or other redox catalysts — are easily available commercially. It suffices to put them into aqueous solution, and then to mix them in appropriate concentrations in order to observe the famous oscillations. The addition of a colored oxydo-reduction indicator permits the observation of the regular change in color: with ferroïn, for example, the solution alternates between red to blue. Here is the composition, expressed in moles/liter (M), of an oscillating mixture at room temperature:

H_2SO_4 : 1,5 M; $CH_2(COOH)_2$: 0.002 M
$NaBrO_3$: 0.08 M; $Ce_2(SO_4)_3$: 0.002 M
ferroïn : $3 \cdot 10^{-4}$ M.

Quantitative measurement imposes additional requirements. First of all, characterization of the dynamical regime requires that the regime be maintained for a sufficiently long time. Concretely, this means that an open reactor must be used,

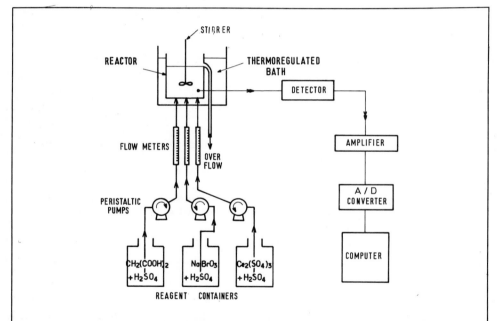

Figure V.9 Diagram of experimental set-up used to study the dynamical behavior of a chemical reaction.
A dynamical regime (stationary, periodic, or aperiodic) is maintained as long as necessary in the open reactor, using a supply system which maintains a constant flux of reactants in the appropriate proportions.
The instantaneous state of the reaction is observed by spectrophotometry or potentiometry. The experiment is controlled by a computer, which also samples the signal in real time and stores it for further numerical processing (e.g., calculation of Fourier transforms, Poincaré sections). The reagents indicated are those of the B.Z. reaction.
From C. Vidal, S. Bachelart and A. Rossi.

continually supplied with reactants and provided with an overflow pipe. To keep the concentrations of the species constant and uniform, the mixture must be stirred rapidly and intensely. Finally, since the speed of these chemical reactions is sensitive to the temperature, thermoregulation of the reactor is necessary. The progress of the reaction is followed by measuring one property of the medium.

Figure V.9 shows a diagram of the apparatus for a B.Z. experiment. Three separate pathways transport the B.Z. reactants from reservoirs to the reactor. The instantaneous state of the medium is measured by potentiometry or spectrophotometry. The signal produced by the detectors — electrodes or spectrophotometer — is amplified, then digitized at regular time intervals. A computer controls the experiment and stores the successive measurements in memory. The data file can eventually contain up to hundreds of thousands of values of $X(t)$, which can be stored for later data analysis: Fourier transform, phase portrait, Poincaré section, etc.

The dynamical variables here are the instantaneous concentrations of certain species which can be measured by potentiometry or spectrophotometry. Within the domain of validity of the Nernst equation, the potential difference between reference and measurement electrodes is proportional to the logarithm of the concentration, while according to the Beer-Lambert law, the optical density is proportional to the concentration itself. To avoid "mixed" information, reflecting the variation in concentration of several species, one often uses an electrode measuring a specific concentration (in general, that of the ion Br^-). Similarly, one measures the optical density at 340 nm, the wavelength at which only Ce^{4+} shows significant light absorption in this reactive medium.

V.4.3 EVOLUTION EQUATIONS

A chemical reaction can be considered as a sequence of elementary steps, symbolized by:

$$v_{ir} X_i \xrightarrow{k_r} v'_{jr} X'_j$$

$i, j = 1, ..., L$ index of the L chemical species
$r = 1, ..., R$ index of the R elementary steps
X_i: reactants; X'_j: products
v_{ir}: molecularity of species i in step r
k_r: rate constant of step r.

This is a concise way to summarize the contents of Table V.2, which describes the B.Z. reaction.

In a homogeneous medium at constant volume, the variations in molar concentration $[X_\ell]$ of a species resulting from the reaction above are given by:

$$\frac{d[X_\ell]}{dt} = \sum_{r=1}^{R} (v'_{\ell r} - v_{\ell r}) k_r \prod_i [X_i]^{v_{ir}}, \quad \ell = 1, ..., L.$$

When the reaction takes place in an open reactor of volume V, fed by a volume flux J, each reactant enters with concentration $[X]_0$ and is evacuated through the overflow with its instantaneous concentration $[X]$. So we must add to the preceding equation the contribution due to transport:

$$[\dot{X}_\ell] = \sum_{r=1}^{R} (v'_{\ell r} - v_{\ell r})k_r \prod_i [X_i]^{v_{ir}} + \frac{J}{V}([X_\ell]_0 - [X_\ell]), \quad \ell = 1, ..., L \qquad (14)$$

This system of differential equations bears some resemblance to the equation of the R.B. instability. They are nonlinear equations, first order in time. Here the nonlinearity comes from interactions between chemical species $\left(\prod_i [X_i]^{v_{ir}}\right)$. Several control parameters can be used, of which one does not appear explicitly in (14) — that is the temperature on which the rate constants k_r depend. Other easily modified parameters are the concentrations at entry $[X_\ell]_0$ of any of the reactants. Finally, and most important, the ratio $\mu = (J/V)$ can be changed since the volume flux J is controlled by adjustable pumps. The physical significance of the parameter μ is that it is the inverse of the refill time of the reactor container, which is also the mean residence time of the species in the reactor. Mathematically, μ enters in an identical way in each of the L equations, as a multiplicative factor of the linear transport term. By changing μ, we can vary at will the relative importance of linear (transport) and nonlinear (chemical) effects. Now we see why μ is frequently chosen as the control parameter to search for bifurcations between different dynamical regimes. When μ is sufficiently large for the chemical reaction terms to be dominated by the transport term, the flow (14) becomes practically linear and has a trivial stationary solution:

$$[X_\ell] \simeq [X_\ell]_0 \quad \ell = 1, ..., L$$

whose physical significance is obvious: at high flux the chemical species stay in the

Table V.2. FKN mechanism of the B.Z. reaction.

$HOBr + Br^- + H^+$	$\rightleftarrows Br_2 + H_2O$
$HBrO_2 + Br^- + H^+$	$\rightarrow 2HBrOr$
$BrO_3^- + Br^- + 2H^+$	$\rightarrow HBrO_2 + HOBr$
$2HBrO_2$	$\rightarrow BrO_3^- + HOBr + H^+$
$BrO_3^- + HBrO_2 + H^+$	$\rightleftarrows 2BrO_2^\cdot + H_2O$
$BRO_2^\cdot + Ce^{3+} + H^+$	$\rightleftarrows HBrO_2 + Ce^{4+}$
$Br_2 + CH_2(COOH)_2$	$\rightarrow BrCH(COOH)_2 + Br^- + H^+$
$6Ce^{4+} + CH_2(COOH)_2 + 2H_2O$	$\rightarrow 6Ce^{3+} + HCOOH + 2CO_2 + 6H^+$
$4Ce^{4+} + BrCH(COOH)_2 + 2H_2O$	$\rightarrow 4Ce^{3+} + HCOOH + 2CO_2 + 5H^+ + Br^-$
$Br_2 + HCOOH$	$\rightarrow 2Br^- + CO_2 + 2H^+$

reactor too short a time to permit them to react with each other. Nonstationary regimes can only be observed for moderate values of the flux J.

We could envisage going further and writing down explicitly Equation (14) for the B.Z. case. But this proves to be of limited utility; knowing the structure of the equations is sufficient from the dynamical point of view. In addition, any reaction mechanism is intrinsically the result of conjectures, a large number of which can never be directly verified. The mechanism proposed in 1972 by Field, Körös and Noyes, which contains fourteen species interacting in thirteen steps as in table V.2, is no exception. A simplified version, with four species and five steps, has been used to show that it is possible to obtain various dynamical regimes, periodic and chaotic, by numerical simulation of the corresponding differential equations.

References for Chapter V

P. Bergé, "Experiments on hydrodynamics instabilities and the transition to turbulence", *Lecture Notes in Physics*, **104**, p. 189 (1979).

P. Bergé, M. Dubois, "Study of Rayleigh-Bénard convection properties through optical measurements", in *Scattering Techniques Applied to Supramolecular and Nonequilibrium Systems*, edited by S. H. Chen, B. Chu, R. Nossal, Plenum Publishing Corporation, p. 493 (1981).

V. Croquette, "Déterminisme et chaos", *Pour la science*, **62**, p. 62 (1982).

V. Croquette, C. Poitou, "Cascades de dédoublements de période et stochasticité à grande échelle des mouvements d'une boussole", *Compte Rendus de l'Académie des Sciences de Paris*, **C292**, p. 1353 (1981).

M. Dubois, *Experimental study of transition to turbulence in convective systems*, cours de l'*Institut von Karman de dynamique des fluides*, série 1982-07 (Bruxelles).

M. Dubois, P. Bergé, "Experimental evidence for the oscillators in a convective biperiodic regime", *Physics Letters*, **76A**, p. 53 (1980).

C. Vidal, J.-C. Roux, "Comment naît la turbulence", *Pour la Science*, **39**, p. 50 (1981).

C. Vidal, S. Bachelart, A. Rossi, "Bifurcations en cascade conduisant à la turbulence dans la réaction de Belousov-Zhabotinsky", *Journal de Physique*, **43**, p. 7 (1982).

PART TWO

...To chaos

PART TWO

INTRODUCTION

Temporal chaos in dissipative systems

I Asymptotic behavior of a dissipative system

Most of the ideas necessary for describing and understanding the time dependent behavior of dynamical systems have been introduced in the first five chapters. Let us summarize the main elements. We wish to analyze the evolution of arbitrary dissipative systems. The systems are assumed to be deterministic, since they are described either by a continuous autonomous flow:

$$\frac{d}{dt}\vec{X}(t) = F(\vec{X}(t))$$

or by a discrete time mapping:

$$\vec{x}_{k+1} = f(\vec{x}_k)$$

where $\vec{X}$ is a vector in $\mathbb{R}^m$ ($m \geq 1$). The function F contains one or several control parameters μ which express the constraints imposed by the external world on the system under investigation. The constraint can be, for example, the intensity of a magnetic field, the Rayleigh number, or the mean residence time in a reactor. By changing a control parameter, it is possible to alter the behavior of the system. During the observation, the control parameter is kept (or assumed to be) constant.

What interests us primarily is the long-term behavior of dissipative systems. We assume that certain technical conditions are met (which can actually be rather difficult to satisfy, even in a simple model like the Van der Pol equation), insuring that the trajectories stay in a bounded volume of phase space.

If the flow is dissipative, then in this finite volume the trajectories converge towards an attractor, a compact set in the phase space which is invariant under the flow or mapping. This attractor has zero volume but its basin of attraction has a finite (or even infinite) volume. The basin of attraction is defined to be the set of initial conditions giving rise to trajectories converging towards the attractor. A flow can have several different attractors, each with its own basin of attraction. Mappings have even been constructed which have a (countably!) infinite number of attractors. We will limit our treatment to the case of a single attractor.

Studying attractors is a substantial simplification, since we can neglect transient effects to concentrate on the asymptotic regime. Such a procedure is of course inapplicable to the analysis of Hamiltonian systems: free of all friction by definition, the concept of attraction is alien to them[1]. Therefore, the ideas developed in this second part concern exclusively dissipative systems. It is important not to lose sight of this point.

Our analysis will consist of two parts:
— defining the different types of attractors
— cataloging the modes of transition between attractors.

The ultimate goal is to understand the origin and characteristics of all kinds of time-evolution encountered, including those which at first seem totally disordered.

II The simplest attractors

We recall briefly the three kinds of attractors already described in the first part, and the dominant properties of the corresponding dynamical regimes.

The point attractor is a solution which is independent of time — that is, a steady state. This situation is completely understood, since the system does not evolve.

A solution which is periodic in time is a limit cycle, characterized by its amplitude and period. Its Fourier spectrum contains a single fundamental frequency and possibly a certain number of harmonics, depending on the form of the oscillations. The solution to the flow can always be expressed as a Fourier series: if the state of the system is known at a given time, one can predict its state at all later times.

A third type of attractor, again relatively simple, is the torus T^r ($r \geq 2$) which corresponds to a quasiperiodic regime with r independent fundamental frequencies[2]. Here, too, the Fourier spectrum is composed of a set of lines, whose frequencies are linear combinations of the fundamental frequencies. While the solution to the flow

---

1. This of course does not imply that the study of Hamiltonian systems is without interest, but only that they have no attractors. We could make an analogous remark about stability, which is closely tied to attraction: in the absence of damping, a small fluctuation never diminishes. This is true, for example, of the stability of non-dissipative (non-viscous) fluid flows: the linear stability of these flows does not imply very much about their true long-term stability against fluctuations of finite amplitude, even small ones. This important fact is often glossed over.

2. While tori T^r with $r > 3$ can certainly exist in theory, there is doubt as to their structural stability, without which they cannot be accessible experimentally. We call a property structurally stable when it "resists" small variations in the parameters of the flow (ordinary stability concerns instead the resistance to small variations in the initial conditions). The connection between observability and structural stability of a property is nevertheless not absolute. A "structurally unstable" property (like the absence of frequency locking in a biperiodic regime) can still be observable if it has a finite probability of occurring for an arbitrary choice of parameters (see Appendix C).

cannot generally be put into the form of an ordinary Fourier series, it is still possible to calculate the state of the system starting from an initial condition.

III Pseudo-definition of chaos

Given that no precise scientific definition exists for the noun "chaos" or for the adjective "chaotic", we will consider these words to be synonymous with certain typical properties[3]. We will say that a dynamical regime is chaotic if its power spectrum contains a continuous part — a broad band — regardless of the possible presence of peaks. Or else we may use the criterion that the autocorrelation function of the time signal has finite support, i.e. that it goes to zero in a finite time. In either case, the same concept is involved: the loss of memory of the signal with respect to itself. Consequently, knowledge of the state of the system for an arbitrarily long time does not enable us to predict its later evolution. Essentially, this means that we are making unpredictability[4] the quality which defines chaos. The characterization is pragmatic; it lacks rigor and contains unavoidable ambiguities. The boundary separating predictability[4] from unpredictability is unclear, leaving open questions such as:
 — On what time scale must the flow be predictable?
 — How precise must the prediction be?
 — Can we allow a statistical prediction?

The distinction between the theoretical and practical impossibility of prediction is also problematic. But given the current state of knowledge, it would be premature to try to do better.

IV Strange attractors

What kind of attractor is associated with a regime whose Fourier spectrum contains a broad band? The Soviet physicist Landau first tried to answer this question in 1944. In trying to explain how a fluid makes a transition from laminar to turbulent flow when the Reynolds number Re (the control parameter) is increased, he advanced the following idea, now known as the Landau theory of turbulence. Above the first instability threshold, the fluid velocity, which previously was constant in time (laminar flow), becomes periodic: it is modulated with a certain frequency f_1. Then if we

---

3. This is also the case for "turbulence" and "turbulent". We will use the word chaos, because space does not enter into the problems considered here, and turbulence inevitably evokes the spatio-temporal behavior of irregular unsteady hydrodynamic flows.

4. More precisely, "unpredictability" is used in the sense that precise knowledge of the past evolution of a system over an arbitrarily long time does not aid in predicting its subsequent evolution past a limited time range.

continue to increase Re, the periodic regime in turn loses its stability. A quasiperiodic regime sets in via the appearance of a second frequency f_2, incommensurate with f_1. Continuing this chain of reasoning, we expect to see other frequencies f_3, f_4, ..., f_r appear in succession, by destabilization of more and more oscillatory modes. For r large enough, turbulent flow would then be obtained[5]. And since the distinction between a Fourier spectrum composed of many neighboring peaks, and a continuous spectrum is, in practice, academic, the attractor of a turbulent flow would therefore be a torus T^r of sufficiently high dimension r.

The corollary to this seductively simple hypothesis is that only systems with a large number of degrees of freedom can display chaotic behavior, since the dimension of the phase space must be greater than that of the attractor. Yet, we know today, mainly from numerical simulations of three-dimensional flows, that this conclusion is false : *three degrees of freedom suffice to give rise to a chaotic regime*, in the sense described above.

It is to Ruelle and Takens that we owe the breakthrough in 1971 of having introduced[6] attractors topologically different from the torus, which they named strange attractors, and of having shown that such attractors could be important in physical problems, including hydrodynamic turbulence. The name "strange attractor" refers to their unusual properties, the most crucial being sensitivity to initial conditions (S.I.C.). By virtue of S.I.C., any two initially close trajectories on the attractor eventually diverge from one another. Their divergence (averaged over short time intervals) even increases exponentially with time.

Figure 1 illustrates schematically the property of S.I.C.. We see that two trajectories, at first almost indistinguishable, later have nothing in common after a finite time has passed. The vanishing of the autocorrelation function, the broad band of the Fourier spectrum, and the intrinsic unpredictability of the system are all

---

5. By identifying the arbitrary phase entering in with each characteristic frequency as a degree of freedom, and by using the Kolmogoroff theory of turbulence to count the degrees of freedom of a turbulent flow, Landau showed that the limit $r \longrightarrow \infty$ of the continuous spectrum corresponds to the limit of infinite Reynolds number Re (i.e. vanishing dissipation). Even though the Landau theory of turbulence is no longer considered to be exact, the determination of the number of degrees of freedom of a turbulent flow (at large Re) remains an unsolved problem. In this approach, "number of degrees of freedom" has the following specific meaning. Let $\lambda_1 \geq \lambda_2 \geq \lambda_3 \geq ...$ be, in decreasing algebraic order, the Lyapunov exponents (see Appendix B) of the dynamical system being considered. The number of degrees of freedom is the maximum number n of exponents such that:

$$\sum_{i=1}^{n} \lambda_i \geq 0.$$

6. In fact, articles describing at least two such attractors had already been published by Rikitake (1958) and by Lorenz (1963) before Ruelle and Takens' article, but had been largely ignored. Appendix D is devoted to a detailed description of the Lorenz model. Rikitake and Allan proposed a model to explain the flipping of the terrestrial magnetic dipole moment over geological time scales. This simple model showed that the seemingly random flips could well be the result of a deterministic process. It challenged other explanations that related the flips to factors external to the dynamics of the earth's magnetic field, such as cataclysms, abnormal solar activity, erratic convective currents inside the earth, etc. Currently, the origin of the flips remains a mystery, although surprising regularities in the trajectory of the magnetic poles have recently been recognized.

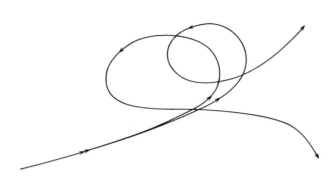

Figure 1 Sensitivity to initial conditions (S.I.C.).
On a strange attractor, two neighboring phase trajectories always diverge, regardless of their initial proximity. Therefore the trajectory actually followed by the system depends crucially on its starting point.

consequences of S.I.C.. In particular, it is clear that the slightest error or imprecision in specifying the initial condition will prevent us from deciding which will be the trajectory followed by the system, and therefore from making other than a statistical prediction on the long-term future of the system. Thus we have arrived at a highly nontrivial result: the impossibility of predicting the behavior of certain deterministic flows with only three degrees of freedom!

V Ways in which chaos appears

The introduction of strange attractors has revived and revolutionized the study of chaos and turbulence. The innumerable articles, books, and conferences devoted to this subject since 1971 bear ample witness to this fact. However, the concept does not by itself solve the problem, because another question immediately arises: what are the origins of the strange attractor, or, if we prefer, how is a chaotic regime established? What are the routes that lead a dynamical system from regular behavior to chaos? To answer this question, we must list the different possible transitions between attractors, a task as indispensable as identifying the attractors.

In the first part, we have emphasized the tendency of many dynamical systems to adopt periodic behavior, regardless of their other characteristics. This fact is not at all obvious[7]. Although periodicity does not exhaust the range of possibilities, a natural first step towards addressing the questions posed above is to examine the conditions under which a periodic regime loses its stability. Two elements will be brought into play: first, Floquet theory of linear stability of periodic solutions, and second, analysis of nonlinear effects that can limit the growth of an instability.

VI Floquet theory

Here we will review the basic elements of Floquet theory that were presented in Section IV.3.1. Consider a nonlinear autonomous flow in an m-dimensional phase space which has a periodic solution of period T :

$$\vec{X}(t + T) = \vec{X}(t).$$

To find out if this solution is stable or not, it suffices to look at what happens to a small initial displacement $\delta\vec{X}$ away from the solution. Linearizing the flow about the periodic trajectory, we find that an initial condition $\vec{X}_0 + \delta\vec{X}$ ($\delta\vec{X}$ infinitesimal) is mapped at the end of the period T into $\vec{X}_0 + M\,\delta\vec{X}$, where M is an $m \times m$ matrix called the Floquet matrix.

The problem of the linear stability of a periodic solution has been reduced in this way to the study of the eigenvalues of M. We first note that this matrix always has an eigenvalue equal to one: this corresponds to a displacement $\delta\vec{X}$ along the trajectory. The significance of this is simply that if we stay on the limit cycle, at the end of a period we arrive exactly at the point of departure: a trivial conclusion for a periodic solution, from which we learn nothing about its stability. Instead, we must study what happens in the directions perpendicular to the trajectory, as shown in the diagram of Figure 2. We see intuitively that, while the eigenvalues of M depend on the form of the limit cycle, they are independent of the reference point $\vec{X}_0$ chosen along it. Since over one period $\vec{X}_0 + \delta\vec{X}$ is mapped into $\vec{X}_0 + M\,\delta\vec{X}$, the solution is linearly stable if all of the eigenvalues of M are located inside the unit circle D of the complex plane. Then, all the components of the vector $\delta\vec{X}$ which are perpendicular to the limit cycle are reduced with each period. On the other hand, if (at least) one of the eigenvalues of M is outside of D,

---

7. To explain the relative universality of periodic regimes, we present two stability arguments. First, Floquet theory shows that, for a given flow, the stability of a periodic trajectory with respect to small variations of the initial conditions depends on certain inequalities, and is therefore not an exceptional situation. But we also have recourse to an argument of structural stability: if a flow has a linearly stable periodic trajectory (i.e. a limit cycle), any neighboring flow (obtained by variation of the control parameters) will also have a limit cycle. We therefore have a good chance of finding periodic behavior in a given dissipative dynamical system. However, this leaves completely open the question of what happens when the periodic trajectory becomes unstable in the usual sense. We will see that the answer is far from unique. It depends in part on nonlinear effects governing the amplitude of fluctuations tending to displace the trajectory from a limit cycle that still exists, but has become unstable.

VII TRANSITIONS RESULTING FROM THE LOSS OF LINEAR STABILITY

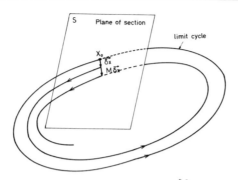

Figure 2 Basis of Floquet theory.
To determine the stability of a periodic solution (limit cycle) we look at what becomes of a small initial displacement $\overrightarrow{\delta X}$ after one period. According to whether the displacement decreases in all directions, or whether it grows in at least one direction, we conclude that the limit cycle is stable or, as shown in the figure, unstable. Here we have chosen an initial displacement $\overrightarrow{\delta X}$ along one of the eigendirections in the plane of section S.

$\overrightarrow{\delta X}$ grows continually in at least one direction: the trajectory moves further and further away from the limit cycle, which is therefore unstable.

By continuous variation of a parameter μ, the periodic solution gradually changes; the same is true of the matrix M and of its eigenvalues. Each of the eigenvalues can be represented in the complex plane by a curve parametrized by μ. Loss of stability of the periodic solution, accompanied by a bifurcation, occurs when one of these curves exits from the unit circle as μ is varied. There exist three[8] generic[9] ways in which to cross the unit circle D, as indicated on Figure 3: at $(+1)$, at (-1) and at two complex conjugate eigenvalues $(\alpha \pm i\beta)$. Aside from the loss of stability, each of these crossing types has different consequences on the later behavior of the system, which depend on the nonlinearities and are closely related to the bifurcations involved.

VII Transitions resulting from the loss of linear stability

When $(+1)$ is crossed, a saddle-node bifurcation occurs (see Appendix A). The periodic solution does not merely become unstable: it disappears entirely. In a

---

8. For a flow in $\mathbb{R}^2$, only crossing at $(+1)$ can occur since a trajectory never intersects itself. The two other crossing types can be manifested only if m is greater than or equal to three (see Chapter VI).

9. Generic means that if we modify the solution a little — by, for example, changing a second control parameter μ' — the crossing remains of the same type.

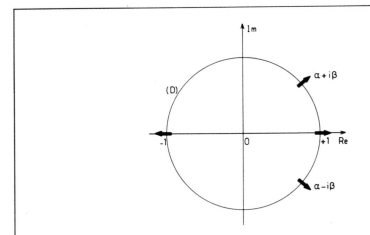

Figure 3 The three possibilities for traversal of the unit circle D by eigenvalues of the Floquet matrix.

parameter region slightly above the bifurcation threshold, the system enters a regime called Type I intermittency. It is characterized by phases of regular, almost periodic behavior (laminar phases), interrupted from time to time by phases of apparently anarchical behavior (turbulent bursts).

If the circle is traversed at (-1), the bifurcation is called subharmonic, and may be either supercritical (normal) or subcritical (inverse). In the case of a supercritical subharmonic bifurcation (see Chapter VIII), a new stable periodic solution, whose period is twice as long, replaces the solution which has become unstable. Period-doubling is repeated for each of the periodic solutions obtained, resulting in an infinite sequence of bifurcations called a subharmonic cascade and ending in chaos. A subcritical bifurcation, on the other hand, leads to Type III intermittency, which qualitatively resembles Type I intermittency: long phases of almost periodic behavior are interrupted from time to time by chaotic bursts. However, Type III is characterized by progressive increase of the amplitude of the subharmonic during the almost periodic phase, the reason being that, here, nonlinear effects amplify the subharmonic instability of the limit cycle. The amplitude increases with each successive oscillation; when it exceeds a critical value, the laminar phase is interrupted.

Finally, a third mode of instability takes place when two complex conjugate eigenvalues $(\alpha \pm i\beta)$ simultaneously cross the unit circle: this is called a Hopf bifurcation[10]. If the Hopf bifurcation is supercritical, it leads to a stable attractor, close to

---

10. Arnol'd calls this a Poincaré-Andronov bifurcation. We will continue to use the term Hopf bifurcation, more current in the Western literature.

the limit cycle which is now unstable (but which still exists). This attractor is a torus T^2 on the surface of which is inscribed the new solution corresponding to a quasiperiodic regime. A second supercritical Hopf bifurcation can then generate a transition $T^2 \longrightarrow T^3$ and a strange attractor appears as the outcome of a third bifurcation (possibly merged with the second). If the bifurcation is subcritical, we can encounter another phenomenon, called Type II intermittency. Type II intermittency has the same global qualitative features as Types I and III, except for the fact that the instabilities which develop during the laminar phases have a frequency which is unrelated to the fundamental frequency of the original cycle. This new frequency is instead related to the ratio α/β where $(\alpha \pm i\beta)$ are the eigenvalues crossing the unit circle at the bifurcation. The dynamical process which initiates a new laminar phase after a turbulent burst is particularly complex to describe in this case, as it involves a flow in a six-dimensional phase space, which means a Poincaré section in $\mathbb{R}^5$.

The following table summarizes the different situations we have described, as well as the chapters in which they are analyzed.

Crossing	Bifurcation		Phenomenon	Chapter
$(+1)$	Saddle-node		Type I intermittency	IX
(-1)	Subharmonic	normal	Subharmonic cascade	VIII
		inverse	Type III intermittency	IX
$(\alpha \pm i\beta)$	Hopf	normal	Quasiperiodicity	VII
		inverse	Type II intermittency	IX

VIII Current limits of the theory

The study of transition through loss of linear stability is a convenient way to classify a number of processes through which chaotic behavior is established. We must nevertheless make two cautionary remarks.

In the first place, as we have already indicated, periodic evolution is not a necessary condition for the appearance of chaos, but merely a favorable circumstance. That it is not obligatory is clearly demonstrated by the example of the Lorenz model.

Secondly, we must be aware of the limits of the theory in its current state. Three phenomena (quasiperiodicity, subharmonic cascade, and intermittency) have been catalogued and analyzed as leading to chaos. But nothing has been specified about the necessary and/or sufficient conditions for them to take place, based on the particular structure of the flow. In this regard, the symphony of dynamical systems remains unfinished.

CHAPTER VI

Strange attractors

VI.1 Dissipation and attractors

VI.1.1 THE PHENOMENON OF ATTRACTION

Dissipative dynamical systems, the only ones considered here, are characterized by the attraction of all trajectories passing through a certain domain of phase space towards a geometric object called an attractor. We illustrate this fact with a simple example: the driven pendulum. The energy supplied to the pendulum from outside compensates for the energy lost and so is dissipated by the system. When the oscillations of the pendulum attain an amplitude such that the energy supplied per cycle is exactly equal to that dissipated per cycle, a steady state is reached. The regime is periodic: the oscillation amplitude is constant and the trajectory in phase space is a limit cycle C (see fig. VI.1). To illustrate its attracting nature, it suffices to displace the system from the limit cycle. For example, we perturb the pendulum by a shock, sending it to amplitude θ_1 and velocity $\dot{\theta}_1$ significantly higher than the maxima θ_m and $\dot{\theta}_m$ of the limit cycle. As time passes, dissipation causes the trajectory to converge rapidly towards the limit cycle C, where the dissipation and the energy supplied are equal. Similarly, temporary braking of the forced pendulum will reduce its amplitude and velocity to θ_2 and $\dot{\theta}_2$ (see fig. VI.1), but the trajectory later evolves back towards C[11].

What we have said about a limit cycle can be generalized in the following way. In phase space, the solutions to a set of n ordinary differential equations:

$$\frac{d}{dt}\vec{X}(t) = F(\vec{X}(t)) \qquad \vec{X} \in \mathbb{R}^n$$

constitute a flow ϕ which, for a dissipative system, has an attractor. An attractor A is defined to be a compact set in phase space with these properties:
- A is invariant under the action of the flow, i.e. $\phi A \equiv A$.
- A has zero volume in the n-dimensional phase space (see next section).

---

11. More specifically, this happens for any trajectory starting in a neighborhood of C constituting what is called its basin of attraction.

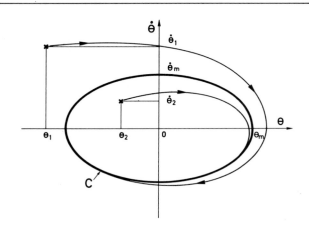

Figure VI.1 Attraction towards a limit cycle C.
The phase space is defined by the angle θ between the pendulum and the vertical, and the angular velocity $\dot\theta = (d\theta/dt)$. We see two trajectories — emanating from initial conditions off the limit cycle — converge rapidly towards the limit cycle.

— A is contained in a domain B, of nonzero volume, which constitutes its *basin of attraction*. The basin of attraction is defined to be the set of points from which originate trajectories which converge to A as $t \longrightarrow \infty$ (see fig. VI.2).

We see that the attractor A is the asymptotic limit of solutions with initial points in its basin of attraction B. We remark that even if A is a simple geometric object, B can have a very complex form.

VI.1.2 TWO CONSEQUENCES OF THE CONTRACTION OF AREAS

We return to the example of the limit cycle in order to examine two very important characteristics of attraction: the loss of memory of initial conditions, and what that implies for the dimensionality of the attractor.

We consider a set of initial conditions in the $(\theta, \dot\theta)$ phase plane occupying a region of size Γ (see fig. VI.3). Because of dissipation, the flow contracts areas. Therefore, the surface Γ is reduced by the flow to a line segment on the attractor C. It follows that there is a loss of information concerning the relative position of points initially contained in Γ; once the attractor is attained, the information is irretrievably lost. This argument rests entirely on the contraction of areas and on the concomitant existence of an attractor. The conclusion then remains valid regardless of the type of attractor. This is why information about initial conditions is also lost in a biperiodic regime, in

VI.1 DISSIPATION AND ATTRACTORS

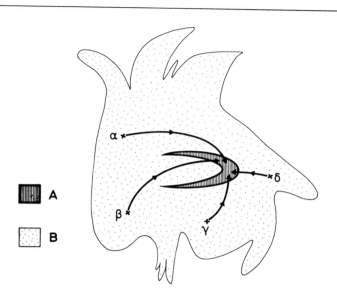

Figure VI.2 Schematic representation of an attractor.
A is the attractor, B its basin of attraction. The trajectories issuing from the initial points α, β, γ and δ are carried to the attractor A by the flow ϕ.

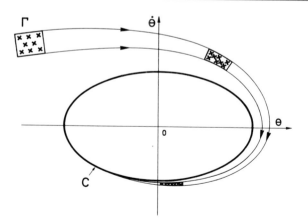

Figure VI.3 Contraction of areas associated with attraction.
The area Γ, delimiting a set of initial conditions, is reduced to a line segment when the attractor C is reached.

which trajectories evolve in a phase space of at least three dimensions and converge towards a torus T^2. The same result holds for aperiodic regimes and the attractors associated with them.

Loss of memory of the exact initial conditions has an important corollary which we will again illustrate using the pendulum. At first — that is, away from the limit cycle — two coordinates θ_i and $\dot{\theta}_i$ are necessary to characterize the state of the dynamical system, so a two-dimensional phase space, or surface, is required. Once the transients have disappeared and the asymptotic regime is attained, only one trajectory remains: the curve C. One curvilinear coordinate along C then suffices to locate a point. This illustrates a general principle: the dimension d of the attractor is less than the dimension n of the phase space, i.e. less than the number of degrees of freedom of the dynamical system[12] :

$$d < n.$$

Let us now evaluate more quantitatively the consequences of the contraction of areas, or, more generally, of volumes. The relative rate of change of a volume V in phase space, under the action of the flow, i.e. $(1/V)(dV/dt)$, is given by the *Lie derivative* :

$$\frac{1}{V}\frac{dV}{dt} = \sum_{i=1}^{n} \frac{\partial \dot{X}_i}{\partial X_i}.$$

where X_i is the i-th component of $\vec{X}$ ($\vec{X} \in \mathbb{R}^n$). In dissipative systems, this quantity, averaged over time, is negative, and measures the rate of contraction[13]. The fact that the Lie derivative is negative implies that at large times, after the trajectories have reached the attractor, any set of initial conditions of volume V has converged to a set of volume zero: the volume of the attractor itself is zero.

In a three-dimensional phase space, the word volume has its usual meaning. Any attractor must then have dimension less than three in order to satisfy the condition $d < n$ stated above.

VI.1.3 NON INTERSECTION OF PHASE TRAJECTORIES

The trajectories of a dynamical system in phase space cannot intersect[14] (except at a singular point, where they converge). This is an unavoidable consequence of the

---

12. We will see later on that the difference between the dimension d and n is not necessarily one, nor even integer.

13. As we have already mentioned, for conservative (or Hamiltonian) systems, the volume is conserved:

$$\sum_{i=1}^{n} \frac{\partial \dot{X}_i}{\partial X_i} \equiv 0.$$

14. This result was already evoked in Section I.3.3 in the context of vector fields.

deterministic nature of the description. Otherwise one initial condition (the point of intersection) could give rise to different trajectories, or else the system could behave differently, in an undetermined fashion, upon traversing the intersection point. This clearly contradicts the hypothesis of a deterministic description of the system as a finite set of ordinary differential equations.

This remark might seem innocuous at first glance, but it actually has very profound consequences, as we will often observe. For now, we can see what it implies for a biperiodic regime. The attractor in this case is a torus T^2 which, for convenience, we will "unfold" onto a plane. This operation is topologically feasible, since the surface of a torus is two-dimensional.

The unfolding of the torus is shown as a three-stage process in Figure VI.4. First, the torus is cut along a small circle (SC) separating regions C and C' and then unfolded to form a cylinder. A second cut along a generator AA' of the cylinder and a second unfolding yields the rectangle $AA'B'B$. Note that AB (or $A'B'$) represents a small circle SC on the torus, whereas AA' (or BB') represents a large circle LC. Let ω_1 and ω_2 denote the angular frequency of rotation of the trajectories about large circles and small circles, respectively. To a complete revolution about a large circle is associated a segment on the rectangle parallel to its larger side AA'; similarly a segment parallel to the smaller side BA of the rectangle corresponds to a rotation about the small circle of

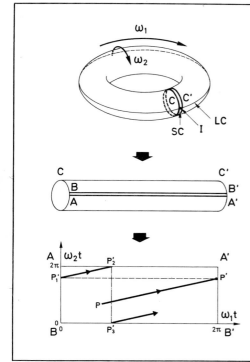

Figure VI.4 Unfolding of the surface of a torus T^2.
The unfolding is carried out in two stages. First, the torus is cut along a small circle C and unfolded: we obtain a tube delimited by the circles C and C' (initially identified on the torus). We then cut again along the line AA' (BB') of the tube; unfolding, we obtain the rectangle $AA'B'B$ (the points A, A', B', and B being originally identified with the single point I on the torus). See text for the way in which a phase trajectory is represented on the rectangle.

the torus. More generally, any phase trajectory on the torus which is some combination of two motions is represented on the rectangle by a line segment such as PP'. Since the line AB and $A'B'$ are identified (i.e. are the same) on the torus, to continue the trajectory[15] beyond P' we must transfer to P'_1 and, from there, draw a line parallel to PP'. Similarly, at P'_2, we transfer to P'_3 and so on. The complete trajectory in this representation then consists of a set of parallel line segments. Figure VI.5 shows an example of the trajectory obtained when the angular frequency ratio ω_1/ω_2 is four. Determinism forbids divergence of trajectories: if two trajectories diverged, they would eventually intersect as in Figure VI.6, and the number of intersection points would increase with time. A phase portrait like that of Figure VI.6 is therefore excluded.

To summarize, on an attracting torus T^2, the only two possibilities are as follows. If the ratio ω_1/ω_2 is irrational, then the line segments densely cover the rectangle (that is, the unfolded torus): the regime is biperiodic, or quasiperiodic. Otherwise ω_1/ω_2 is rational, and a trajectory is composed of a finite number of parallel line segments, traversed sequentially with time, as in Figure VI.5: the regime is periodic. The only kind of structural instability to which a biperiodic regime is subject is frequency locking or synchronization[16], meaning that the ratio of frequencies goes from irrational to rational. As a corollary, a torus T^2 can never be the attractor of an aperiodic regime.

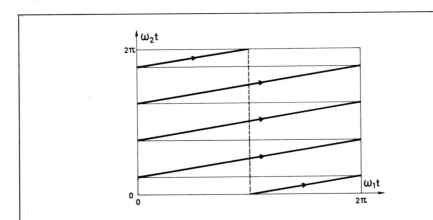

Figure VI.5 Representation of a trajectory on an unfolded torus. Here $\omega_1/\omega_2 = 4$. Because of periodicity, the sides of the rectangle are normalized to the interval $[0, 2\pi]$.

15. Naturally, for t increasing monotonically, the trajectory cannot retrace its steps.
16. Structural stability refers to stability with respect to a change in the control parameters, and not with respect to initial conditions. The quasiperiodic regime is structurally unstable because an infinitesimal modification of a control parameter can always transform it to a periodic regime. For more on this, see Chapter VII and Appendix C.

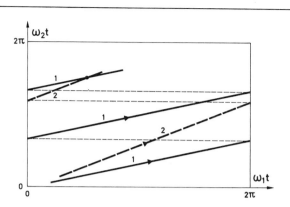

Figure VI.6 Impossibility of divergence of trajectories on T^2. Two divergent trajectories would intersect, which is forbidden.

VI.2 Aperiodic attractors

VI.2.1 CHARACTERISTICS OF A CHAOTIC REGIME

Let us review what was said about chaotic regimes in Chapter III, with a view towards deducing properties of attractors associated with chaotic behavior.

In a chaotic regime, the power spectrum of one of the variables X of the dynamical system contains a continuous part, expressing the fact that the evolution of X is disordered and erratic (see fig. VI.7 a). To estimate the amount of disorder, it is useful to introduce a function measuring the resemblance of X at time t with itself at a later time $t + \tau$. This quantity $C(\tau)$ is obtained by averaging over a large number of products $X(t) \cdot X(t + \tau)$:

$$C(\tau) = \frac{1}{t_2 - t_1} \int_{t_1}^{t_2} X(t) \cdot X(t + \tau) \, dt$$

or in more condensed notation:

$$C(\tau) = \langle X(t) \cdot X(t + \tau) \rangle.$$

By varying the interval τ we construct $C(\tau)$, called the temporal *autocorrelation function*. In Chapter III, we have seen that, according to the Wiener-Khintchine theorem, $C(\tau)$ is the Fourier transform of the power spectrum. If $X(t)$ is constant, periodic, or quasiperiodic, then the power spectrum is composed of distinct peaks. Therefore $C(\tau)$ will remain nonzero as τ tends to infinity.

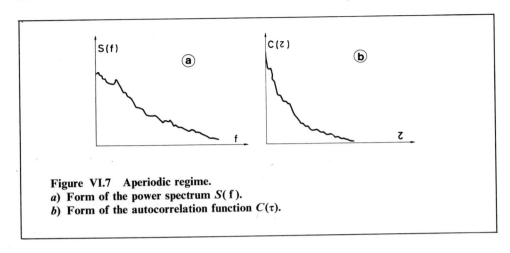

Figure VI.7 Aperiodic regime.
a) Form of the power spectrum $S(f)$.
b) Form of the autocorrelation function $C(\tau)$.

A periodic (or quasiperiodic) signal resembles itself at later times. This means that the behavior of the system is predictable since knowledge of it for a sufficiently long time enables us to construct it at all later times, by simple comparison. On the other hand, in a chaotic regime, where the power spectrum includes a continuous part, $C(\tau)$ necessarily tends to zero as τ increases (see fig. VI.7 *b*). The autocorrelation function has finite support: the resemblance of the signal with itself in time diminishes, and even disappears for times that are sufficiently far apart. It follows that no finite interval of observation of $X(t)$ suffices to predict its future behavior. The chaotic regime is intrinsically unpredictable, by progressive loss of self-similarity. We have already encountered loss of memory of initial conditions in Section VI.1.1; in fact, this reflected only *insensitivity* to initial conditions and an impoverishment of the original information since many initial states *outside of the attractor* merge on the attractor. But when we speak of loss of memory of initial conditions in a chaotic regime, we mean that a set of imperceptibly different initial states *on the attractor* lead, in an unpredictable way, to many final states; in a way, information has not been lost, but rather gained [17]. A crucial consequence on the dynamics is this: two trajectories that are initially very close will diverge, resulting in loss of all resemblance after a finite time (see fig. VI.8). Conversely, if a regime is represented by an attractor on which neighboring trajectories diverge, then the regime is chaotic. This very important property of amplification (in fact, exponential amplification) of errors or of initial uncertainty in a chaotic regime is called *sensitivity to initial conditions* or SIC.

---

17. Unlike the temporary loss of information due to attraction, there is perpetual creation of information on the attractor.

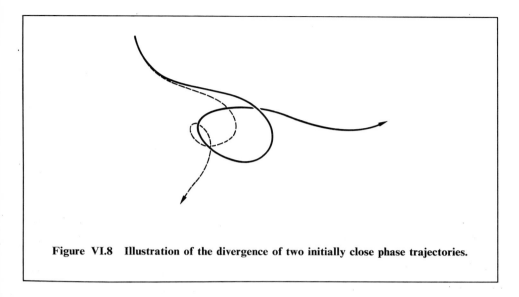

Figure VI.8 Illustration of the divergence of two initially close phase trajectories.

VI.2.2 PROPERTIES OF APERIODIC ATTRACTORS

Landau's idea was to attribute the existence of a continuous Fourier spectrum to the presence of a very large number of independent frequencies[18]. Yet a torus T^r, with r very large, is not required to describe chaos: according to what we have seen, any attractor, even low-dimensional, which exhibits SIC, will correspond to a chaotic regime.

The idea of an attractor exhibiting SIC is highly nonintuitive. This is probably the explanation for the late discovery of chaos with a small number of degrees of freedom: deterministic chaos. The paradox resides in the apparent contradiction between attraction, which implies the convergence of trajectories, and SIC which implies their divergence. But in fact, the divergence of trajectories merely sets a lower bound on the attractor dimension. We have seen in Section VI.1.3 that SIC is impossible on a two-dimensional attractor for topological reasons: it is therefore necessary that the phase trajectories evolve in a space of at least three dimensions. For simplicity, we will from now on consider exclusively the case of a three-dimensional phase space. In contrast to what can occur on the attractor T^2, the diagram[19] of Figure VI.9 demonstrates one of the radically different possibilities which can open the way to a new type of attractor. Trajectories diverge within a plane by spiralling out. They then emerge from the plane, and return again, reinjected into the center of the spiral. The process is repeated *ad infinitum*. This requires two operations: *stretching* due to SIC, followed by *folding*, without which the trajectories could not remain confined in a bounded space.

---

18. In the introduction to the second part, we define this as chaos with a large number of degrees of freedom.
19. This is not the only possibility, but only serves as an example.

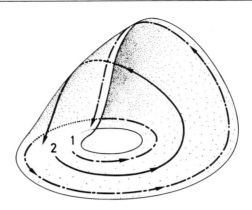

Figure VI.9 Schematic representation of the divergence of two trajectories in a three-dimensional phase space.
The neighboring trajectories 1 and 2 emerge from the horizontal plane, "cross" without intersecting and return to the spiral. These curves are drawn on a simplified view of the Rössler attractor.
From Abraham and Shaw.

To understand the origins of such a flow, imagine a three-dimensional flow, diverging in the xx' direction and converging in the perpendicular direction yy' (see fig. VI.10 a). A set of initial conditions will lead to a quasi-two-dimensional sheet $ABEC$ exhibiting the divergence of trajectories required for SIC (see fig. VI.10 b). After this first stretching operation, we must insure that the flow remains in a bounded three-dimensional space: hence the folding[20]. When the sheet's width has doubled ($CE = 2AB$), CE is folded in two along CDE (see fig. VI.10 c); then, the "end" CD is connected up with the "beginning" AB, (see fig. VI.10 d). We have thus constructed a three-dimensional flow exhibiting SIC in a finite space[21].

We see that in a three-dimensional phase space, the conflicting demands of attraction and of SIC are reconciled via the concept of *hyperbolicity*: attraction takes place in one direction, while divergence occurs in another. In Figure VI.11, we show a plane perpendicular to the mean direction of the flow. The point O is called hyperbolic by geometric analogy. Along the sheet, i.e. in the xx' direction, trajectories diverge from O, while in the perpendicular direction yy', the trajectories approach O. Off the two axes, the points move along curves which, when projected onto the plane of

---

20. We will see later on that it is the repeated foldings of the flow that give rise to the characteristic layered structure of strange attractors.
21. This object created by stretching and folding has essentially the same topology as the Rössler attractor illustrated in Figure VI.9.

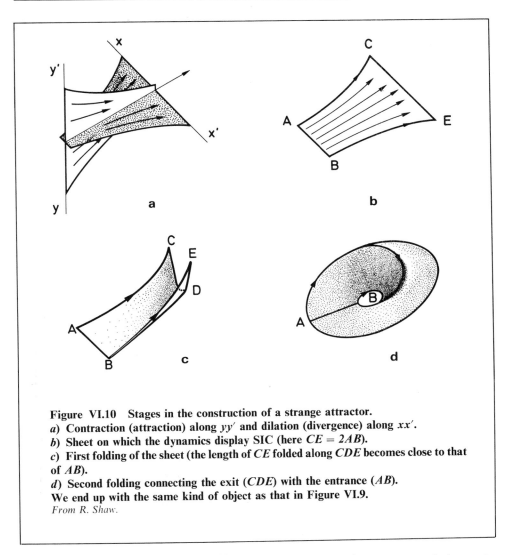

Figure VI.10 Stages in the construction of a strange attractor.
a) Contraction (attraction) along yy' and dilation (divergence) along xx'.
b) Sheet on which the dynamics display SIC (here $CE = 2AB$).
c) First folding of the sheet (the length of CE folded along CDE becomes close to that of AB).
d) Second folding connecting the exit (CDE) with the entrance (AB).
We end up with the same kind of object as that in Figure VI.9.
From R. Shaw.

Figure VI.11 resemble hyperbolas[22]. Thus there now exists a counterbalance to attraction, which destroys information: divergence, which creates it. Having concluded

---

22. In this way we can define two different lines "parallel" to the same "line" (or rather, geodesic). One possible method of constructing a line parallel to a line D passing through a point P not on D uses a running point M on D. The line parallel to D through P is the limit of the straight line MP when M goes to infinity on D. Euclid's fifth postulate is equivalent to saying that this limit is the same for M going in either direction on D. However, in a hyperbolic plane — that is, not satisfying the fifth postulate — this is not so: there exist two different parallel lines, depending on the direction on D. This remark is the basis of the non-Euclidean geometries of Gauss-Bolyai-Lobatchevsky. Some of the ideas used to study chaotic systems in a finite phase space have their roots in hyperbolic geometries. For example, the Lyapunov numbers studied later (cf. Appendix B) are related to the idea of intrinsic curvature in the hyperbolic plane.

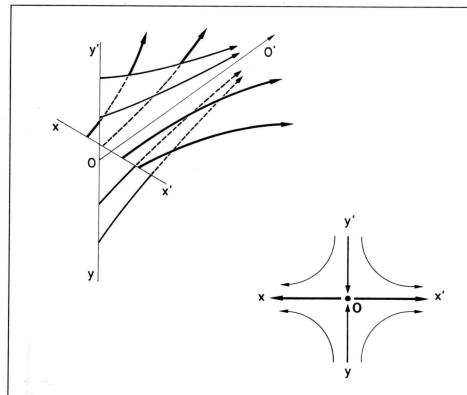

Figure VI.11 Illustration of hyperbolicity.
The bold curves on the leftmost diagram are the divergent trajectories and the finer curves, the convergent trajectories. OO' is the direction of the mean flow, xx' the dilating direction, and yy' the contracting direction. The figure at bottom right shows a projection of the flow onto a plane perpendicular to the direction OO'. The surface containing the convergent trajectories — here the plane $YY'OO'$ — is called the stable manifold, while the surface containing the divergent trajectories — here the plane $XX'OO'$ — is called the unstable manifold.

that SIC requires that the dimension d of the attractor satisfy $d < 2$, we might at this point wonder whether it is possible for a three-dimensional flow to exhibit SIC. The question arises because in a dissipative system (i.e. a system in which an attractor exists), volumes in phase space contract as time increases (see Chapter I). The volume of the attractor must therefore be zero, which, in a three-dimensional phase space, implies $d < 3$. An attractor[23] capable of representing a chaotic regime (thus displaying SIC) must be such that $2 < d < 3$. This condition would seem to preclude the existence

---

23. Recall that we are in a three-dimensional phase space.

of such attractors, if we refer to the Euclidean dimension, which is necessarily an integer. Nevertheless we do encounter attractors verifying the condition $2 < d < 3$. They have, among other curious properties, a noninteger dimension, called a *fractal dimension*[24] (see Section VI.4.2). Because of their peculiar properties, they are called *strange attractors*.

To summarize, a dissipative dynamical system can become chaotic if the phase space has dimension greater than, or equal to, three. This chaos — with a small number of degrees of freedom — is due to the SIC of trajectories on strange attractors. The essential properties of a strange attractor are:
1) Phase trajectories are attracted towards it (at least those originating close by).
2) Pairs of neighboring trajectories diverge on it (SIC).
3) Its dimension d is fractal.

VI.3 Examples of strange attractors

VI.3.1 THE LORENZ ATTRACTOR

We refer the reader to Appendix D for an introduction to this model of turbulence and its main properties. Making some radical simplifying hypotheses, the dynamical behavior of a convecting fluid is reduced to three ordinary differential equations[25] which define a flow in a three-dimensional phase space with variables X, Y and Z:

$$\frac{dX}{dt} = \Pr Y - \Pr X$$

$$\frac{dY}{dt} = -XZ + rX - Y$$

$$\frac{dZ}{dt} = XY - bZ.$$

The values of Pr and b are usually fixed (often at $\Pr = 10$ and $b = 8/3$, as they will be here), leaving r as the control parameter. This system of coupled nonlinear equations looks simple but nonetheless is non-integrable (analytically) in the general case. However, it can be solved numerically by starting with initial values $(X(0), Y(0), Z(0))$, and using a calculator to compute the flow $X(t)$, $Y(t)$, and $Z(t)$ step by step.

---

24. Unlike the usual Euclidean dimension, the fractal dimension is defined in several different ways which are not formally equivalent and, indeed, yield different numerical values in concrete examples. Simplifying a great deal, some definitions of the fractal dimension are more sensitive to the intrinsic connectivity of the object, others to the way in which the object is embedded in the ambient space. These questions, which are now under careful study, enter into the description of phenomena as varied as percolation, turbulence, and polymers. Later on we will discuss the Hausdorff dimension, also called the Hausdorff-Besicovitch dimension (to be defined in Section VI.4), which is one of the possible fractal dimensions.

25. Ordinary differential equations, unlike partial differential equations, involve functions of only one variable: here the time.

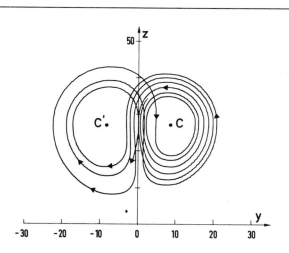

Figure VI.12 The Lorenz attractor.

Projection of part of a trajectory onto the plane (Y, Z) of the phase space for $r = 28$. C and C' are unstable fixed points.

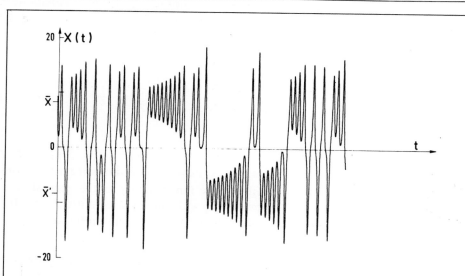

Figure VI.13 Graph of $X(t)$ for $r = 28$.

Note the disordered evolution, sometimes about a mean value $\overline{X}$, and sometimes about another value $\overline{X}'$. ($\overline{X}$ and $\overline{X}'$ are the X coordinates of the unstable fixed points C and C').

VI.3 EXAMPLES OF STRANGE ATTRACTORS

For low values of r, the stable solutions are stationary[26] (see Appendix D). When r exceeds 24.74 (for example, for $r = 28$), the trajectories projected onto the (Y, Z) plane become irregular orbits about points C and C' (see fig. VI.12), which are unstable fixed points of the flow. Figure VI.13 shows that the time evolution of, for example, $X(t)$ exhibits the irregular appearance of chaotic behavior.

For initial values $(X(0), Y(0), Z(0))$ taken at random, the corresponding trajectories are rapidly attracted towards a region A composed of the set of trajectories rotating about C and C' (see fig. VI.14). The attractor A — associated with a chaotic regime — is a strange attractor. It was in fact the first strange attractor[27] to be discovered and studied. Using the Lie derivative:

$$\frac{\partial \dot X}{\partial X} + \frac{\partial \dot Y}{\partial Y} + \frac{\partial \dot Z}{\partial Z} = -(\mathrm{Pr} + b + 1) = -\frac{41}{3}$$

we can calculate the contraction of volumes by the flow. After one unit of time (which here is roughly one rotation about C or C'), a volume is reduced by a factor $e^{-41/3} \sim 10^{-6}$. The Lorenz model is thus highly dissipative.

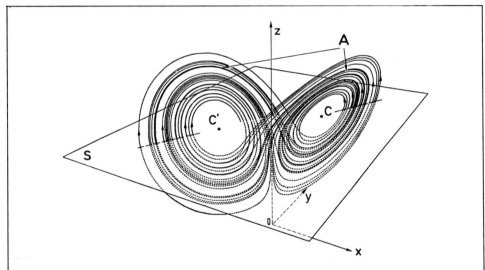

Figure VI.14 View of the Lorenz attractor for $r = 28$.
The apparent asymmetry of the attractor is due only to the perspective of the drawing. The two dashed lines show the intersections of the trajectories with the plane of section S (defined by $Z = r - 1 = 27$) with the condition $\dot Z > 0$, i.e. the Poincaré section of the attractor with this plane.
From O. Lanford.

26. Recall that throughout this section, $\mathrm{Pr} = 10$ and $b = 8/3$.
27. When Lorenz studied his model, the term "strange attractors" did not yet exist, nor had their characteristic properties been explored. Previously, the work of Rikitake and Allan had led to results which can today be said to relate to the properties of strange attractors.

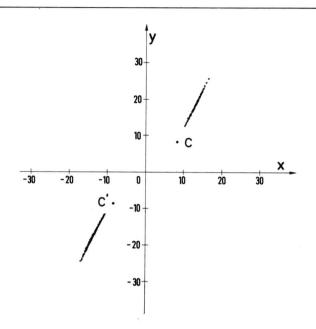

Figure VI.15 Poincaré section of the Lorenz attractor. The plane of section is the plane S of Figure VI.14. It contains the unstable fixed points C and C'. The Poincaré section appears to consist of two line segments.

The Poincaré section of the trajectories with the plane[28] $Z = r - 1$ containing C and C' is shown on Figure VI.15. At first glance, the section seems to consist of two segments, each corresponding to rotation about one of the unstable fixed points C and C' in a given direction. This implies that the trajectories can almost be inscribed on a surface and that the attractor has dimension two. In fact, though, if we "pierce" the apparently two-dimensional sheet we find a complex structure consisting of a large number of closely packed sheets. Thus, the Lorenz attractor is not quite a surface, but neither does it have any volume (the sheets have no transverse extent and are separated by empty space). Indeed, if we calculate its Hausdorff-Besicovitch dimension, which we will define in Section VI.4.2, we find a number very close to two, but nonetheless greater:

$$d = 2.06.$$

This result confirms the idea that the Lorenz attractor is not a simple surface. The fact

---

28. The plane $Z = r - 1$ contains the unstable fixed points C and C'.

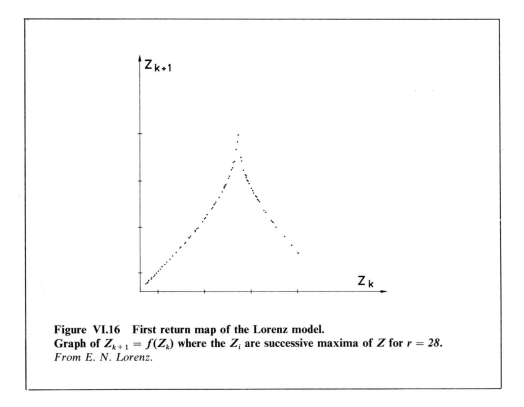

Figure VI.16 First return map of the Lorenz model.
Graph of $Z_{k+1} = f(Z_k)$ where the Z_i are successive maxima of Z for $r = 28$.
From E. N. Lorenz.

that its fractal dimension is very close to two is due to the strong volume contraction noted above. Also due to strong volume contraction is the fact that the Poincaré map can be described by a one-variable first return map $x_{k+1} = f(x_k)$ (see Section IV.4). Lorenz himself had defined such a mapping by graphing successive maximum values of Z, i.e. Z_{k+1}, as a function of the preceding maximum Z_k. These values Z_{max} are the Z coordinates of the points on the Poincaré section, with the surface $XY - bZ = 0$ (so that $dZ/dt = 0$, by the third equation of the Lorenz model). On Figure VI.16 we can see that, to a good approximation[29], the Poincaré map of the Lorenz attractor can be reduced to a first return map $x \mapsto f(x)$. The essential dynamical properties of the model are easily found using this first return map. It suffices to analytically approximate the curve represented in Figure VI.16, and then to carry out iterations of $x_{k+1} = f(x_k)$ according to the procedure outlined in Section VI.4.1.

Using this example we can illustrate one of the advantages of the reduction of a flow to a one-dimensional mapping. According to what we have seen in Chapter IV, if the slope of the curve $x_{k+1} = f(x_k)$ is everywhere greater than one (on the interval of the

---

29. The approximation is the same as that which identifies the Poincaré section of the attractor with two line segments.

mapping), then there will be sustained chaotic behavior, for any initial value x_0. For example, for $r = 28$ (the value chosen in numerous studies of the Lorenz attractor) the slope of the curve $Z_{k+1} = f(Z_k)$ is indeed everywhere greater than one in the interval under consideration. Therefore, no matter what initial value of Z_0 is taken, the

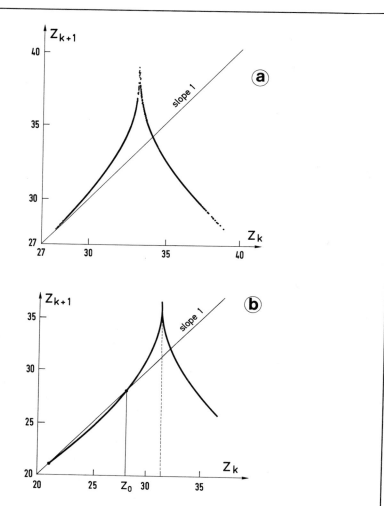

Figure VI.17 First return map of the Lorenz model.
a) $r = 24.06$.
b) $r < 24.06$.
Figure *a*) shows the limiting case for which the curve has a slope whose absolute value is everywhere greater than one.
From J. A. Yorke, E. D. Yorke.

behavior will be chaotic. This situation changes when r is decreased. In particular, one very interesting case occurs when we take values of r slightly under 24.06, such that the fixed points C and C' are stable. For such values, the first return map looks like Figure VI.17 b: one part of the curve has slope less than one ($Z < Z_0$) which implies the existence of an unstable fixed point Z_0 (intersection of the first return map with the identity map). Trajectories originating from initial values less than Z_0 immediately approach the stable fixed point below Z_0. On the other hand, trajectories originating from $Z > Z_0$ evolve chaotically for a while before arriving in the zone $Z < \check{Z}_0$, where they converge rapidly to the stable fixed point. This kind of chaos, called metastable, has some interesting properties which can be deduced from the form of the first return map.

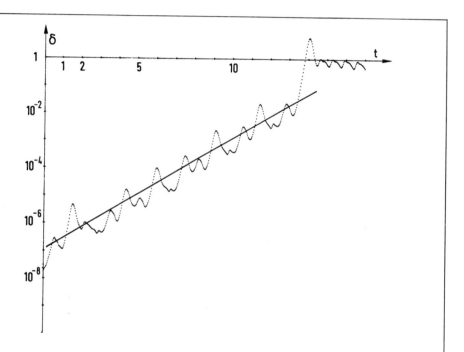

Figure VI.18 Evolution of a separation δ introduced in the initial values of the Lorenz equations.

The distance δ separates trajectories originating from two points located near the attractor and separated by only $\delta_0 = 10^{-8}$. On the average, the distance δ between the two trajectories increases with time t like $\delta = \delta_0 \exp(\lambda_1 t)$ where λ_1 is the largest Lyapunov exponent. Plotted logarithmically, an estimate of λ_1 is given by the average slope of the curve above, i.e. $\lambda_1 \sim 0.9$, in reasonable agreement with more detailed numerical calculations. The maximum value of the distance δ is, of course, limited to the size of the attractor, which explains the saturation observed for long times.

Finally, we illustrate the sensitivity to initial conditions displayed by the Lorenz attractor for the value $r = 28$. We calculate the flow starting from two very close initial values and watch the distance between the two trajectories evolve in time. Figure VI.18 shows that, on the average, this distance δ increases exponentially, as predicted by the general theory.

VI.3.2 THE HÈNON ATTRACTOR

We could simplify further than did Lorenz in his model of thermal convection by transforming the continuous-time equation to discrete-time iterations. A system of three differential equations, analogous to the Lorenz model, is then replaced by a two-dimensional mapping[30]. The successive points obtained in $\mathbb{R}^2$ are to be considered as belonging to the Poincaré section of the flow described by the system of three differential equations.

It is with this idea in mind that Hénon suggested the following mapping of the plane onto itself:

$$X_{k+1} = Y_k + 1 - \alpha X_k^2$$
$$Y_{k+1} = \beta X_k$$

where the constant α is used to control the nonlinearity of the mapping and β the role of dissipation. The values usually chosen for α and β, which we will use for the rest of this section, are $\alpha = 1.4$ and $\beta = 0.3$. Starting from a point (X_0, Y_0), we calculate (X_1, Y_1) and the succeeding points by iteration. The transformation is invertible[31], so that from P_k, we can find its antecedents P_{k-1}, P_{k-2} etc.

For all initial conditions in the basin of attraction (whose structure is extremely complex), the successive iterates converge rapidly towards the attractor represented in Figure VI.19. Once on the attractor, successive points appear in an irregular, unpredictable, and random fashion: the dynamics on the Hénon attractor are chaotic.

Examination of the evolution of the distance between two initially very close points confirms the diagnosis of SIC. Figure VI.20 shows the distance growing exponentially, on the average. Based on numerical calculations, we conclude that the Hénon attractor (for $\alpha = 1.4$ and $\beta = 0.3$) is indeed a strange attractor[32].

Let us use the Hénon mapping to illustrate some of the ideas mentioned previously. First, we return to the contraction of areas (analogous to volume contraction in a three-dimensional dissipative system). For a discrete-time mapping,

---

30. One welcome consequence is the reduction by an order of magnitude of the computational power needed to construct the attractor.

31. This is to be contrasted with the behavior of quadratic mappings of the interval into itself (see Chapter VIII).

32. In fact there exists no rigorous proof of this, despite the extreme simplicity of the original equations.

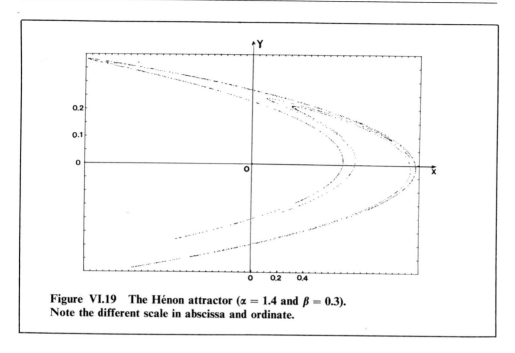

Figure VI.19 The Hénon attractor ($\alpha = 1.4$ and $\beta = 0.3$). Note the different scale in abscissa and ordinate.

the Jacobian plays the role of the Lie derivative (see Appendix E). For the Hénon mapping it is:

$$J = \begin{vmatrix} \dfrac{\partial X_{k+1}}{\partial X_k} & \dfrac{\partial X_{k+1}}{\partial Y_k} \\ \dfrac{\partial Y_{k+1}}{\partial X_k} & \dfrac{\partial Y_{k+1}}{\partial Y_k} \end{vmatrix} = \begin{vmatrix} -2\alpha X_k & 1 \\ \beta & 0 \end{vmatrix} = -\beta.$$

Therefore each iteration multiplies areas by a factor $|\beta|$. Areas are contracted if $|\beta| < 1$. For $\beta = 0.3$ as in the case under consideration, a unit area at iteration k is mapped into a set of points of area 0.3 at iteration $k + 1$.

This relatively weak contraction is to be contrasted with that occurring in the Lorenz model where the volume reduction rate is on the order of 10^{-6} over each time interval between points in a Poincaré section (for the Hénon mapping the corresponding time is that separating consecutive iterates). This very rapid contraction prevents us from discerning the fractal structure of the Poincaré section of the Lorenz model. The more moderate contraction rate of the Hénon attractor permits observation of its fractal structure, using graphs at different scales of magnification. In Figure VI.21 *a* we note the presence of three parallel lines, one of which appears thicker than the others. An enlargement, Figure VI.21 *b*, shows that this uppermost line is in fact composed of three lines. A second enlargement, Figure VI.21 *c*, reveals the

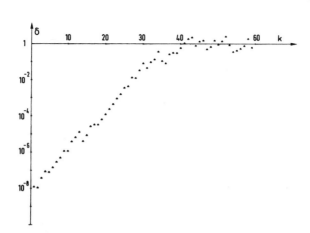

Figure VI.20 Evolution of a separation δ introduced in the initial value of Hénon mapping.

The distance δ separates trajectories originating from two points located near the attractor and separated by only $\delta_0 = 10^{-8}$. On the average, the distance δ between the two trajectories increases with the number k of iterations like $\delta = \delta_0 \exp(\lambda_1 k)$ where λ_1 is the largest Lyapunov exponent. Plotted logarithmically, an estimate of λ_1 is given by the average slope of the curve above, i.e. $\lambda_1 \sim 0.50$ in reasonable agreement with more detailed numerical calculations. Saturation of the slope occurs for the reason given in Figure VI.18. See also Figure VIII.9.

uppermost line of Figure VI.21 b to be in turn composed of three lines, and so on. That is, the structure of the attractor is repeated at successive scales [33] of observation. Such a structure, characteristic of a fractal object [34], can be considered as a section through a highly layered set, such as a strange attractor. The Hénon attractor, an object somewhere between a line and a surface, has a Hausdorff dimension of $d = 1.26$.

The Hénon attractor can be used to illustrate the coexistence of area contraction with SIC. Recall that the contraction takes place at a constant rate of $|\beta| = 0.3$ per iteration. This contraction is not isotropic: in a first approximation, a circle (of radius r small compared to the size of the attractor) becomes an ellipse, increasingly elongated in one direction, and compressed in the other (see fig. VI.22). The lengths of the principal axes are, after k iterations, $r\Lambda_1^k$ and $r\Lambda_2^k$, respectively, where Λ_1 and Λ_2 are

---

33. In some cases, this is scale invariance.
34. The simplest and most famous example of an object of this type is the Cantor set described in Section VI.4.2.

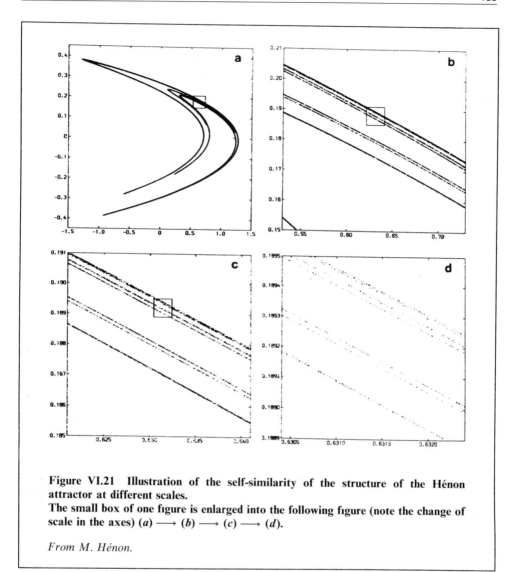

Figure VI.21 Illustration of the self-similarity of the structure of the Hénon attractor at different scales.
The small box of one figure is enlarged into the following figure (note the change of scale in the axes) (a) ⟶ (b) ⟶ (c) ⟶ (d).

From M. Hénon.

two constants[35] which can be calculated only numerically. With $\Lambda_1 > 1$ and $\Lambda_2 < 1$ (and $\Lambda_1 \Lambda_2 = |\beta| < 1$ since there is global contraction) lengths are stretched in one direction and contracted in the perpendicular direction (see fig. VI.22 c). Here we again encounter the idea of hyperbolicity, implying convergence of trajectories in one direction and divergence in another.

35. We do not know how to calculate Λ_1 and Λ_2 analytically, but only that $\Lambda_1 \Lambda_2 = |\beta|$.

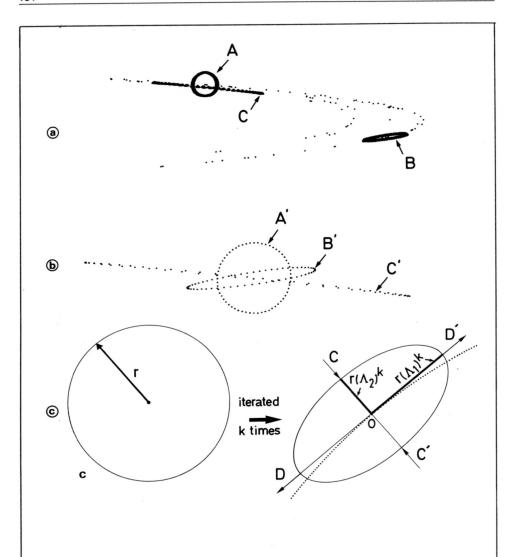

Figure VI.22 Deformation of a circle by iteration of the Hénon mapping.
a) We consider a set of points *A* located on the circumference of a circle whose center is on the attractor. After a first iteration, we obtain the set *B*. Note the effect of stretching in the direction tangent to the attractor, and of contraction in a direction approximately perpendicular to it. A second iteration transforms *B* into *C*.
b) This transformation of a circle into ellipses and the corresponding contraction of areas is made more visible by enlargement and recentering of *A*, *B*, *C* into *A'*, *B'*, *C'*.
c) Illustration of hyperbolicity and of the role of Lyapunov coefficients using as an example the transformation of a circle into ellipses.

VI.3 EXAMPLES OF STRANGE ATTRACTORS

The Hénon attractor can be considered as being the Poincaré section of the three-dimensional flow of a chaotic regime, so that each point represents the intersection of the plane on which the Hénon mapping is defined with a trajectory which is locally perpendicular to it. For a small ellipse located around the point O, the direction of divergence corresponds to the tangent DD' (see fig. VI.22 c), while the direction of contraction is along CC', perpendicular to DD'. The transformation of a circle into increasingly elongated ellipses provides a concrete illustration of *Lyapunov coefficients* (cf. Appendix B): in the limit of r small and k large, Λ_1 and Λ_2 are the two Lyapunov coefficients of the Hénon mapping. For the other extreme, of r on the order of the attractor size, Figure VI.23 shows that here, too, evolution on the attractor involves the dual operations of stretching and folding.

VI.3.3 EXPERIMENTAL ILLUSTRATIONS OF STRANGE ATTRACTORS

The examples that follow come from experiments carried out on the three dynamical systems described in Chapter V.

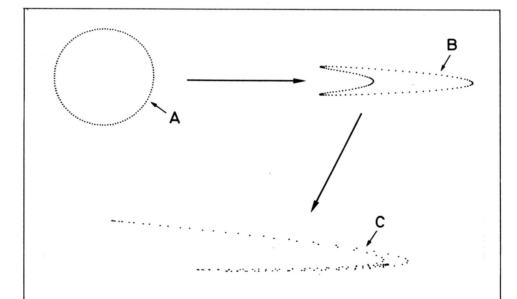

Figure VI.23 Illustration of folding in the Hénon mapping.
The procedure is the same as that described in Figure VI.22, but here, the original circle A is comparable in size to the attractor. The first and second iterates B and C illustrate the effect of stretching combined with folding.

We begin with the compass which offers the advantage of an adjustable dissipation rate. By adopting moderate values for the fluid friction, we can observe the beautifully layered structure of an experimental strange attractor. The compass, as described in Chapter V, is a magnet subject to the combined influence of a stationary magnetic field with induction B_1 and another magnetic field with induction B_0 rotating with angular velocity ω. The viscous friction is produced by the oil contained in one of the bearings of the magnet. It is easy to obtain a Poincaré section of the phase trajectories, since we are dealing with a forced oscillator, for which the period $2\pi/\omega$ of the rotating field is a natural sampling time.

It therefore suffices to record a variable $x(t)$ proportional to the velocity of the magnet, and its derivative $\dot{x}(t)$, each time B_0 rotates by 2π. Carried out on a chaotic regime, this operation results in the graph of Figure VI.24. We observe the

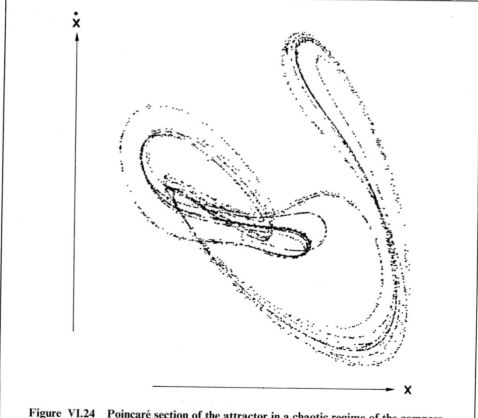

Figure VI.24 Poincaré section of the attractor in a chaotic regime of the compass. The angular velocity of the magnet is proportional to x.
From V. Croquette.

VI.3 EXAMPLES OF STRANGE ATTRACTORS

characteristic structure of a strange attractor, with a great deal of folding and a complex structure recalling the Hénon attractor.

Another, perhaps even more striking, resemblance between the Hénon attractor and experiment can be found in the R.B. instability. The experimental procedure is slightly different from that utilized for the compass; since R.B. convection is not forced periodically, the calculation of Poincaré sections is more complicated. Recalling what was said in Chapter V, two variables[36] (ΔT) and $(\Delta T)'$ are measured in two different locations of the convecting fluid (see fig. V.8). A three-dimensional phase space is defined by the coordinates $(\Delta T), (\Delta \dot T)$ and $(\Delta T)'$. If it is possible to detect a signal $(\Delta T)'$ representing essentially one of the thermoconvective oscillators, then its period can be used as a time interval for stroboscoping trajectories projected onto the $(\Delta T, \Delta \dot T)$ plane. It is, however, generally very difficult to obtain signals $(\Delta T)'$ that are of sufficient spectral purity, even after they have been filtered. The Poincaré section is then obtained as the intersection of trajectories in the $(\Delta T), (\Delta \dot T), (\Delta T)'$ phase space with a plane $(\Delta T)' = K$. To do this, one plots the coordinates $(\Delta T), (\Delta \dot T)$ each time $(\Delta T)'$ attains a given value K (with a condition on the sign of $d(\Delta T)'/dt$, e.g. $d(\Delta T)'/dt > 0$).

This is how the Poincaré section of Figure VI.25 was obtained, in a chaotic convective regime for which the Fourier spectrum contains not only discrete lines but also substantial broad-band noise. We see (fig. VI.25) that the points in the Poincaré section are arranged in a complex but well-defined structure[37].

We now turn to the chemical B.Z. reaction to study an aspect directly related to dynamics. The variable measured is the concentration of bromide ions in the reactor. Recordings of the time series $X(t_i)$ containing several tens of thousands of points have been made for a fixed value of the reactant flux (the control parameter μ). Phase portraits are obtained by the method of time delays (see Section IV.5) which gives a representation of the attractor in a three-dimensional phase space with coordinates $X(t_i)$, $X(t_i + \tau)$, and $X(t_i + 2\tau)$.

The simplest illustration that can be given of this procedure is presented in Figure VI.26. For certain values of the reactant flux, the B.Z. reaction is periodic and the phase portrait is then a limit cycle, representable in a space of two dimensions. The situation changes radically for other flux values, where chaotic regimes can be found, as the phase portrait projection of Figure VI.27 testifies. The coordinates chosen are:

$$X = X(t_i)$$
$$Y = X(t_i + \tau)$$

with $\tau = 8.8$ seconds[38].

---

36. Generally (ΔT) and $(\Delta T)'$ are measured by shining narrow light beams horizontally through the fluid. The beams are deflected proportionately to the temperature gradient.

37. If we had a nondeterministic chaotic regime, we would see a cloud of points without any fine structure.

38. This time is to be compared with the time of traversal of the attractor, which is of the order of a hundred seconds.

138　　　　　　　　　　　　　　　　　　　　　CHAPTER VI　STRANGE ATTRACTORS

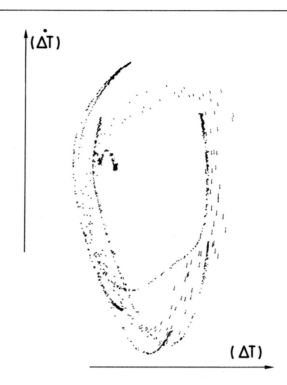

Figure VI.25 Poincaré section of the attractor in a chaotic regime of R.B. convection.
(ΔT) is approximately proportional to the vertical temperature gradient at a given point in the fluid layer. Here, the Poincaré section has been obtained by stroboscoping the signals at the frequency of one of the two thermoconvective oscillators.
From M. Dubois.

Apart from its aesthetic interest, this projection tells us little about the dynamical state of the system. It would be risky to conclude on the sole evidence of Figure VI.27 that we have a strange attractor. The same uncertainty in interpretation could exist even if we represented the attractor in a three-dimensional phase space. A decisive step is taken, however, when we make a Poincaré section through a three-dimensional representation of the attractor, the third coordinate being $X(t_i + 2\tau)$. The plane of section is perpendicular to the plane of Figure VI.27, on which it is represented by a dashed line. Figure VI.28 shows the Poincaré section thus obtained. We note that its quasi-linear appearance is entirely fortuitous: another choice of τ and another plane of intersection would have given a different curve (as we will see below). However the narrowness of this quasi-linear object is reminiscent of the Lorenz attractor. The

VI.3 EXAMPLES OF STRANGE ATTRACTORS

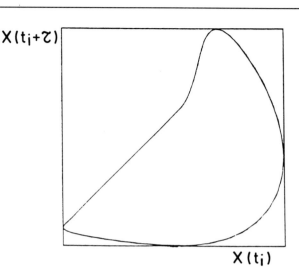

Figure VI.26 Limit cycle obtained for a periodic regime of the B.Z. reaction. The signal measured — the potential of a specific electrode — is proportional to the logarithm of the concentration of Br^- ions. The representation uses what is called the method of delays: the signal at a time $t + \tau$ is plotted as a function of its value at time t (here the delay time is equal to 8.8 sec).

From J. C. Roux, R. Simoyi, and H. Swinney.

reason is the same: the B.Z. system — under the conditions studied here — is highly dissipative; hence the very small width of the attractor. Later, we will see further confirmation of this strong contraction.

The deterministic nature of the chaotic behavior corresponding to the attractor can be indisputably confirmed by the first return map, which can be constructed from the Poincaré map when the dissipation rate is as great as it is here. By graphing the ordinate X_{k+1} of a point on Figure VI.28 as a function of the ordinate X_k of its antecedent, we see on Figure VI.29 *a* that all of the points are located on a well-defined curve. Thus, knowledge of one point determines that of the following point. This fact illustrates the *order* governing the trajectories on the attractor.

To again illustrate the two crucial properties of a strange attractor let us demonstrate the effects of attraction and of SIC. The first phenomenon can be illustrated by introducing a perturbation: for example, an air bubble carried by the flux of one of the reactants temporarily perturbs the reaction from its operating regime. Figure VI.29 *b* shows the first return map obtained, as well as the continuous curve from Figure VI.29 *a*. After the point designated by 4, the arrival of a bubble has temporarily altered the state of the reactor. The effect of this perturbation can be seen

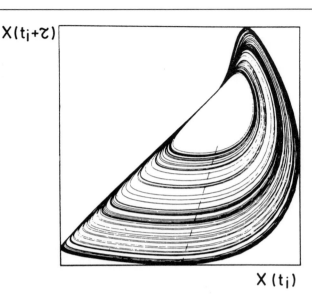

Figure VI.27 Aperiodic regime of the B.Z. reaction.
Same type of diagram as Figure VI.26, but for a chaotic regime. The loops are projections of trajectories onto the $(X(t_i), X(t_i + \tau))$. plane.

From J. C. Roux, H. Swinney.

on point 5, which is visibly off the curve. The next points 6, 7, and 8 approach the curve — and thus the attractor — and after point 9, the measurement is not sufficiently precise for us to detect any displacement from the curve. In this way, the strength of attraction brings a "dissident" point back to the attractor in four orbits.

As for sensitivity to initial conditions — i.e. (exponential) amplification of displacements — let us return to the projection of the phase portrait. Figure VI.27 is obtained by graphing a large number of coordinate pairs $(X(t_i), X(t_i + \tau))$, where the t'_is are separated by an interval Δt (here 0.88 second) of about a hundredth of a cycle. Instead of systematically plotting all points, we now proceed in the following way: each time a point passes near O (see fig. VI.30) we follow its evolution by plotting its position at later times. In Figure VI.30 a, a set of points is shown at a time $t = 80\ \Delta t$, where $t = 0$ is the time of passage near O. In Figure VI.30 b, the time is $105\,\Delta t$, and in Figure VI.30 c it is $210\,\Delta t$. We see that at $t = 0$, there exists a set of points so close together as to be practically indistinguishable from O. At $t = 80\,\Delta t$, we notice that these points begin to separate (fig. VI.30 a, zone A). The separation increases rapidly with time: by $t = 210\,\Delta t$ (about the time necessary to cycle twice around the attractor) the points are scattered almost throughout the attractor (fig. VI.30 c, zone C). This means that memory of the initial condition (all the points are practically

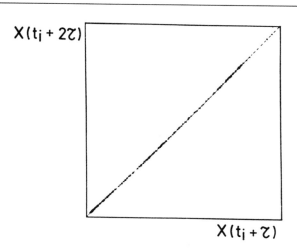

Figure VI.28 Poincaré section.
This is a section through the trajectories in the $X(t_i)$, $X(t_i + \tau)$, $X(t_i + 2\tau)$ space, of which Figure VI.27 is a projection. The plane of section is perpendicular to that of Figure VI.27, where their intersection is marked by a dashed line. We remark that the quasi-linear appearance of the section is completely fortuitous.

From J. C. Roux, H. Swinney.

indistinguishable from O) is lost due to the divergence of the initial displacements from O. This is indeed SIC.

We can even show explicitly how the B.Z. reaction displays the dual operations of stretching and folding common to all strange attractors. By changing the delay time τ from that adopted in Figure VI.27, the appearance of the phase portrait changes radically, as we can see in Figure VI.31, which shows the trajectories projected onto the $(X(t_i), X(t_i + \tau'))$ plane with $\tau' = 53$ seconds[39]. To follow the evolution of the corresponding flow, we examine Poincaré sections with planes perpendicular to that of Figure VI.31, each denoted by a number ①, ②, ③, ..., ⑨.

Figure VI.32 shows how the Poincaré section gradually changes during an average cycle around the attractor of Figure VI.31. We see that folding takes place between ② and ⑧. Stretching (which is less visible since the scale varies greatly from one figure to the next) occurs mainly as we go from ⑨ to ① . This is why the apparent width of the section is much smaller in ① than in ⑨.

---

39. We emphasize that the experimental data — the values of $X(t_i)$ — used in Figures VI.31 and VI.32 are the same as those used in Figures VI.27 and VI.28. This confirms the fortuitousness of the quasi-linear shape of the Poincaré section of Figure VI.28.

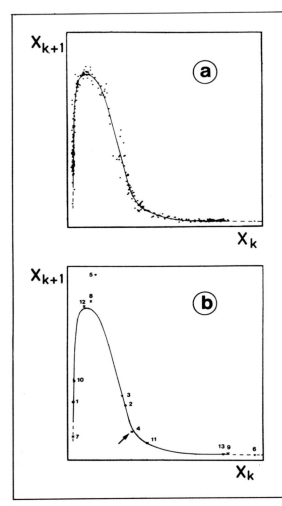

Figure VI.29 First return map.
X_k is the ordinate of a point of the section of Figure VI.28. To construct the first return map, we plot the ordinate X_{k+1} as a function of the ordinate X_k of the immediately preceding point.

a) Set of points obtained for a chaotic regime of the attractor. The continuous curve is drawn merely to guide the eye.

b) The preceding continuous curve is redrawn. During the recording of a series of points-numbered 1 to 13, a perturbation intervenes approximately at the moment that point 4 is recorded. The subsequent points return rapidly towards the asymptotic curve, illustrating the effect of attraction.

Figure VI.30 Illustration of SIC.
We recognize the dashed lines as being a few of the chaotic trajectories of Figure VI.27. ▶
The position of the points at $t = 0$ is O.
a) Position of the points at $t = 80 \, \Delta t$.
b) Position of the points at $t = 105 \, \Delta t$.
c) Position of the points at $t = 210 \, \Delta t$.
($\Delta t = 0.88$ second. The avarege time required to complete a loop is $120 \, \Delta t$).
From J. C. Roux, H. Swinney.

VI.3 EXAMPLES OF STRANGE ATTRACTORS

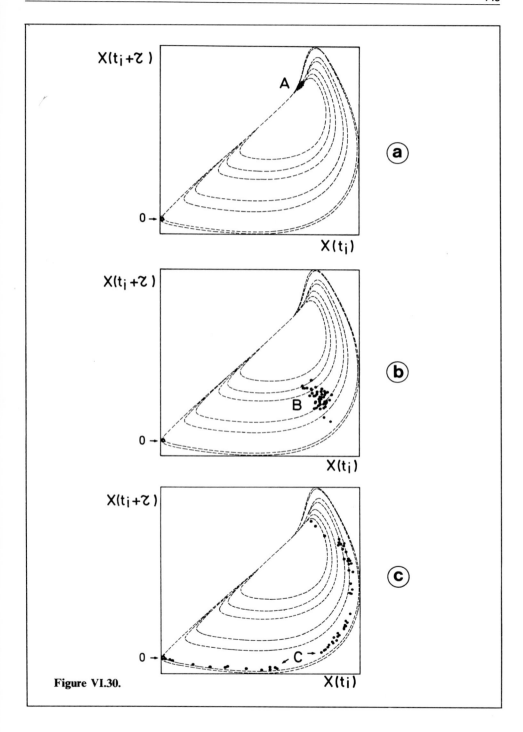

Figure VI.30.

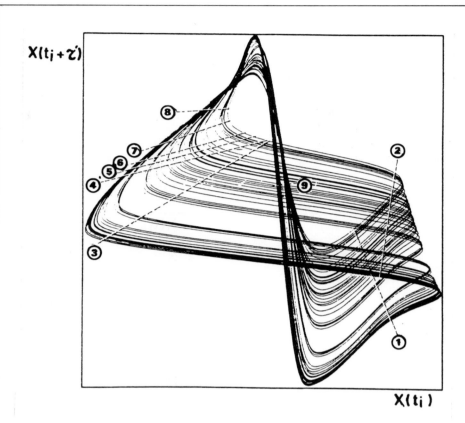

Figure VI.31 Projection of the attractor for a different delay time τ'. The trajectories are taken from the same chaotic regime as in Figure VI.27, but here, $\tau' = 53$ seconds (instead of 8.8 seconds). The dashed lines indicate different planes of section, all perpendicular to that of the figure used for the Poincaré sections presented in Figure VI.32.

From J. C. Roux, R. Simoyi, H. Swinney.

VI.4 Measuring the dimension of strange attractors

VI.4.1 PROBLEMS OF CHARACTERIZATION

A dynamical regime can be characterized by Fourier analysis. However Fourier analysis does not distinguish between chaos involving a small number of degrees of

VI.4 MEASURING THE DIMENSION OF STRANGE ATTRACTORS

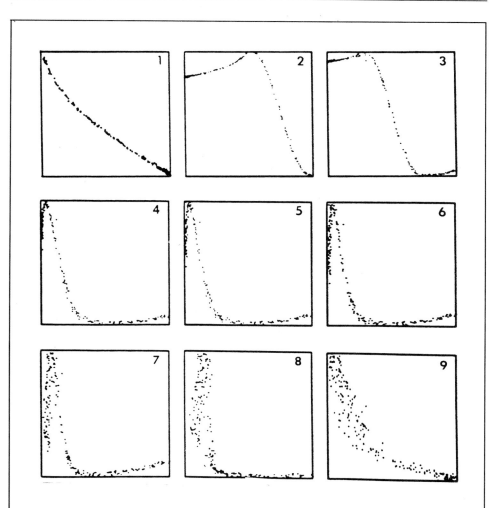

Figure VI.32 Poincaré sections.

Each Poincaré section is identified by a number (N) and corresponds to a different plane of section, whose projection in Figure VI.31 is indicated by the same number. The scale varies from one section to the next. For each section (N), the relative scales for the abscissa EX and ordinate EY, are listed below as $(N) EX \times EY$:
(1) 0.3×0.2; (2) 1.1×0.9; (3) 1.0×1.0; (4) 0.5×0.9; (5) 0.5×0.9;
(6) 0.5×0.8; (7) 0.4×0.5; (8) 0.4×0.2; (9) 0.4×0.08.

From J. C. Roux, R. Simoyi, H. Swinney.

freedom (deterministic chaos) and white noise[40]. In this respect, the study of phase trajectories by means of Poincaré sections offers a significant advantage. However, only qualitative information can be obtained. In addition, the use of Poincaré sections is limited in practice to three-dimensional phase spaces.

A more quantitative characterization of a chaotic regime comes from determination of the largest Lyapunov exponent (see Appendix B). We will not discuss the algorithms for determining the largest Lyapunov exponent from experimental data, since they are not yet totally operational. Another interesting characterization of a chaotic regime is given by the fractal dimension of the attractor, to which this section is devoted. Since the dimension of a strange attractor is generally noninteger, let us develop this idea in more detail.

VI.4.2 FRACTAL DIMENSIONS

Consider a set of points in a p-dimensional space. We seek to cover this set by (hyper)cubes of linear dimension ε. Let $N(\varepsilon)$ be the smallest number of cubes necessary to accomplish this (see fig. VI.33). The Hausdorff (also called Hausdorff-Besicovitch) dimension D is defined to be the limit, if it exists, of the ratio $\ln N(\varepsilon)/\ln (1/\varepsilon)$ as the length ε of the hypercubes tends to zero. That is:

$$D = \lim_{\varepsilon \to 0} \frac{\ln N(\varepsilon)}{\ln (1/\varepsilon)}$$

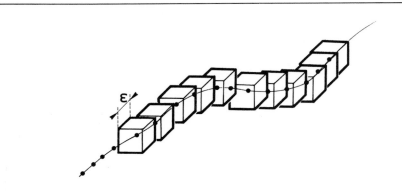

Figure VI.33 Illustration of the covering of an object (a set of points) by cubes of linear dimension ε.

---

40. We designate by "white noise" a noise which — unlike deterministic chaos — originates in the existence of a very large number of independent modes or degrees of freedom (see Section III.3.3).

VI.4 MEASURING THE DIMENSION OF STRANGE ATTRACTORS

Another way of saying this is that the minimum number $N(\varepsilon)$ of cubes necessary to cover the set of points varies as ε^{-D} with ε. When the set is merely a single point, we have:

$$N(\varepsilon) = \text{constant} = 1.$$

Thus $D = 0$ is the Hausdorff dimension of the point, whose Euclidean dimension is also 0. If the set is a line segment of length L, then:

$$N(\varepsilon) = L\varepsilon^{-1}$$

so that $D = 1$, while for a surface of area S:

$$N(\varepsilon) = S\varepsilon^{-2}$$

leading to $D = 2$.

Up to this point the definition of Hausdorff dimension has not added anything to that of Euclidean dimension. However let us now consider the Cantor set, obtained by an iterative process. First, the central third of the unit segment is removed. Then, the central third of each of the two remaining segments is removed. The operation is repeated indefinitely, as illustrated in Figure VI.34. In this way we obtain an infinite set of disconnected points[41] whose dimensionality is between 0 and 1. The Hausdorff

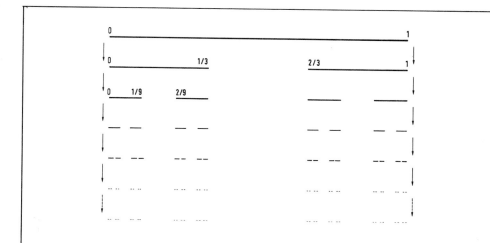

Figure VI.34 First steps in the construction of the Cantor set.

41. Note that the structure obtained is invariant under changes in scales. "Seen from afar", the Cantor set seems to be an interval missing its central third, but seen from closer up, we find the same internal structure at all scales. The trace of this scale invariance is present in the organization of strange attractors into sheets.

dimension of the Cantor set can be easily deduced from its construction. For $\varepsilon = 1/3$, the number of elements (here, the hypercubes are merely segments) necessary to cover the set is $N(1/3) = 2$. Similarly, for $\varepsilon = 1/9$, $N(\varepsilon) = 4$, and more generally, for $\varepsilon = (1/3)^m$, $N(\varepsilon) = 2^m$. From the definition of Hausdorff dimension:

$$D = \lim_{m \to \infty} \frac{\ln 2^m}{\ln 3^m} \simeq 0.63.$$

With this example, we see that the Hausdorff dimension is a generalization of the usual geometric dimension, permitting characterization of "fractal" objects.

We consider a second classical fractal object, called the snowflake. It has the curious property — shared by Poincaré sections of many strange attractors — of having an infinite perimeter while occupying only a bounded region of the plane. Take an equilateral triangle, divide each side into thirds, and attach smaller equilateral triangles onto each of the three central thirds (see fig. VI.35). By iterating this

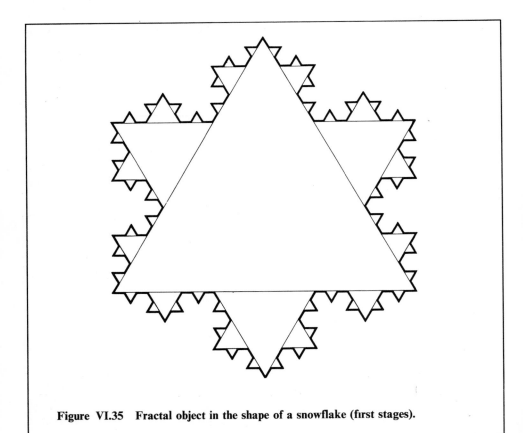

Figure VI.35 Fractal object in the shape of a snowflake (first stages).

construction an infinite number of times, we obtain a fractal object[42], sometimes called *Koch's curve*. Using the same reasoning as before, we find its Hausdorff dimension to be:

$$D = \frac{\ln 4}{\ln 3} \simeq 1.26.$$

Unfortunately, in many other cases of practical interest, the computation of the Hausdorff dimension directly from its definition as a limit converges too slowly when the dimension p of the phase space is greater than two. We must therefore use another fractal dimension v which can be calculated more rapidly. Although v, called the correlation dimension, is in general not exactly equal to the Hausdorff dimension D, it does bound it from below:

$$v \leqslant D.$$

We illustrate this second approach with examples from plane geometry. Consider a set of points on the plane representing, for example, the Poincaré section of a flow in $\mathbb{R}^3$, and let $N(r)$ be the number of points of the set located inside a circle of radius r. The correlation dimension v is determined from the variation of $N(r)$ with r. For a discrete set of points uniformly distributed on a curve (dimension one), we have, for r sufficiently small:

$$N(r) \sim r$$

that is, $N(r) \sim r^v$ with $v = 1$ (see fig. VI.36 b). If, on the other hand, the points are uniformly distributed on a surface (dimension two; see fig. VI.36 c):

$$N(r) \sim r^2, \quad v = 2.$$

We can now consider general objects of arbitrary dimension, such as the Cantor set described above (see fig. VI.36 d). The number of points $N(r)$ located inside a circle will grow, on the average, more slowly than the radius r. Setting $N(r) \sim r^v$, it can be calculated that $v \approx 0.63$, which is equal to the Hausdorff dimension calculated previously. Similarly, for the Hénon attractor (see Section VI.3.2), we obtain an average dependence:

$$N(r) \sim r^v, \text{ with } : v \simeq 1.26.$$

The method is generalized to p-dimensional spaces by defining $N(r)$ to be the number of points contained in a p-dimensional hypersphere of radius r[43].

---

42. Like the Cantor set, this object displays scale invariance.
43. The correlation dimension v depends on the density of points on the attractor, since it involves counting points. In contrast, the Hausdorff dimension D of a set depends only on its geometric composition; by definition, it is a "maximal" dimension of the object. D is dominated by the piece requiring the largest number of spheres to cover it. This is the most "folded" part, if the object is a strange attractor obtained by stretching and folding. If the density of points on the attractor is low in just this region, it will not contribute appreciably to v. This is a qualitative justification of the fact that v is bounded from above by D. In constructions such as those of the Cantor set or of the Koch snowflake, the two dimensions are equal when the density of points used to calculate v is uniform over the object.

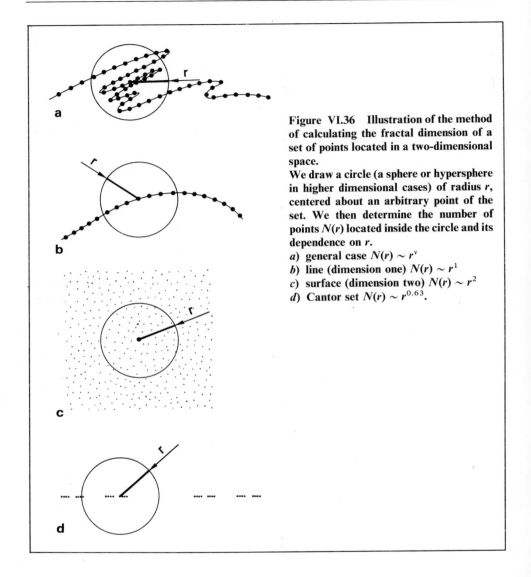

Figure VI.36 Illustration of the method of calculating the fractal dimension of a set of points located in a two-dimensional space.
We draw a circle (a sphere or hypersphere in higher dimensional cases) of radius r, centered about an arbitrary point of the set. We then determine the number of points $N(r)$ located inside the circle and its dependence on r.
a) general case $N(r) \sim r^\nu$
b) line (dimension one) $N(r) \sim r^1$
c) surface (dimension two) $N(r) \sim r^2$
d) Cantor set $N(r) \sim r^{0.63}$.

VI.4.3 GEOMETRIC CHARACTERIZATION OF AN ATTRACTOR

Inspired by the approach described above, combined with the ability to reconstruct phase space trajectories by time delays, an efficient method of geometric characterization has recently been proposed.

Recall that starting with one time-dependent variable $X(t)$, we can reconstruct a trajectory in a p-dimensional phase space (see Section IV.5) by taking as coordinates

VI.4 MEASURING THE DIMENSION OF STRANGE ATTRACTORS

$X(t), X(t + \tau), X(t + 2\tau) \ldots X(t + (p - 1)\tau)$ where τ is an appropriate[44] delay time. In practice the time t is discretized, so that we obtain a series of p-dimensional vectors representing the phase portrait of the dynamical system (see fig. VI.37).

In a chaotic regime (corresponding to a strange attractor), the positions of two points along the same trajectory but far apart in time are, by definition, uncorrelated[45], due to SIC. However, since all the points are on the attractor, there exists a spatial correlation that we can seek to characterize through some function. We can write:

$$C(r) = \lim_{m \to \infty} 1/m^2 \times [\text{number of pairs } i, j \text{ whose distance } |\vec{x}_i - \vec{x}_j| < r]$$

where i and j are indices ordering the points along a trajectory containing a total of m points (see fig. VI.37). This function can be defined in a more formal manner:

$$C(r) = \lim_{m \to \infty} \frac{1}{m^2} \sum_{i,j=1}^{m} H(r - |\vec{x}_i - \vec{x}_j|)$$

where H is the Heaviside function defined by $H(x) = 1$ for positive x, 0 otherwise. The number of pairs of points i, j whose distance is less than r is also the sum of all $N(r)$ (see

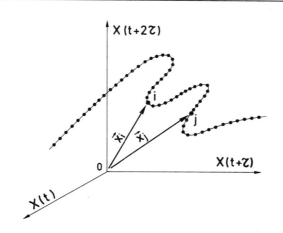

Figure VI.37 Schematic representation of a trajectory reconstructed point by point from a simple variable $X(t)$ using the method of delays.
Here we have used a three-dimensional space with coordinates $X(t)$, $X(t + \tau)$, $X(t + 2\tau)$.

---

44. The optimal choice of τ depends a great deal on the particular form of $X(t)$, and is determined in part empirically.
45. S.I.C. is again the culprit here.

above), counting up the points in hyperspheres centered on *every*[46] point of the attractor. $C(r)$ is therefore proportional to $N(r)$:

$$C(r) \sim r^\nu.$$

Note that the validity of this power law is limited to values of r reasonably small compared to the dimensions of the attractor; as r is increased, $N(r)$ necessarily saturates as r attains values comparable to the attractor size. On the other hand, for very small values of r, the number of pairs i, j whose distance is less than r becomes small, since the number of points on the attractor is finite, and the statistics become poor. In addition, the relative contribution of instrumental uncertainty then becomes dominant. In practice, therefore, it is only in a limited range of r that the power law $C(r) \sim r^\nu$ is satisfied and can be used to determine ν, the correlation dimension of the attractor.

VI.4.4 IMPLEMENTATION

Starting from the discrete values $X(t_i)$ obtained experimentally, we reconstruct the trajectory in a p-dimensional space, as described above, for increasing integer values of p:

$$p = 2, 3, 4, 5, \ldots$$

For each value of p, we calculate $C(r)$ and determine the slope of the function f defined by $\log C(r) = f(\log r)$, arriving at an exponent ν. For a periodic regime, whose phase portrait is a limit cycle, the dependence of $C(r)$ on r is strictly linear (up to size effects). Contrast this with the case of white noise[47]. The signal can be considered to be a superposition of an infinite number of independent oscillatory modes. Such a regime can therefore be described by an attractor T^n, with n very large. The trajectories will densely cover any phase space of dimension:

$$p \leqslant n.$$

Indeed, Figure VI.38 shows that the characteristic functions $C(r)$ obtained from a white noise have slopes on a log-log plot which continue to increase with p: we find $\nu \sim p$. This result can be extended: as long as the calculated value of ν is equal to p (or continues to grow with p), we know that the dimension of the space used for the calculation is smaller than (or comparable to) that of the corresponding attractor. If, on the other hand, the dimension ν calculated for a chaotic regime becomes independent of p, then the chaos is deterministic and the corresponding attractor strange.

---

46. This summation over all pairs i, j has the beneficial effect of increasing the statistical accuracy of $C(r)$. On the other hand, the procedure is very costly in computation time. In practice, experience shows that a very reasonable average is already obtained after a hundred or so points ($m \simeq 100$), or about 10^4 operations, resulting in a considerable economy of computation time.

47. For the definition of white noise, see Section III.3.3.

VI.4 MEASURING THE DIMENSION OF STRANGE ATTRACTORS

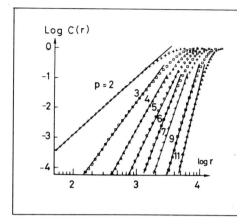

Figure VI.38 Log-log plot of $C(r)$ for white noise.
Note that the slope (which measures the exponent v) continues to grow as the dimension p of the representation space is increased.
From B. Malraison, P. Atten, P. Bergé, M. Dubois.

To illustrate this method of calculating v for strange attractors, we consider the Hénon attractor defined in Section VI.3.2. Figure VI.39 shows the function $C(r)$ on a log-log plot. The value of v that can be deduced from it is:

$$v = 1.25 \pm 0.05$$

which is consistent with the value of the Hausdorff dimension calculated from its definition:

$$D = 1.26...$$

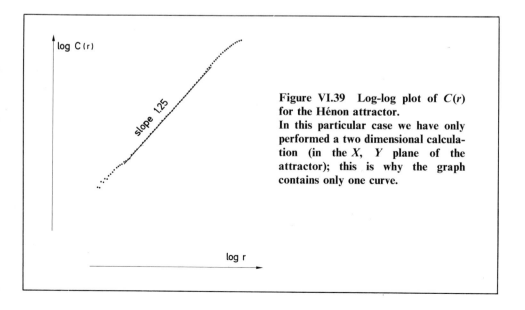

Figure VI.39 Log-log plot of $C(r)$ for the Hénon attractor.
In this particular case we have only performed a two dimensional calculation (in the X, Y plane of the attractor); this is why the graph contains only one curve.

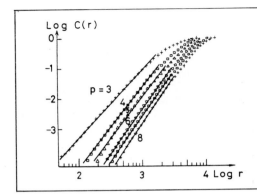

Figure VI.40 Log-log plot of $C(r)$ for turbulent R.B. convection.
We see that, unlike for the case of white noise presented in Figure VI.38, the slope of the curves stops increasing beyond the spatial dimension $p = 4$.
From B. Malraison, P. Atten, P. Bergé, M. Dubois.

Finally, we study an example from the R.B. instability in a confined geometry (cf. Section V.3). The fluid is silicone oil with Prandtl number 40, contained in a cell whose aspect ratio (width/height) is 2. The convective structure consists of two rolls. By increasing the Rayleigh number, we go from the periodic regime to chaos by a subharmonic cascade (see Chapter VIII). The variable monitored is again the deflection of a narrow light beam shone through the cell, measuring the vertical temperature gradient. The data analyzed here come from a chaotic regime at the end of an inverse cascade (again, see Chapter VIII). Figure VI.40 shows that the slopes of the functions $C(r)$ on a log-log plot are reasonably well-defined and independent of the dimension p of the phase space as soon as p exceeds 3. The saturation of v as p is increased is better illustrated by Figure VI.41, and contrasted with the linear dependence of v on p for white noise (or random signal). This type of data analysis demonstrates the deterministic nature of the chaotic behavior in this regime and, in addition, determines a lower bound on the number of degrees of freedom excited.

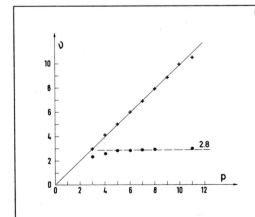

Figure VI.41 Variation of the exponent v as a function of the dimension p.
We deduce v from the slopes of the characteristics presented in Figures VI.38 and VI.40. For white noise, v increases linearly with p. In contrast we see a distinct saturation of v when the data is taken from turbulent R.B. convection.
From B. Malraison, P. Atten, P. Bergé, M. Dubois.

VI.5 The horseshoe attractor

Throughout this chapter we have described the various properties of strange attractors. A theoretical model which illustrates these properties is a mapping of the plane into itself proposed by Smale, enlarging on previous work by Levinson.

The basic operations necessary for creating a strange attractor are *stretching*, followed by *folding*, so that the attractor will only occupy a bounded portion of the phase space. This twofold operation is the starting point of the model. We take a rectangle $ABCD$ (see fig. VI.42) and stretch it by a factor of two (for example) in the x direction, while contracting it by a factor of 2η in the y direction; we thus obtain the rectangle $A_1 B_1 C_1 D_1$. For $\eta > 1$, this transformation *contracts areas*, as is necessary for a dissipative system. We then fold $ABCD$ back on itself and fit the image back inside the original rectangle $ABCD$. The shape of the folded rectangle leads to the name of horseshoe for the model. Repeating the transformation — stretching again in the x direction and folding in the y direction — we obtain as the image of the horseshoe a kind of double hairpin[48] (see fig. VI.43). Iterating an infinite number of times, we obtain a complex layered structure A which has all the properties of a strange attractor[49], as we will show.

The fact that the structure A is an attractor is guaranteed by the value $\eta > 1$ which entails the contraction of areas: every point on the initial rectangle $ABCD$ has its image inside A. Furthermore, this attractor is strange, since it displays SIC; each iteration doubles the x component of the distance between two points, thus separating them exponentially.

Smale's transformation gives us a particularly simple example of *hyperbolicity*. The dilating direction is x (one also says that x corresponds to the unstable manifold), while the contracting direction is y (corresponding to the stable manifold). To these two directions x and y (which are perpendicular for the horseshoe attractor) are associated the two *Lyapunov coefficients* of the mapping. The larger coefficient (along x) is $\Lambda_1 = 2$ and the other Lyapunov coefficient is $\Lambda_2 = 1/2\eta$. Their product $\Lambda_1 \cdot \Lambda_2 = 1/\eta$ measures the rate of contraction of areas (see also fig. VI.22), and the Lyapunov exponents are $\lambda_1 = \ln \Lambda_1$ and $\lambda_2 = \ln \Lambda_2$.

Finally, let us verify that the attractor is a fractal object. Figure VI.44 shows the evolution of a cross section OO' through the rectangle under repeated iteration.

---

48. Smale's model and the baker's transformation that we will encounter in Chapter IX are closely related. Note, however, two differences: the baker's transformation is discontinuous and also conserves areas.

49. Rigorously, A is the intersection of all of the iterates of the initial rectangle R. If f denotes the Smale mapping, then:

$$\mathscr{A} \equiv \bigcap_{k=0}^{\infty} f^k(R).$$

In what follows, we assume that the dominant effect of the transformation is the stretching of horizontal branches, neglecting its effect on the bent part of the horsehoe. Equivalently, we can assume that the length of the rectangle $ABCD$ along x is much greater than its width.

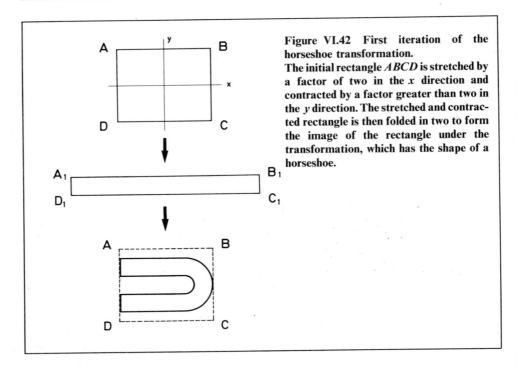

Figure VI.42 First iteration of the horseshoe transformation.
The initial rectangle *ABCD* is stretched by a factor of two in the *x* direction and contracted by a factor greater than two in the *y* direction. The stretched and contracted rectangle is then folded in two to form the image of the rectangle under the transformation, which has the shape of a horseshoe.

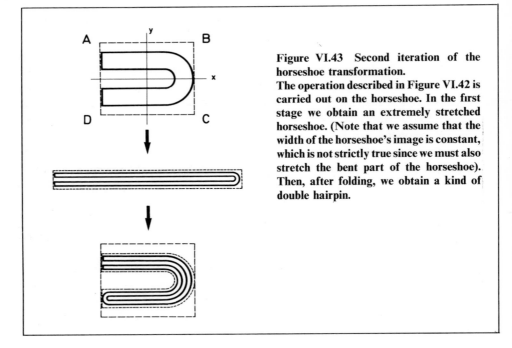

Figure VI.43 Second iteration of the horseshoe transformation.
The operation described in Figure VI.42 is carried out on the horseshoe. In the first stage we obtain an extremely stretched horseshoe. (Note that we assume that the width of the horseshoe's image is constant, which is not strictly true since we must also stretch the bent part of the horseshoe). Then, after folding, we obtain a kind of double hairpin.

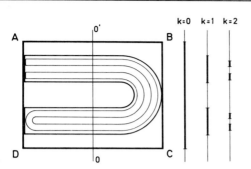

Figure VI.44 Cross-sections through the first few images of the rectangle under the horseshoe transformation.
Cross-sections along the OO' axis are taken through the initial rectangle ($k = 0$), its iterate (the horseshoe, $k = 1$), and its second iterate (the double hairpin, $k = 2$). We recognize, by comparing with Figure VI.34, the principle of construction of Cantor set.

Initially ($k = 0$) we have the rectangle $ABCD$, and the cross section OO' through it is a line segment. After one iteration ($k = 1$), we obtain the horseshoe and the cross section through it consists of two segments. After the second iteration, the cross section through the "double hairpin" consists of four segments, and so on. Here we recognize the principle of construction of a Cantor set, as described in the previous section. We can calculate, as we did for the Cantor set, the Hausdorff dimension of the cross section OO' through the attractor:

$$D = \frac{\ln 2}{\ln 2\eta}.$$

Hence we obtain a *fractal dimension* for the attractor A:

$$D' = 1 + \frac{\ln 2}{\ln 2\eta}$$

which indeed characterizes an object intermediate between a line and a surface[50].

We remark that in the limiting case $\eta \longrightarrow 1$ (negligible dissipation, almost no contraction of areas), D' approaches two, so that the iterates of the points of the rectangle $ABCD$ fill practically its entire surface. In the opposite limit of strong dissipation, with $\eta \ll 1$, D' approaches one, and the attractor A is almost a line (like the Poincaré section of the Lorenz attractor).

The attractor A provides a relatively simple illustration of the properties of strange attractors. Beyond its pedagogical interest, the essential properties of the horseshoe

---

50. Like the Hénon attractor, this attractor can be considered to be the Poincaré section of a chaotic three-dimensional flow.

attractor can be found in many Poincaré sections of chaotic flows. For example, numerical studies have revealed that, for certain values of the parameter r, the Poincaré section of the Lorenz attractor results from repeated iteration of the horseshoe mapping. The Hénon mapping was in fact constructed based on this remark: it consists of a mapping which stretches, contracts, and folds a given shape at each iteration. In Figure VI.23 we have already illustrated results from the Hénon mapping which greatly resemble those from the Smale mapping.

References for Chapter VI

A. Brandstäter, J. Swift, H. L. Swinney, A. Wolf, J. D. Farmer, E. Jen, J. P. Crutchfield, "Low dimensional chaos in a system", *Physical Review Letters*, **51**, p. 1442 (1983).
V. Croquette, C. Poitou, "Cascade de dédoublements de période et stochasticité à grande échelle des mouvements d'une boussole", *Compte-rendus de l'Académie des Sciences de Paris*, **C292**, p. 1353 (1981).
V. Croquette, "Déterminisme et chaos", *Pour la Science*, **62**, p. 62 (1982).
M. Dubois, "Experimental aspects of the transition to turbulence in Rayleigh-Bénard convection", *Lecture Notes in Physics*, **164**, p. 117 (1982).
P. Grassberger, I. Procaccia, "Characterization of strange attractors", *Physical Review Letters*, **50**, p. 346 (1983).
P. Grassberger, I. Procaccia, "Measuring the strangeness of strange attractors", *Physica* **9D**, p. 189 (1983).
J. Guckenheimer, G. Buzina, "Dimension measurements of geostrophic turbulence", *Physical Review Letters*, **51**, p. 1438 (1983).
M. Hénon, Y. Pomeau, "Two strange attractors with a simple structure", *Lecture Notes in Mathematics*, **565**, p. 29 (1976).
M. Hénon, "A two-dimensional mapping with a strange attractor", *Communications in Mathematical Physics*, **50**, p. 69 (1976).
E. N. Lorenz, "Deterministic non-periodic flow", *Journal of Atmospheric Sciences*, **20**, p. 130 (1963).
B. Malraison, P. Atten, P. Bergé, M. Dubois, "Dimension d'attracteurs étranges : une détermination expérimentale en régime chaotique de deux systèmes convectifs", *Comptes-rendus de l'Académie des Sciences de Paris*, **C297**, p. 209 (1983).
J.-C. Roux, R. H. Simoyi, H. L. Swinney, "Observation of a strange attractor", *Physica*, **8D**, p. 257 (1983).
J.-C. Roux, H. L. Swinney, "Topology of chaos in a chemical reaction", dans *Non Linear Phenomena in Chemical Dynamics*, ed. C. Vidal, A. Pacault, Springer-Verlag, p. 38 (1981).
C. Simó, "On the Hénon-Pomeau attractor", *Journal of Statistical Physics*, **21**, p. 465 (1979).
C. Tresser, P. Coullet, A. Arneodo, "Topological horseshoe and numerically observed chaotic behavior in the Hénon mapping". *Le Journal de Physique Lettres*, **13**, p. L123 (1980).
J. Yorke, E. Yorke, "The transition to sustained chaotic behavior in the Lorenz model", *Journal of Statistical Physics*, **21**, p. 263 (1979).

CHAPTER VII

Quasiperiodicity

VII.1 Hopf bifurcation from a limit cycle

We have seen in Chapter VI that strange attractors are characteristic of chaotic behavior in a dissipative dynamical system with a small number of degrees of freedom. The question addressed in this chapter and those which follow is: by what mechanisms does the transition between periodic and strange attractor occur? It is to answer this question that "*routes*" or "*scenarios*" towards chaos — characteristic series of a limited number of bifurcations — have been proposed. These are classified into three major routes according to the way in which the periodic regime loses its stability. Analysis of the Poincaré section establishes that a limit cycle is linearly unstable when the successive intersections $x_1, x_2, \ldots$ of a trajectory initially close to a limit cycle with a plane S (see fig. VII.1) are such that:

$$|\overrightarrow{x_0 x_1}| < |\overrightarrow{x_0 x_2}| < \ldots$$

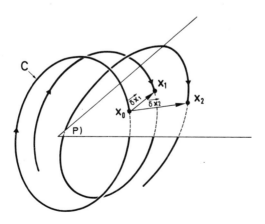

Figure VII.1 Poincaré section of the phase trajectories near a limit cycle C.

Defining $\overrightarrow{\delta x_1} = \overrightarrow{x_0 x_1}$ and $\overrightarrow{\delta x_2} = \overrightarrow{x_0 x_2}$, we know that $\overrightarrow{\delta x_2} = M \overrightarrow{\delta x_1}$ where M is the Floquet matrix. The condition for instability of the limit cycle is that the modulus of at least one of the eigenvalues λ of M be greater than one. If $\lambda = 1 + \varepsilon$, each $\overrightarrow{\delta x_i}$ is amplified in the same direction with every cycle. If $\lambda = -(1 + \varepsilon)$, the successive $\overrightarrow{\delta x_i}$ are amplified in modulus, but alternately in one direction and in the opposite direction. Finally, if there are two complex conjugate eigenvalues, the $\overrightarrow{\delta x_i}$'s rotate by an angle γ with each cycle, while their lengths increase (see fig. VII.2). If the bifurcation accompanying the last type of instability is supercritical, nonlinear effects — not included in the reasoning above — limit the amplification of $|\overrightarrow{\delta x_i}|$ to a finite value. The limit cycle, represented on the Poincaré section by the point x_0, is then transformed by a Hopf bifurcation into a torus T^2 whose Poincaré section consists of the set of points $P_i, P_{i+1}, ...$ (see fig. VII.3). This chapter treats the events which can then lead from the torus T^2 — a quasiperiodic regime — to a strange attractor corresponding to a chaotic regime.

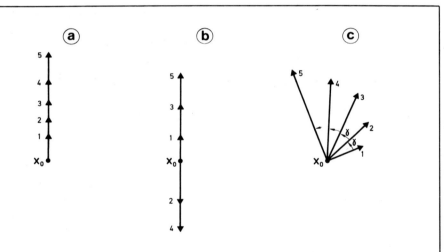

Figure VII.2 Représentation in the plane of the Poincaré section of the vectors $\overrightarrow{\delta x_i} = \overrightarrow{x_0 x_i}$ for increasing values of i.
There are three ways for the eigenvalue λ of the Floquet matrix to exit from the unit circle:
a) $\lambda > 1$
b) $\lambda < -1$
c) $\lambda = \alpha \pm i\beta$ with $|\lambda| > 1$.
We have shown only the dynamics in the direction of the eigenvector associated with the eigenvalue of greatest modulus.

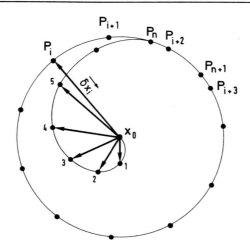

Figure VII.3 Evolution of $\vec{\delta x_i}$ in the plane of the Poincaré section after a Hopf bifurcation.
The point x_0 is the section of the now unstable limit cycle and the points P_i belong to the torus T^2.

VII.2 The theory of Ruelle-Takens (R.T.)

VII.2.1 DESCRIPTION

This theory, put forward in 1971 and elaborated in 1978 by Ruelle, Takens and Newhouse (R.T.N.) is of great historical importance insofar as it fundamentally challenged, for the first time, the mechanism proposed by Landau, which required an infinite number of Hopf bifurcations in order to generate turbulence. Ruelle and Takens (R.T.) made an epistemological break with tradition in advancing this revolutionary idea: a very small number of bifurcations would then suffice to produce chaotic behavior. Without reproducing the rather abstract mathematical statement they proved, the approach of R.T.N. can be schematically presented in the following way. Consider a dynamical system in a steady state, such as the laminar flow of a viscous fluid. Suppose that, when the control parameter (Reynolds number) is increased, the regime loses its stability and begins to oscillate with frequency f_1. Now suppose that the same kind of process is repeated two more times, so that a total of three successive Hopf bifurcations have occurred, producing three independent frequencies f_1, f_2, and f_3. Then according to R.T.N., the corresponding torus T^3 can, under fairly general conditions, become unstable and be replaced by a strange attractor. The time-dependent behavior is no longer quasiperiodic with three frequencies (on the torus T^3), but distinctly chaotic (see fig. VII.4).

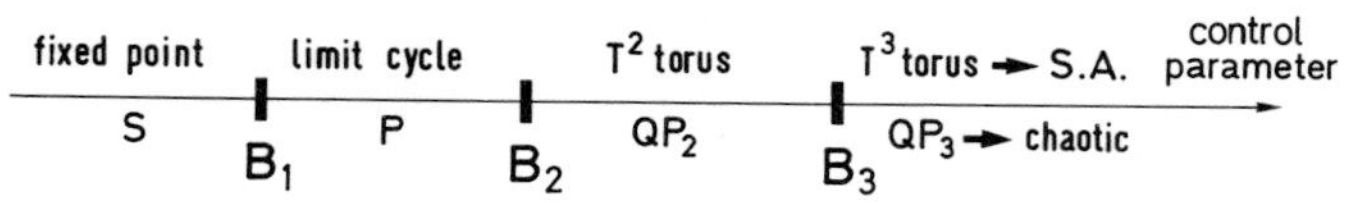

> ```
>     fixed point   limit cycle   T² torus      T³ torus → S.A.   control
>                                                                 parameter
>         S              P           QP₂           QP₃ → chaotic
>         •──────────────•────────────•─────────────•──────────────────→
>                       B₁           B₂            B₃
> ```
>
> **Figure VII.4** Schematic representation of the successive bifurcations B_1, B_2 and B_3 leading to chaos in the Ruelle-Takens theory.
> S = steady state, P = periodic state
> QP_2 = quasiperiodic regime with two frequencies
> QP_3 = quasiperiodic regime with three frequencies
> SA = strange attractor.

What R.T.N. have shown is that certain perturbations are capable of destroying the torus T^3 and transforming it into a strange attractor. But, unlike the torus, the strange attractor is stable, or robust, with respect to perturbations acting on the system. We then expect the power spectrum of the dynamical system to evolve, as a function of the control parameter, in the following manner. There will be a spectrum consisting of one frequency (f_1), then two (f_1 and f_2), and sometimes three frequencies ($f_1, f_2,$ and f_3). As soon as the third frequency arrives, the broad-band noise characteristic of chaos should start to appear (see fig. VII.5 a, b, c). The third frequency may or may not be detectable in the spectrum before chaos is identified. We again emphasize the fundamental difference between this and the Landau mechanism: three frequencies suffice instead of an infinite number. This tells us that a system with a small number of degrees of freedom can engender a chaotic regime. Let us now attempt to justify the necessity of having a quasiperiodic regime with three frequencies before a strange attractor can appear by this route.

VII.2.2 TOPOLOGICAL INTERPRETATION

We have seen in the previous chapter that a phase space of at least three dimensions[51] is necessary for the existence of a strange attractor. We return to the sequence of three successive Hopf bifurcations. The first leads from the initial steady state (fixed point, dimension zero) to a periodic state (limit cycle, dimension one). The second Hopf bifurcation transforms the periodic regime (frequency f_1) into a quasiperiodic regime (frequencies f_1, f_2 with f_1/f_2 irrational). The corresponding attractor is a torus T^2 whose dimension is two. On T^2, the only way the trajectories can change topologically is by synchronization, meaning that f_1/f_2 becomes rational (see Section VI.1.3 and its footnotes). The third Hopf bifurcation, if supercritical,

---

51. We emphasize again that the dimension of the attractor must not be confused with the dimension of the phase space in which it is found.

VII.2 THE THEORY OF RUELLE-TAKENS (R.T.)

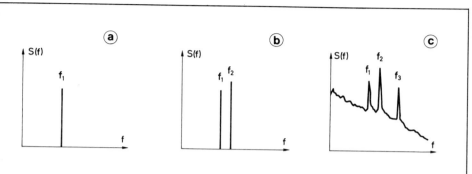

Figure VII.5 Schematic evolution of the power spectrum according to the Ruelle-Takens theory.
a) periodic regime
b) quasiperiodic regime with two frequencies
c) chaotic regime
$S(f)$ is the power spectrum defined in chapter III (designated there by $|\dot{x}_k|^2$).

causes a transition — in a linear approximation — from the quasiperiodic regime with two frequencies to a quasiperiodic regime on the torus T^3 with three frequencies $(f_1, f_2,$ and $f_3)$. Just as T^2 is a two-dimensional attractor, the torus T^3 is a three-dimensional attractor; T^3 can be "unfolded" in $\mathbb{R}^3$ in the same way as T^2 can be unfolded in $\mathbb{R}^2$ (see Section VI.1.3). On T^2, the trajectories are constrained to describe line segments on a rectangle (the rectangle which is the unfolded surface of T^2; see Section VI.2.2). However the trajectories on T^3 live in a cube [52] (see fig. VII.6) so that they can very well not be parallel and yet not intersect each other (cf. Section VI.2.2). Therefore the additional degree of freedom [53], revealed by the appearance of a third frequency, brings a radically new element into the picture. It allows the development of instabilities (see fig. VII.7) different from synchronization, the only possibility that exists for T^2 as we have shown in Section VI.1.3. It is by these elementary topological considerations that we can try to interpret the *necessity* of having at least three degrees of freedom [54] in order to obtain chaos described by a strange attractor. But is it sufficient? This is the question we will discuss in the next section.

---

52. Just as the trajectories on a torus T^2 are conveniently represented on a rectangle (cf. section VI.1.3), so too can trajectories on a torus T^3 be represented inside a parallelepiped whose opposite faces are identified. (The corresponding phase space must be at least four-dimensional; see fig. VII.6 and caption).
53. Note that the appearance of a new frequency in the regime requires that we take into account another dimension in phase space; to each new frequency is associated a new degree of freedom. Landau's interpretation is that to each frequency there corresponds a phase whose value depends on the exact process of onset of the instability. This "free" phase is the degree of freedom.
54. This implies three successive Hopf bifurcations.

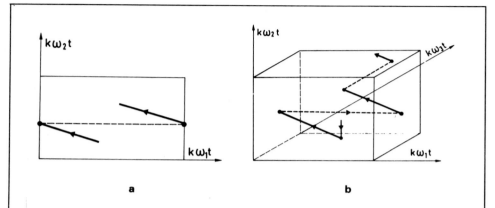

Figure VII.6 Unfolding of the tori T^2 and T^3.
a) In the unfolding of T^2 to a rectangle, when a trajectory intersects a side, it jumps to the opposite side to continue in the same direction.
b) For the torus T^3 we identify each face of the parallelepiped to the opposite face. The angular frequencies of the quasiperiodic regime are ω_1, ω_2, and ω_3.

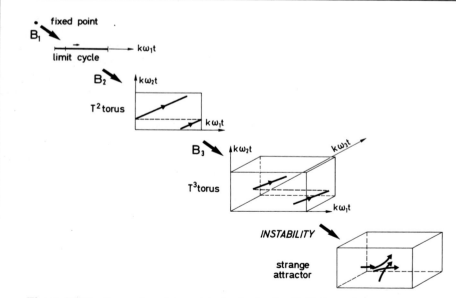

Figure VII.7 Successive phase portraits in the Ruelle-Takens theory. Each Hopf bifurcation brings another dimension into the phase space.

VII.2.3 PRACTICAL SIGNIFICANCE: A NUMERICAL SIMULATION

In the framework discussed here, three independent frequencies are required for a flow in phase space to have a strange attractor. Our goal is to determine whether small perturbations are sufficient to destabilize a torus T^3 and transform it into a strange attractor. The existence of stable experimental regimes which are quasiperiodic with three frequencies, and free from chaos, proves at least that *all* perturbations do not destabilize T^3.

These considerations, as well as the abstract mathematical nature of the R.T.N. theorem, prompt us to illustrate the theorem with numerical experiments. We consider a quasiperiodic function $X(t)$ with three frequencies f_1, f_2, f_3:

$$X(t) = F(f_1 t, f_2 t, f_3 t).$$

The equation of the flow is perturbed in a well-controlled way, so as to analyze its robustness with respect to these perturbations. The stability of $X(t)$ is studied by means of the Poincaré section of the flow. Specifically, we choose a surface of section such that $X(t)$ can be examined at multiples of the period $T_1 = 1/f_1$, i.e. at the discrete times[55] $t_k = k/f_1$.

The Poincaré section of the attractor T^3 corresponding to $X(t)$ is a torus T^2 which can be described by two angles θ and ϕ. The successive intersections with the surface of section are governed by the equations[56]:

$$\begin{aligned} \theta_{k+1} &= [\theta_k + \omega_1] \bmod 1 \\ \phi_{k+1} &= [\phi_k + \omega_2] \bmod 1 \end{aligned} \quad \text{(VII.1)}$$

We determine the stability of $X(t)$ by perturbing the Poincaré mapping defined by the system (VII.1). Let $\varepsilon \cdot P(\theta, \phi)$ be the perturbation[57] applied to evaluate the robustness of the solution. Its amplitude is adjusted through ε, $\varepsilon = 1$ signifying a perturbation of 100 %.

The perturbed Poincaré mapping of T^3 becomes:

$$\begin{aligned} \theta_{k+1} &= [\theta_k + \omega_1 + \varepsilon P_1(\theta_k, \phi_k)] \bmod 1 \\ \phi_{k+1} &= [\phi_k + \omega_2 + \varepsilon P_2(\theta_k, \phi_k)] \bmod 1 \end{aligned} \quad \text{(VII.2)}$$

Many numerical experiments are carried out on the system (VII.2) by taking P_1 and P_2 at random, for increasing values of ε. Conclusions about the perturbed system (VII.2) are drawn by calculating the Lyapunov exponents of the mapping (see Appendix B).

---

55. Here again is the stroboscopic technique described in the previous chapter for calculating Poincaré sections in thermoconvection experiments.
56. We again note the advantage of the Poincaré method in reducing a quasiperiodic flow with three frequencies to an iterated mapping in two dimensions.
57. $P(\theta, \phi)$ is a periodic function in θ and ϕ whose amplitude and phase are chosen at random. The average amplitude of the perturbation is controlled through ε.

We can classify the regimes obtained into four types, according to the signs of the Lyapunov exponents:
- quasiperiodic regimes with three frequencies $(0, 0, 0, -)$
- quasiperiodic regimes with two frequencies $(0, 0, -, -)$
- periodic regimes with one frequency $(0, -, -, -)$
- chaotic regimes $(+, 0, -, -)$.

The statistical results of many numerical experiments corresponding to different kinds of perturbations $P(\theta, \phi)$ and to various frequency ratios are displayed in the piecharts[58] of Figure VII.8 for three values of ε (i.e. for three perturbation amplitudes). These results clearly confirm the concrete implications of the R.T.N. theorem: a torus T^3 is unstable with respect to certain (sufficiently large) perturbations, by which it is transformed into a strange attractor shown to be robust by further numerical experiments.

However, it seems that in the limit $\varepsilon \longrightarrow 0$, the probability of obtaining a chaotic regime becomes zero. We can make an analogy between the structural instability of trajectories on T^2 and on T^3. The only instability of the trajectories on T^2 leads to the phenomenon of synchronization studied in detail in Appendix C; there we can see that synchronization becomes infrequent as we approach the threshold of the Hopf bifurcation at which the attractor T^2 first appears. Analogously, we should not be surprised to find that instability on T^3 leading to chaos also becomes infrequent — for weak perturbations — for an arbitrary choice of the bifurcation parameter. This point,

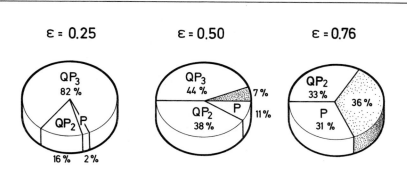

Figure VII.8 Percentage of different dynamical regimes observed numerically for three different amplitudes ε of the perturbations applied to a quasiperiodic regime with three frequencies.
P: periodic, QP_2: quasiperiodic with two frequencies,
QP_3: quasiperiodic with three frequencies, gray: chaotic regime.

---

58. The designation of 7 % chaotic regime, for example, means that out of N numerical experiments carried out on the system (VII.2), $7 \times (N/100)$ led to a chaotic regime.

VII.2 THE THEORY OF RUELLE-TAKENS (R.T.)

already remarked by Arnol'd, is therefore well in accord with the results of the numerical model presented here.

VII.2.4 EXPERIMENTAL ILLUSTRATION

We illustrate the R.T. theory using the R.B. instability. The convective fluid is water whose average temperature is such that its Prandtl number is equal to 5. The aspect ratio of the cell is $\Gamma_x = 3.5$, so that three convective rolls are formed. The control parameter is Ra/Ra_c[59].

As the ratio Ra/Ra_c is increased, stationary convection loses stability to periodic behavior starting at $Ra/Ra_c = 30$: the first Hopf bifurcation. At $Ra/Ra_c = 39.5$, the periodic regime in turn loses its stability, to be replaced by a quasiperiodic regime with two frequencies: the second Hopf bifurcation. Finally, at $Ra/Ra_c = 41.5$, a third Hopf bifurcation takes place, marking the appearance of the quasiperiodic regime with three frequencies whose spectrum is shown in Figure VII.9.

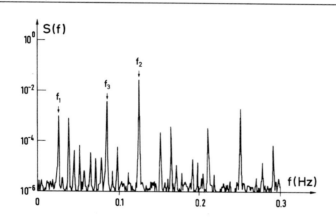

Figure VII.9 Power spectrum of a quasiperiodic regime with three incommensurate frequencies observed in R.B. instability.
The fluid is water, with $Ra/Ra_c = 42.3$. The spectral density is plotted logarithmically as a function of the frequency in Hertz.
From J. Gollub, S. Benson.

---

59. Ra/Ra_c is the ratio of the Rayleigh number in an experiment to the critical Rayleigh number Ra_c for onset of convection in an infinite geometry.

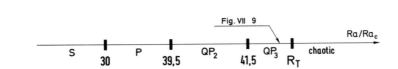

Figure VII.10 The stages of a route to chaos via a quasiperiodic regime with three frequencies. The observations were made on the R.B. instability in water.

We remark that identification of such a spectrum, containing a large number of peaks, with a regime having three incommensurate frequencies is highly nontrivial. It is necessary to show that the frequency of each of the lines *can* be indexed by $f = m_1 f_1 + m_2 f_2 + m_3 f_3$ with m_1, m_2, m_3 integers (which we can limit to a maximum value of 20) and — crucially — that they *cannot* be indexed by $f = m'_1 f'_1 + m'_2 f'_2$. To establish this, numerous adjustments by computer are required. Incommensurability is verified by making sure that the ratios f_1/f_2, f_1/f_3, and f_2/f_3 vary continuously with Ra/Ra$_c$. If there were synchronization (f_1/f_2 rational, for example), it would be manifested by a frequency-locking interval — i.e. a finite interval of Ra/Ra$_c$ over which f_1/f_2 would remain constant (see Appendix C on synchronization).

After a quasiperiodic regime with three frequencies has been clearly identified, experiments show that in a small domain of Ra/Ra$_c$, the corresponding spectra are — up to measurement accuracy — free from noise. This supports one of the conclusions drawn from the numerical model of the previous section: under some circumstances, the three-frequency regime is stable. However, above a certain value $R_T =$ Ra/Ra$_c \sim 43$, broad-band noise becomes noticeable, with accompanying widening of the peaks: the regime has become chaotic. Figure VII.10 shows the succession of different regimes leading to chaos in this particular experiment.

It is quite difficult to know if the appearance of chaos has a threshold R_T truly distinct from that of quasiperiodicity with three frequencies, or if these two thresholds are actually the same. There may exist a significant noise level of instrumental origin: when the threshold R_T is exceeded we can only say the intrinsic noise level of the system has surpassed the instrumental noise level. Analyzing only the experimental data, we cannot exclude the possibility that, for negligible instrumental boise, R_T and the threshold of appearance of the third frequency might be the same.

VII.3 Transition to chaos from a torus T^2

VII.3.1 CURRY-YORKE MODEL

We have described a route to chaos through which a torus T^3 is transformed into a strange attractor. We will now examine another route, also via quasiperiodicity, but

VII.3 TRANSITION TO CHAOS FROM A TORUS T^2

where chaos appears directly from a quasiperiodic regime with two frequencies (i.e. without the appearance of a third frequency). In other words, this route corresponds to the destabilization, or rather the destruction, of a torus T^2. There is no incompatibility between destabilization of T^2 and the reasoning set forth above tending to demonstrate that T^3 has the minimum dimension required for a strange attractor to appear. Indeed, in the case we are about to study, if chaos arises starting from T^2, it is due to the manifestation of another degree of freedom, not in the form of a third frequency, but by the gradual departure of the trajectories from T^2, which amounts to destruction of the torus.

The proof of the existence of such a route does not follow from a rigorously demonstrated theory, but rather from numerical experiments on an iterated two-dimensional mapping representing the Poincaré map of a three-dimensional flow.

The mapping must satisfy several criteria:
— it must be nonlinear
— it must be contracting (i.e. correspond to a dissipative system)
— it must exhibit a Hopf bifurcation.

As the control parameter is increased, a fixed point of the mapping must lose its stability via a Hopf bifurcation to a limit cycle. For the three-dimensional flow whose Poincaré map is being represented this Hopf bifurcation corresponds to a bifurcation: limit cycle $\longrightarrow T^2$.

Curry and Yorke have constructed a model satisfying these conditions. Consider the composition of two homeomorphisms (see Appendix E) on $\mathbb{R}^2$:

$$\psi = \psi_1 \circ \psi_2$$

where $\circ$ denotes composition of mappings[60].

* The mapping ψ_1 defines, in polar coordinates (ρ, θ), the $(k+1)^{st}$ iterate as a function of the kth by:

$$\psi_1 \begin{cases} \rho_{k+1} = \varepsilon \, \log(1 + \rho_k) \\ \theta_{k+1} = \theta_k + \theta_0 \end{cases}$$

where ε is the control parameter ($\varepsilon \geq 1$) and $\theta_0 > 0$ is fixed for a given series of experiments.

* The mapping ψ_2 is given by the following relation between the Cartesian coordinates (X, Y) of two consecutive iterates:

$$\psi_2 \begin{cases} X_{k+1} = X_k \\ Y_{k+1} = Y_k + X_k^2. \end{cases}$$

---

60. The composition $\psi = \psi_1 \circ \psi_2$ could also be put in a form analogous that of the Hénon model:

$$\psi \begin{cases} X_{k+1} = f(X_k, Y_k) \\ Y_{k+1} = g(X_k, Y_k). \end{cases}$$

However the final formulation is very awkward and is in fact not adapted to intuitive comprehension.

We can verify that ψ_1, ψ_1^{-1}, ψ_2 and ψ_2^{-1} are continuous, so that ψ is indeed a homeomorphism.

We begin by considering the effects of ψ_1. The result of the transformation $\rho_{k+1} = \varepsilon \log(1 + \rho_k)$ is illustrated by the graphs in Figure VII.11. The slope of the curve at the origin is ε. Therefore, if $\varepsilon < 1$, the origin is a stable fixed point: all initial values ρ tend to zero under iteration (fig. VII.11 a). For $\varepsilon > 1$, there exists another fixed point in addition to the origin: under iteration, all initial values ρ tend towards a limit ρ_L, which is a positive, nonzero solution to $\rho = \varepsilon \log(1 + \rho)$ (fig. VII.11 b).

The second part of the ψ_1 mapping, $\theta_{k+1} = \theta_k + \theta_0$ a simple rotation by the angle θ_0. Figure VII.11 c combines the two types of iterated graphs of ψ_1. For

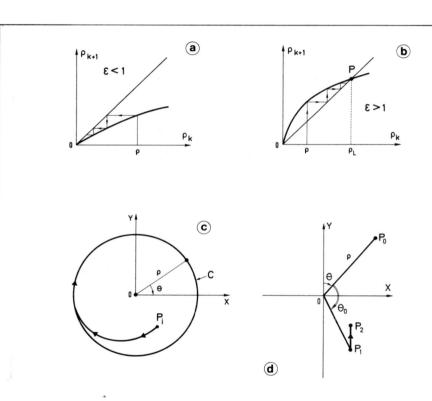

Figure VII.11 Curry-Yorke model.
a) Graph of mapping $\rho_{k+1} = \varepsilon \log(1 + \rho_k)$ for $\varepsilon < 1$.
b) Graph for $\varepsilon > 1$.
c) Iteration of the mapping ψ_1 alone for $\varepsilon > 1$. Any initial point P_i is mapped into the attracting circle C in several iterations.
d) Representation of the composed mapping $\psi = \psi_1 \circ \psi_2$ in the Cartesian plane. We go from P_0 to P_1 via ψ_1, and from P_1 to P_2 via ψ_2.

VII.3 TRANSITION TO CHAOS FROM A TORUS T^2

$\varepsilon \leqslant 1$, ths origin o is the only attracting point, while for $\varepsilon > 1$, ψ_1 has an attracting circle[61] C of radius ρ_L increasing with ε. We see that the conditions set down initially (dissipation and existence of a Hopf bifurcation) are satisfied.

The mapping ψ_1 does in fact describe a dissipative system. In the corresponding three-dimensional flow (whose Poincaré map is represented by the graph of ψ_1), there is a bifurcation from a limit cycle to an invariant torus T^2 at $\varepsilon = 1$.

The composed mapping ψ is nonlinear by virtue of, first, the logarithmic term in ψ_1, and second, the quadratic term in ψ_2:

$$Y_{k+1} = Y_k + X_k^2.$$

The effect[62] of ψ_2 is to couple θ ("angle") and ρ ("action"). As long as angle and action are decoupled, only circles can be attractors of the transformation. Furthermore, we can show that, in the composition $\psi = \psi_1 \circ \psi_2$, the nonlinear effects increase with ε.

Figure VII.11 d shows the effect of the transformation ψ in the (X, Y) plane. Starting from a point P_0 with coordinates (X_0, Y_0), we calculate its image P_1 under the transformation ψ_1, and then the image of P_1 under ψ_2 gives the point P_2. The procedure is repeated by applying the transformation ψ on P_2, and can be iterated indefinitely. After transient effects have been eliminated (which is accomplished in practice by discarding the first hundred iterates), the points are located on the attractor.

VII.3.2 BEHAVIOR OF THE MODEL AS A FUNCTION OF ε

We set $\theta_0 = 2$ in radians. This value being close to $2\pi/3$, at each iteration $\theta_{k+1} = \theta_k + \theta_0$ we rotate approximately a third of the way around the circle. Every three iterations, we find ourselves near the point of departure (see fig. VII.12 a where we have represented the consecutive iterates a, b, c, a', b', c'). We have said that for $\varepsilon = 1$, the origin is the only attracting point of the mapping. For $\varepsilon = 1_+$, the attractor is a circle centered about the origin (see fig. VII.12 b). When ε is increased, the radius of the circle grows rapidly. Nonlinear effects deform the curve and give rise to a kind of symmetry of order three in the closed loop constituting the attractor (which nonetheless remains topologically equivalent[63] to a circle). Note the presence of three zones (fig. VII.13) whose density is well above that of the rest of the attractor. This foreshadows the arrival of frequency locking of order three, but at present, the regime is definitely quasiperiodic with two incommensurate frequencies. The attractor is a torus T^2 whose Poincaré section is the closed curve represented in

---

61. We can draw a parallel between the iterated map of ψ_1 for $\varepsilon > 1$ and the phase diagram of the Van der Pol oscillator. In both cases there is dilation in a neighborhood of the origin and contraction far away from the origin; hence an attractor exists at a finite distance (see Section II.1.1).
62. Note that, unlike ψ_1, ψ_2 preserves areas (Jacobian = 1).
63. To be mathematically more precise we could say "homeomorphic to a circle".

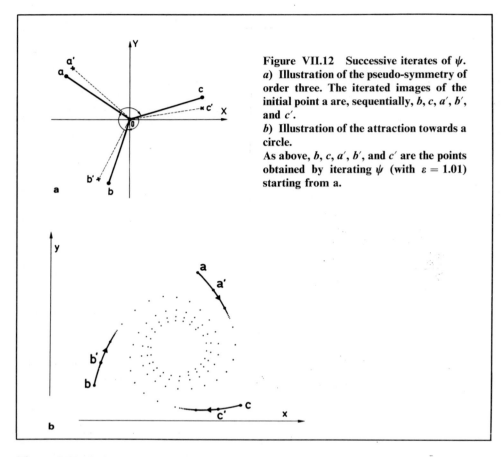

Figure VII.12 Successive iterates of ψ.
a) Illustration of the pseudo-symmetry of order three. The iterated images of the initial point a are, sequentially, b, c, a', b', and c'.
b) Illustration of the attraction towards a circle.
As above, b, c, a', b', and c' are the points obtained by iterating ψ (with $\varepsilon = 1.01$) starting from a.

Figure VII.13. As we continue to increase ε, making nonlinear effects more important, what began as a tendency towards locking becomes true frequency locking of order three. That is, the graph consists of only three points, and the orbit is of period three. From a physical point of view, this means that the two frequencies f_1 and f_2 of the quasiperiodic regime studied have become commensurate: $f_1/f_2 = 3$.

Frequency locking subsists for $1.28 < \varepsilon < 1.3953$. Just above $\varepsilon_c = 1.3953$, a closed loop reappears. It bears a strong resemblance to that obtained for ε slightly under 1.28. But the loop is no longer homoemorphic to a circle. A detailed examination (see fig. VII.14 for $\varepsilon = 1.48$) reveals a qualitatively new phenomenon: small "bumps" — an infinite number, in fact — have appeared on the loop[64], so that it now has a certain "thickness"[65]. Its dimension is no longer one (that of a line or a circle) but rather has become fractal (close to one, but greater). The graph is therefore a section through a

---

64. This structure, barely detectable just above ε_c, becomes more and more visible as ε is increased.
65. A section perpendicular to the loop would reveal it to be made up of an infinite number of sheets with the structure of a Cantor set.

VII.3 TRANSITION TO CHAOS FROM A TORUS T^2

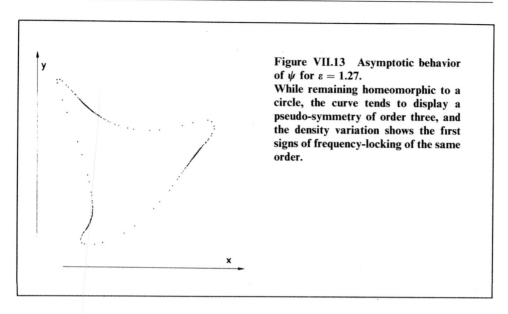

Figure VII.13 Asymptotic behavior of ψ for $\varepsilon = 1.27$.
While remaining homeomorphic to a circle, the curve tends to display a pseudo-symmetry of order three, and the density variation shows the first signs of frequency-locking of the same order.

strange attractor and the ordering of the points on the loop is now slightly chaotic. That is, the torus T^2 has been destroyed; above $\varepsilon = \varepsilon_c$, the attractor has a certain width in the form of wrinkles which corrugate the surface of the torus.

Let us return to the transformation ψ to illustrate the origin of this chaotic behavior. Consider two points P_1 and P_2 initially very close on the Poincaré section

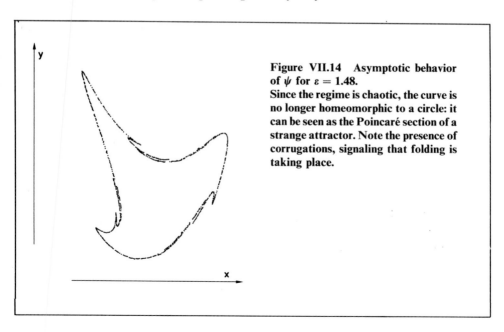

Figure VII.14 Asymptotic behavior of ψ for $\varepsilon = 1.48$.
Since the regime is chaotic, the curve is no longer homeomorphic to a circle: it can be seen as the Poincaré section of a strange attractor. Note the presence of corrugations, signaling that folding is taking place.

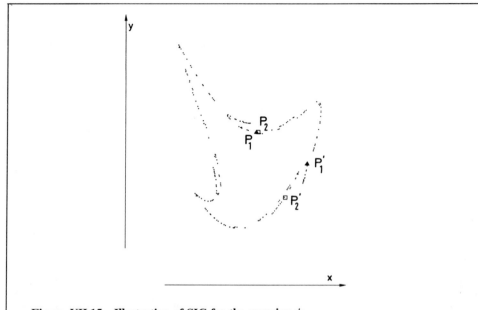

Figure VII.15 Illustration of SIC for the mapping ψ.
P_1 and P_2 are two points on the attractor, initially very close. While the separation between P_1 and P_2 is hardly visible, after forty iterations, their images P'_1 and P'_2 are clearly separated.

(see fig. VII.15) — i.e. two neighboring trajectories of the three-dimensional flow. As we iterate (that is, as time passes), the images $P'_1 = \psi^k(P)$ and $P'_2 = \psi^k(P_2)$ move apart until they are actually on opposite sides of the loop. The separation illustrates the divergence of neighboring trajectories (S.I.C.) typical of strange attractors. After having attained their maximum separation, the images P'_1 and P'_2 are drawn back together by a folding process. Neighbors once again, they begin to diverge, and so forth. We have already encountered several times this combination of stretching and folding, which is a necessary and sufficient condition for the existence of chaos[66].

Figure VII.16 summarizes the different routes to chaos followed by the model of Curry and Yorke. We note that in the region $\varepsilon > \varepsilon_c$, corresponding to the chaotic regime, "windows" occur having periodic behavior. For specific values of ε, the attractor is no longer strange and it becomes periodic. We will also encounter this situation for the route to chaos by period doubling (see Chapter VIII).

---

66. We will see that this twofold phenomenon induces "mixing", one of the essential characteristics of chaos.

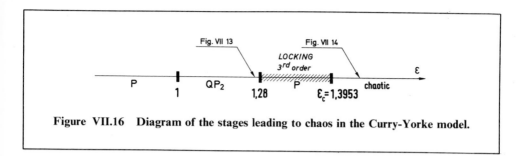

Figure VII.16 Diagram of the stages leading to chaos in the Curry-Yorke model.

VII.3.3 SOME EXPERIMENTAL ILLUSTRATIONS

Certain experimental observations of the onset of chaos starting from a quasiperiodic regime could almost be mistaken for the behavior of the numerical Curry-Yorke model.

The illustrations which follow are for the most part taken from the R.B. instability in a confined geometry, which is very rich in interesting behavior[67] in this domain. The control parameter Ra/Ra_c has already been defined. Knowledge of Ra/Ra_c is useless if we do not, in addition, specify the convective structure[68] present in the cell. We therefore designate each structure symbolically by a letter A, B, or C without entering into detail about the structure itself. We can also use this occasion to illustrate the variety of manifestations of the R.B. instability in high Prandtl number fluids. If the fluid is a silicone oil, its transparency, combined with the substantial thermal variation of its index of refraction, suggests the use of local or demi-local optical measurements. We can then constantly oversee the type and quality of the convective structure. In particular, it is possible using optical techniques to localize — and thus to study separately — the thermoconvective oscillators responsible for the regimes observed.

The spectrum taken from structure A, shown in Figure VII.17 a for $Ra/Ra_c = 569$, indicates quasiperiodicity with two frequencies such that $f_1/f_2 = 2.91$. The two thermoconvective oscillators I and II have been identified and localized. That is, if we place the sights of the optical device in the immediate neighborhood of the site of oscillator I, it is essentially[69] the frequency f_1 which is detected (and similarly for II

---

67. Maintaining the same convective structure throughout the route to chaos is a constraint very difficult to satisfy experimentally. The slightest modification of the convective structure changes the dynamical behavior. Often, when observations seem incomplete, it is not because the experimentalist has neglected to look beyond the domain presented but, rather, that the convective structure has ceased to be stable.
68. A description of the convective structure should include the number of convective rolls, their symmetry, their direction of rotation, the defects in the structure, etc.
69. The two oscillators I and II are, of course, coupled by the dynamics of the fluid. In I we find a bit of the frequency f_2 characterizing the oscillator II, and vice versa (see caption of fig. V.8).

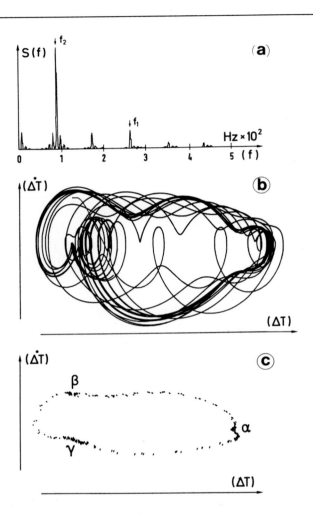

Figure VII.17 R.B. Convection, structure A, $Ra/Ra_c = 569$.
a) Linearly scaled power spectrum of the temperature gradient. The two characteristic frequencies are designated by f_1 and f_2, the other peaks being linear combinations $m_1 f_1 + m_2 f_2$. Here $f_1/f_2 = 2.91$.
b) Phase trajectories in the $[(\Delta T), (\dot{\Delta T})]$ plane. (ΔT) is the temperature gradient and $(\dot{\Delta T})$ its time derivative.
c) Poincaré section of trajectories shown in *b)* obtained by stroboscoping the signals at the frequency of one of the two thermoconvective oscillators. The zones of greatest density, designated by α, β and γ, foreshadow frequency-locking of order three.

From M. Dubois, P. Bergé, V. Croquette.

and f_2). This is clearly very valuable in understanding what is going on and identifying unambiguously the principal frequencies of the spectrum.

Being able to separate the contributions of the two oscillators also facilitates the construction of Poincaré sections. On one detector we collect the contribution due primarily to oscillator I, which we designate by (ΔT). Simultaneously we differentiate (ΔT) analogically (via an RC circuit) to obtain $(\dot{\Delta T})$. The signals (ΔT) and $(\dot{\Delta T})$ are sent to the two channels of an X-Y plotter (see fig. V.8 and section VI.3.3). The phase trajectories in the $(\Delta T, \dot{\Delta T})$ plane can be graphed: Figure VII.17 b shows how complex these can be. If we now, with a second detector, collect a second signal dominated by oscillator II, we can use this information to stroboscope[70] the trajectories at a fixed phase of oscillation of II. The result of this procedure is shown in Figure VII.17 c. By comparing Figure VII.17 b and c, we can appreciate the simplification resulting from the technique of Poincaré sections. The tangled trajectories of Figure VII.17 b are impossible to decipher, while Figure VII.17 c is easily recognized as the Poincaré section of a torus. We can even discern something more subtle: the greater density of points at α, β and γ. This is due to the fact that, since f_1/f_2 is 2.91, we are close to simple frequency locking of order three. Oscillators I and II interact strongly and therefore tend to remain temporarily frequency locked in this ratio.

This remarkable tendency towards temporary[71] frequency locking is reinforced at higher values of the control parameter Ra/Ra_c. For $Ra/Ra_c = 590$, the ratio f_1/f_2 has become 2.99, very close to 3. We can see on Figure VII.18 a that three zones contain almost all of the points of the Poincaré section. This appears even more clearly on Figure VII.18 b, which represents $(\dot{\Delta T})$ as a function of time. We remark that the relative phase varies very slowly[72] with time before changing abruptly by $2\pi/3$. The alternation of periods of slow and rapid variation corresponds to alternation between frequency locking and unlocking. The sequence is strictly periodic of period $1/(f_1 - 3f_2)$, where $f_1 - 3f_2$ is the departure from frequency locking. By increasing Ra/Ra_c still further we arrive, at $Ra/Ra_c = 593$, at true permanent frequency locking $f_1/f_2 = 3$, and the Poincaré section then consists of only three points. Figure VII.19 summarizes the different stages we have described.

What happens when we increase Ra/Ra_c above the synchronization values at which $f_1/f_2 = 3$? Experiments show that structure A disappears. The dynamical behavior of the structure which replaces it has no relation to the route studied up till now. It is now necessary to consider another structure which, although exhibiting less coupling between oscillators, does reach a chaotic regime. Let B be this new structure. For $Ra/Ra_c = 520$, the dynamical regime of structure B is quasiperiodic as attested to by

---

70. In practice, instead of leaving the plotter pen down as was done to obtain Figure VII.17 b, we lower it only when oscillator II has a given phase.
71. Note the astonishing similarity with the frequency-locking in the numerical model of Section VII.3.2.
72. In the absence of coupling, the relative phase would vary linearly with time.

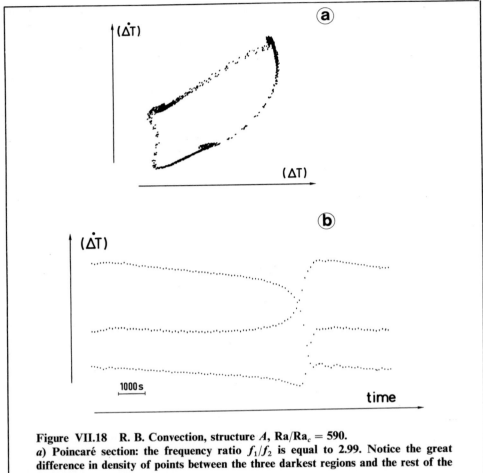

Figure VII.18 R. B. Convection, structure A, $Ra/Ra_c = 590$.
a) Poincaré section: the frequency ratio f_1/f_2 is equal to 2.99. Notice the great difference in density of points between the three darkest regions and the rest of the curve.
b) $(\dot{\Delta}T)$ as a function of time: a long period (note the time scale) of slow evolution of the phase is followed by a sudden rotation by $2\pi/3$.
From M. Dubois, P. Bergé, V. Croquette.

the Poincaré section[73] on Figure VII.20. By increasing the control parameter Ra/Ra_c, the Poincaré section begins to display the characteristic structure seen on Figure VII.21 where $Ra/Ra_c = 563$. The wrinkling of the surface of the attractor is characteristic of the twofold operation leading to a strange attractor[74]: stretching and folding.

---

73. We remark that the curve on this figure is closed, but intersects itself. This is due merely to the presence of a large second harmonic in the signal used to construct the Poincaré section.

74. Since $Ra/Ra_c = 563$ corresponds to the beginning of corrugation of the torus, chaos is not yet very pronounced. The power spectrum shows only the peaks of the quasiperiodic regime, as the noise level remains low.

VII.3 TRANSITION TO CHAOS FROM A TORUS T^2

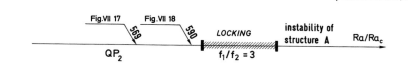

Figure VII.19 Diagram of the dynamical regimes of structure A as a function of Ra/Ra_c.

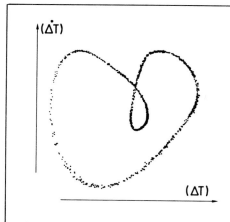

Fig. VII.20 R.B. Convection; structure B; $Ra/Ra_c \sim 520$.
Poincaré section taken from the quasi-periodic regime with two frequencies. The intersection of the curve with itself is due to the presence of a large second-order harmonic in the signal.

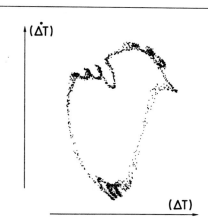

Figure VII.21 R.B. Convection; structure B; $Ra/Ra_c \sim 563$.
The Poincaré section shows the characteristic corrugations of a strange attractor.

The resemblance between this attractor and that of the Curry-Yorke model is obvious in Figure VII.22, where the two Poincaré sections are displayed together. In both cases, corrugations appear in the vicinity of frequency locking of order three. But, for structure B, corrugation immediately precedes the frequency locking which occurs

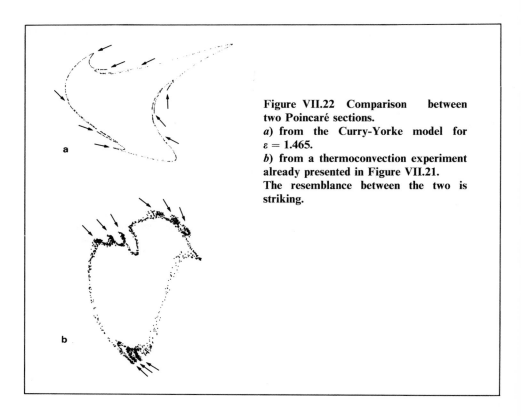

Figure VII.22 Comparison between two Poincaré sections.
a) from the Curry-Yorke model for $\varepsilon = 1.465$.
b) from a thermoconvection experiment already presented in Figure VII.21.
The resemblance between the two is striking.

Figure VII.23 Diagram of the dynamical regimes of structure B as a function of Ra/Ra_c.
We cannot exclude the possibility of "windows" of periodicity in the chaotic domain, despite our current inability to identify them experimentally.

for $570 < (Ra/Ra_c) < 590$ (see fig. VII.23), in contrast to the Curry Yorke model in which corrugation follows frequency-locking.

Finally we mention a case of transition towards chaos via quasiperiodicity where, unlike those we have discussed above, the study of Poincaré sections yields no information. The measurements are taken on a third type of structure, called C, obtained by using a silicone oil of lower Prandtl number than in the preceding

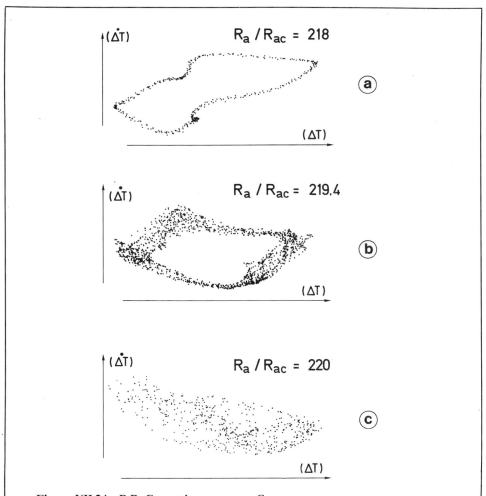

Figure VII.24 R.B. Convection; structure C.
Poincaré sections for three very close values of Ra/Ra_c.
a) quasiperiodic regime.
b) and *c*) chaotic regimes. The increasing dispersal of the points may reflect the presence of a strange attractor of dimension greater than (or very close to) 3.
After M. Dubois and P. Bergé.

examples (Pr = 38 instead of Pr = 130). Figure VII.24 *a* shows that for $Ra/Ra_c = 218$, the regime of structure *C* is quasiperiodic. After an extremely small increase of Ra/Ra_c (less than 1 %) we no longer get corrugation of the section as before, but dispersal of the points (see fig. VII.24 *b*). When Ra/Ra_c reaches 220 (fig. VII.24 *c*) there is only a cloud of points. In this case, there is no reason to conclude that the chaotic regime corresponds to a strange attractor of dimension greater than 3 (or very slightly under 3) rather than to a torus of high dimensionality (non-deterministic chaos). Only the method described in VI.4.4 for determining attractor dimension can settle this question.

VII.4 Elements of a mathematical theory of transition to chaos from a torus T^2

We have seen that a simple mathematical model exhibits a transition from a torus T^2 directly to a strange attractor, without varying the number[75] of degrees of freedom. In what follows we will attempt to situate this phenomenon in a more general framework.

The first return map of a flow on T^2 can be described by a continuous and invertible mapping of the circle T^1 into itself. The mapping $f(\theta)$ (graphed in fig. VII.25) is a monotonically increasing (or decreasing) function, with two fictitious discontinuities due only to the identification of 0 and 1 on the circle. Such a mapping

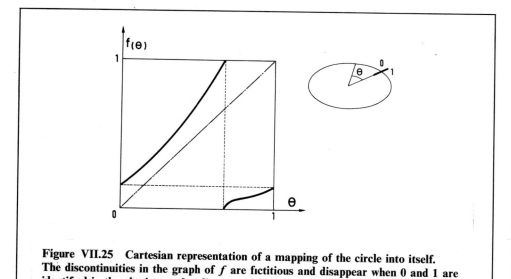

Figure VII.25 Cartesian representation of a mapping of the circle into itself. The discontinuities in the graph of f are fictitious and disappear when 0 and 1 are identified in the abscissa and ordinate.

---

75. Here the number of degrees of freedom has the meaning described in Footnote 53.

on the circle is characterized by its winding number[76]. When the winding number is rational, one says that there is frequency locking. Frequency locking occurs — generically — in an open interval of control parameter values. An irrational winding number corresponds to quasiperiodic behavior. To analyze the phenomenology, it is desirable to introduce a two-parameter set of mapping $f(\theta)$. An example due to Arnol'd is:

$$f : \theta \longmapsto \theta + \theta_0 + \frac{\alpha}{2\pi} \sin(2\pi\theta).$$

The two control parameters are α and θ_0. To each pair of parameters, represented by a point on the Cartesian plane, there corresponds a precise value of the winding number. The plane is therefore densely covered by a set of nonintersecting bands of finite width (but for the most part very narrow), such that inside each band, the winding number is constant and rational. These bands, called frequency-locking bands, are dense in the plane (just as the rationals are dense on the interval). Frequency locking of p_3/q_3, between p_1/q_1 and p_2/q_2 (for example 3/5 between 1/2 and 2/3), will take place in a band (p_3/q_3) located between the bands (p_1/q_1) and (p_2/q_2), where the p_i, q_i are positive integers. The structure of the set of frequency-locking bands is, at least locally, akin to that of a Cantor set: between each pair of frequency-locking bands, we find other bands, and between these, other bands.

We can imagine deforming $f_{\theta_0,\alpha}(\theta)$ in such a way that f acquires a zero derivative on a curve $\theta_0(\alpha)$ in the control-parameter plane (see figs. VII.26 and VII.27). Crossing the curve $\theta_0(\alpha)$ signifies the loss of invertibility of f which can explain the transition $T^2 \longrightarrow$ strange attractor related in this chapter: The preceding construction raises two questions:

(1) What happens when we iterate $\theta \longmapsto f_{\theta_0,\alpha}(\theta)$ near the parameter values $\theta_0(\alpha)$?

(2) How can loss of invertibility take place for a first return map of the plane onto itself?

We now consider these questions in turn.

(1) If we limit ourselves to the region $\theta_0 > \theta_0(\alpha)$ (invertibility preserved) the dominant phenomenon is frequency locking in the neighborhood of $\theta_0(\alpha)$. Indeed, we know that when there is frequency locking, two initially neighboring points converge towards one another under iteration of the mapping $\theta \longmapsto f(\theta)$. Imagine that the curve $f(\theta)$ contains an almost horizontal plateau, i.e. that θ_0 is slightly greater than $\theta_0(\alpha)$.

A simple calculation shows that the distance between two close points is multiplied by $|df/d\theta|$ at each iteration. Therefore if there are values of θ for which the derivative is very small, the mapping will tend to draw together the iterates of neighboring points very efficiently which is, as we have said, typical of frequency locking. In addition, Herman has shown that the probability of having an irrational winding number (and thus no frequency locking) by choosing a point at random on the

---

76. In a mapping like that of Figure VII.25, the successive iterates of a point rotate systematically in the same direction. The winding number measures the average speed of this motion. See also Appendix C.

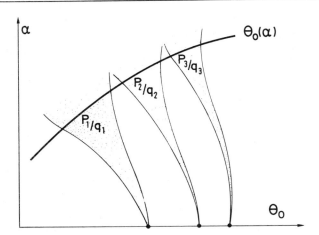

Figure VII.26 Rational frequency-locking bands of a mapping $\theta \mapsto f_{\theta_0,\alpha}(\theta)$ with two parameters θ_0 and α.
Within each band, the winding number is constant and rational. The bands are dense in the parameter plane and do not intersect. The curve $\theta_0(\alpha)$ is the boundary in parameter space between the region where f is invertible and where it is not (see fig. VII.27).

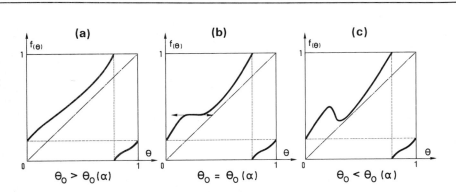

Figure VII.27 Deformation of the graph of $f_{\theta_0,\alpha}(\theta)$.
When θ_0 crosses the curve $\theta_0(\alpha)$ in the parameter plane, the graph of f has a horizontal tangent (b). In (a), the mapping is invertible, while in (c) it is not.

$\theta_0(\alpha)$ curve is zero. We can understand (although it is difficult to prove) that, as we approach $\theta_0(\alpha)$, frequency locking becomes the predominant phenomenon, although the absence of frequency locking subsists for certain values of (θ_0, α) since the winding number must become irrational to pass continuously from one rational number to another [77].

In the region of parameter space for which f is not invertible, we have a very complicated picture which is not yet entirely understood. We can say, however, that frequency locking persists, at least in part. This can be explained by the fact that frequency locking with winding number p/q rational depends only on the existence of solutions to $f^{(q)}(\theta) = \theta$. We should almost always be able to follow this solution by continuity across $\theta_0(\alpha)$, since this equation does not imply the invertibility of f nor that of $f^{(q)}$. But, unlike the case where f is invertible, the frequency-locking bands have a complex internal structure.

How does frequency locking — with period 3, for example — arise in the mapping on the circle $\theta \longrightarrow f(\theta)$? Frequency locking is due to the presence of three fixed points of the mapping $f^{(3)}$, the third iterate of f. The graph of $f^{(3)}$ undulates about the diagonal, which it intersects six times: three intersections comprise a stable cycle of period 3, and the others an unstable cycle of period 3 (see fig. VII.28). The deformation of the graph of $f^{(3)}$ corresponding to frequency locking can be seen as a simple vertical translation, obtained by adding a constant angle θ^* to $f^{(3)}(\theta)$. Let θ_1, θ_2 and θ_3 be the points of the stable cycle, $\theta'_1, \theta'_2, \theta'_3$ those of the unstable cycle. As soon as the cycle appears, the graph of $f^{(3)}$ (dashed curve in fig. VII.28) is just tangent to the diagonal. The two cycles are therefore the same and we have $\theta_1 = \theta'_1, \theta_2 = \theta'_2, \theta_3 = \theta'_3$. When the cycle disappears, i.e. when θ^* is sufficiently large (the topmost curve of fig. VII.28) we have the same collapse of the stable and unstable cycles, but now with $\theta_1 = \theta'_3, \theta_2 = \theta'_1$, and $\theta_3 = \theta'_2$, the cycles and points having been followed by continuity as $f^{(3)}$ is translated. This explains geometrically the process of appearance and disappearance of a cycle on each side of a frequency-locking interval.

Let us now consider what happens when $f(\theta)$ has an extremum [78] and is therefore no longer invertible. By using the relations for differentiation of composition of functions, we find that, if the derivative of f vanishes for one value of θ, then the derivative of $f^{(q)}$ vanishes for at least q different values of θ unless it happened to vanish exactly on one period. In our example of the 3-cycle, $f^{(3)}$ has a zero derivative, at a local maximum or minimum, between each pair of fixed points (fig. VII.29). Examination of the curve in Figure VII.29 shows that, unlike the case where f — and thus $f^{(3)}$ — is

---

77. The concomitant disappearance of the quasiperiodic orbits has been the subject of very detailed study, due in part to its connection with the destruction of magnetic surfaces confining fusion plasmas in toroidal configurations.

78. If f maps [0,1] once onto itself, as we will suppose, i.e. if f traverses the unit interval once when θ traverses it once, then f has an even number of extrema (two in the present case).

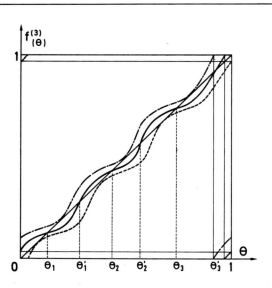

Figure VII.28 Deformation of the graph of $f^3(\theta)$ in a frequency-locking window of period 3.
The two dashed curves, both tangent to the diagonal, mark the limits of the period-3 frequency-locking window. The standard situation inside the window is represented by the solid line, which contains three stable fixed points $(\theta_1, \theta_2, \theta_3)$ and three unstable fixed points $(\theta'_1, \theta'_2, \theta'_3)$.

invertible, we can no longer determine *a priori* the stability of θ_1, say, as a fixed point of $f^{(3)}$. We still have $(df^{(3)}/d\theta|_{\theta'_1} > 1$, so that θ'_1 is unstable, and we still have $(df^{(3)}/d\theta)|_{\theta_1} < 1$. But we also have $(df^{(3)}/d\theta)|_{\theta_1} < 0$ and the point θ_1 will be unstable if $(df^{(3)}/d\theta)|_{\theta_1} < -1$, in which case both 3-cycles of f will be unstable. Therefore, inside the frequency-locking band of the 3-cycle in the (θ_0, α) parameter plane, we find two regions, according to whether $|(df^{(3)}/d\theta)|_{\theta_1}$ is greater or less than 1, i.e. according to whether there are one or two unstable 3-cycles. The region of instability for both cycles forms a kind of band within the frequency-locking band (fig. VII.30). In the same way, every frequency-locking band of winding number p/q is divided into two parts. The higher the denominator q, the closer the dividing line comes to the curve $\theta_0(\alpha)$.

Inside the domain of instability of both cycles, we find a great variety of behavior. As shown by Figure VII.29, the mapping $f^{(3)}(\theta)$ has locally parabolic regions like those to be studied in Chapter VIII. It can generate chaos by the mechanism of a cascade of period doublings, the subject of Chapter VIII. Other more complicated possibilities exist, for $f^{(3)}$ does not map the interval $[\theta'_1, \theta_1]$ into itself. In particular, if the image of the maximum of $f^{(3)}$ located between θ'_1 and θ_1 is located in $[\theta'_2, \theta_2]$, and

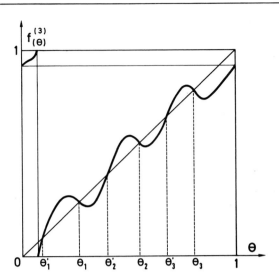

Figure VII.29 Example of a non-invertible mapping f with two unstable 3-cycles. Unlike the (invertible) case shown as the solid line in Figure VII.28, the cycle $(\theta_1, \theta_2, \theta_3)$ is not necessarily stable, since the derivative of $f^{(3)}$ at any of the three points can be less than (-1).

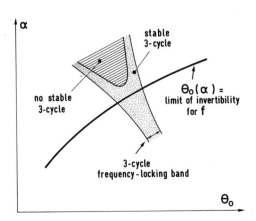

Figure VII.30 Period-3 frequency-locking band in the (θ_0, α) parameter plane. In the region where f is no longer invertible, above and to the left of the $\theta_0(\alpha)$ curve, there exists a domain (hatched region) inside the frequency-locking band in which both 3-cycles are unstable. In the rest of the band, one of the 3-cycles is stable. It is in the hatched region that we can observe chaotic behavior.

if the image of the minimum between θ_1 and θ_2' is located in $[0, \theta_1']$, then iteration of $f^{(3)}$ is accompanied by a phenomenon found neither in the period-doubling cascade nor in invertible mappings of the circle: random diffusion on the circle, superimposed onto the chaotic behavior.

(2) We have already seen that some mappings of the plane have a "circular" attractor, that is, a simple closed curve which can be seen as the Poincaré section of the attracting torus of the three-dimensional flow. On the circle, the mapping is a continuous and invertible transformation characterized by a winding number. It can therefore seem paradoxical to be interested in a transition related to the loss of invertibility of the mapping on the circle. We will show that the loss of invertibility *on the circle* is nevertheless compatible with the invertibility of the original mapping on the plane. The situation can therefore exist for a given dynamical system whose parameters are varied.

The justification of this point is as follows. From Poincaré, we know that a circular attractor for a mapping on the plane results from a dissipative transformation of a ring into itself (fig. VII.31). This transformation can itself be seen as the result of radial contraction followed by rotation about the center of the ring. In this simple case, the attractor is a circle of zero thickness, resulting from the cumulative effect of radial contraction on the finite thickness of the ring. Imagine now that the ring undergoes folding at the same time (fig. VII.32). For the transformation reduced to its effect on the angle, we see that the folding does destroy invertibility, since it introduces a maximum and a minimum in the function $f(\theta)$ defining the operation on the angles. Of course, the mapping of the ring onto itself remains invertible: points whose images share the same angle are distinguished by their different radial positions (fig. VII.32).

Another way of seeing this is the following. In the "standard" case, i.e. before loss of invertibility, the image of the ring resulting from the iterated map is a closed curve of finite length, which can therefore be parametrized by an angle. The increase in the length of this closed curve due to folding[79] is such that at the moment of loss of invertibility (i.e. at every point of the curve $\theta_0(\alpha)$ in parameter space discussed earlier) the length of the curve becomes infinite, excluding any parametrization by an angle.

We have seen that on the curve $\theta_0(\alpha)$ there is a very small probability of finding quasiperiodic behavior. We nevertheless discuss this case, where the iteration describes a curve of infinite length. When the circle is of finite length and the winding number irrational, the different possible values of the angle are visited with equal probability. But, at the moment of loss of invertibility (and provided frequency locking does not occur), the points visited form a Cantor set in values of the angle.

The practical importance of this result is limited, of course, by the fact that periodic behavior becomes dominant as the curve $\theta_0(\alpha)$ is approached. This probably explains the fact that, in experiments, chaos frequently appears via disappearance of frequency locking. This situation occurs far above the threshold for loss of invertibility discussed immediately above.

79. Recall that contraction of areas does not imply contraction of all lengths (cf. Section I.4.2).

VII.4 ELEMENTS OF A MATHEMATICAL THEORY OF TRANSITION

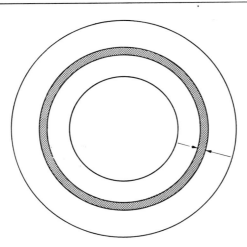

Figure VII.31 An invertible mapping of the plane.
Through contraction of an annulus, the mapping gives rise to a circular attractor. The shaded region is the set of points which are images of the outer annulus.

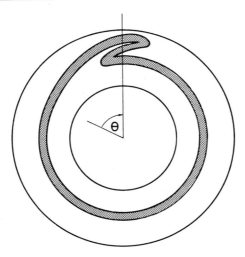

Figure VII.32 A non invertible mapping of the plane.
As in Figure VII.31 the shaded region is the image of the outer annulus. The same angle θ for the image can be obtained from three different values of the angle in the annulus. The mapping on the angles is therefore no longer invertible.

References for Chapter VII

P. Bergé, M. Dubois, "Transient reemergent order in convective spatial chaos", *Physics Letters*, **93A**, p. 365 (1983).

P. Bergé, "Study of the phase-space diagrams through experimental Poincaré sections in prechaotic and chaotic regimes", *Physica Scripta*, **T1**, p. 71 (1982).

J. M. Curry, "A generalized Lorenz system". *Communications in Mathematical Physics*, **60**, p. 193 (1978).

J. Curry, J. A. Yorke, "A transition from Hopf bifurcation to chaos: computer experiments with maps in $\mathbb{R}^2$" in *The structure of attractors in dynamical systems, Springer Notes in Mathematics*, **668**, p. 48, Springer-Verlag (1977).

M. Dubois, "Approach of the turbulence in hydrodynamic instabilities", in *Symmetries and Broken Symmetries in Condensed Matter Physics*, N. Boccara ed., I.D.S.E.T. Paris (1981).

M. Dubois, P. Bergé, V. Croquette, "Étude de régimes convectifs instationnaires à l'aide des diagrammes de Poincaré", *Compte-rendus de l'Académie des Sciences de Paris*, **C293**, p. 409 (1981).

M. Dubois, P. Bergé, "Instabilités de couche limite dans un fluide en convection. Évolution vers la turbulence", *Le Journal de Physique*, **42**, p. 167 (1981).

J. P. Gollub, S. V. Benson, "Many routes to turbulent convection", *Journal of Fluid Mechanics*, **100**, p. 449 (1980).

J. P. Gollub, S. V. Benson, "Phase locking in the oscillations leading to turbulence", in *Pattern Formation*, H. Haken ed., Springer-Verlag, p. 74 (1979).

J. P. Gollub, T. O. Brunner, B. G. Danly, "Periodicity and chaos in coupled nonlinear oscillators", *Science*, **200**, p. 48 (1978).

C. Grebogi, E. Ott, J. A. Yorke, "Are three-frequency quasiperiodic orbits to be expected in typical dynamical systems?", *Physical Review Letters*, **51**, p. 339 (1983).

M. Herman, "Sur la conjugaison différentiable des difféomorphismes du cercle à des rotations", *Publications mathématiques de l'I.H.E.S.*, **49**, p. 5 (1979).

L. Landau, E. Lifschits, *Mécanique des fluides*, Mir, Moscou (1971).

S. Newhouse, D. Ruelle, T. Takens, "Occurrence of strange axiom-A attractors near quasiperiodic flows on T^m, $m \geq 3$", *Communications in Mathematical Physics*, **64**, p. 35 (1978).

S. Ostlund, D. Rand, J. Sethna, E. Siggia, "Universal properties of the transition from quasiperiodicity to chaos in dissipative systems", *Physica*, **8D**, p. 303 (1983).

D. Ruelle, F. Takens, "On the nature of turbulence", *Communications in Mathematical Physics*, **20**, p. 167 (1971).

CHAPTER VIII

The subharmonic cascade

VIII.1 Introduction

VIII.1.1 SUBHARMONIC INSTABILITY

Subharmonic instability was already described in Chapter II apropos of the parametric oscillator. There, we saw the mechanism by which an excitation, even of weak amplitude, succeeds in destabilizing the motion of a pendulum, provided that it occurs at times separated by a whole number of half-periods (cf. Section II.2.4). The excitation is most efficient when it occurs at each half-period: the frequency (period) of the motion is then equal to half (twice) the frequency (period) of excitation, which is described by the adjective subharmonic. This result must be extended and generalized.

Because it is the process of subharmonic instability in and of itself which interests us, we will discuss the subharmonic instability of a general periodic solution, without any further reference to the parametric oscillator. We note that, for similar reasons, the solution with frequency $f/2$ (of period $2T$) could in turn undergo a subharmonic instability, yielding a half frequency $f/4$ (of period $4T$), and so on. In other words, appropriate variation of a control parameter should permit the observation of a whole series of subharmonic instabilities, giving rise to solutions of period $2T, 4T, 8T$, etc. We call this a cascade of subharmonic bifurcations or *subharmonic cascade*, each stage of which is accompanied by halving of the frequency, i.e. doubling of the period.

VIII.1.2 PERIOD-DOUBLING MECHANISM

Floquet theory explains the process by which period-doubling takes place. On Figure VIII.1 are shown the main ingredients necessary for the study of the linear stability of a periodic solution: the Poincaré section of the trajectory (fig. VIII.1 *a*) and the crossing of the unit circle in the complex plane by an eigenvalue of the Floquet matrix (fig. VIII.1 *b*). Let us see what happens when an eigenvalue approaches the critical value -1 and finally attains it. Let x be the eigenvector associated with this eigenvalue and let S be a plane of section containing x. Then without loss of generality, we can consider only the evolution of the successive points of intersection of the phase trajectory with the plane S (fig. VIII.2). Before the bifurcation, any displacement of initial amplitude x_0 decreases with every period, since its value x is multiplied by a negative factor of absolute value smaller than one (fig. VIII.2 *a*). The limit cycle of

192   CHAPTER VIII   THE SUBHARMONIC CASCADE

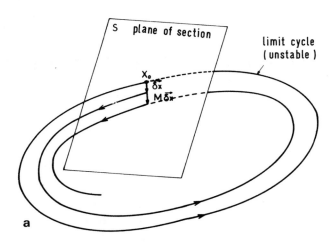

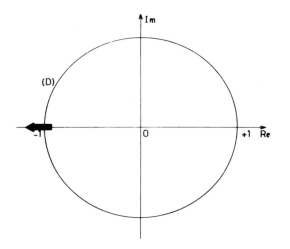

Figure VIII.1   Linear stability of a limit cycle.
a) Using a Poincaré section of the limit cycle, we can determine the evolution of an initial displacement $\vec{\delta X}$ under the action of the flow. Depending on whether $M\vec{\delta X}$ grows larger (case shown) or smaller than $\vec{\delta X}$, the limit cycle is unstable or stable.
b) Marginal stability is attained when an eigenvalue of the matrix M, called the Floquet matrix, crosses the unit circle (D) in the complex plane. For a subharmonic instability, (D) is traversed at -1.

period T is therefore linearly stable; its Poincaré section is merely one point, taken as the origin on Figure VIII.2. But, when the eigenvalue becomes equal to (-1), the situation changes. The distance from the origin does not decrease with time: with each intersection, it changes sign while retaining the same modulus. We return to the initial point x_0 every two intersections (designated by x_2, x_4, etc. on Figure VIII.2 b). We therefore see a new periodic orbit appearing, whose period is twice that of the original orbit. This new orbit depends on the initial displacement x_0 considered. Moreover, we

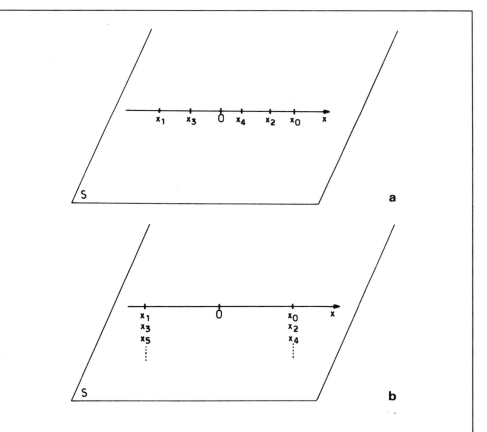

Figure VIII.2 Sequence of iterates of the flow.
We show what happens along the eigenvector x associated with the eigenvalue of the Floquet matrix which crosses -1.
a) Before the bifurcation, the limit cycle is stable. The iterates of an initial point x_0 converge to 0, which is the Poincaré section of the cycle. The eigenvalue being negative, we alternate from one side to the other of 0.
b) At the bifurcation point, the iteration no longer converges. The modulus of any initial displacement x_0 is conserved while its sign changes with each iteration.

can say nothing of its stability while we remain in the linear framework of Floquet theory. Nevertheless it seems plausible that, just above the bifurcation point, a stable limit cycle of period $2T$ will appear, if the bifurcation is supercritical.

VIII.1.3 FIRST RETURN MAP

The preceding paragraph shows yet again the central role played by the critical eigenvalue of the Floquet matrix and by its associated eigenvector. Hence we focus our attention on the Poincaré section and on this eigenvector, more or less setting aside the rest of the trajectory whose importance is secondary. This leads us to study what becomes of an initial condition after discrete time intervals $T, 2T, 3T, \ldots$ i.e. to analyze the properties of the iterated transformation of a point on an axis. The behavior of the flow can be understood through a one-dimensional mapping of the form:

$$x_{k+1} = f(x_k)$$

relating the coordinate x_{k+1} of a point at time $(k+1)$ to the coordinate x_k of its antecedent at time kT, representing the *first return map* of the original flow. We can again see the heuristic advantage of reducing the analysis of a nonlinear flow to a mathematically much simpler problem [80].

Starting in 1918, the study of what we also call iterated maps on the interval had been undertaken by Julia and Fatou. But at that time, the question was considered rather arcane and its practical significance was not perceived. Current developments in the theory of dynamical systems have finally revealed its utility. Since 1975 there has been such an extraordinary proliferation of articles devoted to the subject that it would now be impossible to be exhaustive: this is why we will present only the essential results relative to the appearance of chaos.

Recall that a one-dimensional approach is justified when we are dealing with a highly dissipative system, as we showed in Chapter IV. Because of the draconian contraction of areas resulting from substantial dissipation, only one degree of freedom plays an appreciable role. All useful information is therefore concentrated in one privileged direction in phase space and we can safely ignore the others. This chapter presents the analysis of the subharmonic cascade and the properties of quadratic mappings of the interval into itself.

---

80. The utility of first return maps is, of course, very widespread, and not limited merely to the mapping of the interval onto itself considered in this chapter. Other illustrations will arise when we treat intermittency phenomena in Chapter IX.

VIII.2 The subharmonic cascade

VIII.2.1 QUADRATIC MAPPING OF THE INTERVAL

Consider the continuous function:
$$f(x) = 4\mu x(1-x) \quad x \in [0, 1].$$

For μ positive and less than or equal to one, this function describes a first return map:
$$x_{k+1} = 4\mu x_k(1 - x_k) = f(x_k) \tag{VIII.1}$$

which assigns to every point x_k of the unit interval another point x_{k+1} of the unit interval, called the iterate of x_k. The condition $\mu \leq 1$ is required to insure that $f(x_k)$,

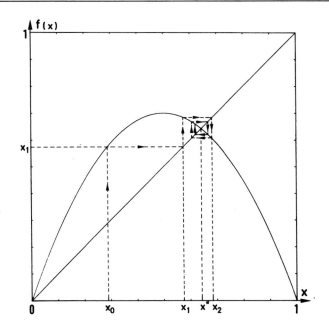

Figure VIII.3 Graph of the mapping $f(x)$ for $\mu = 0.7$.
The origin is an unstable fixed point. The point x^* at which the graph of f intersects the diagonal is a stable fixed point to which the iterates of all initial conditions in $]0, 1[$ converge.

like x_k itself, belongs to the interval [0,1]. In fact, neither the exact form[81] of the function, nor the restriction of x to the unit interval limit the generality of the conclusions we will present. But, in order to develop the conclusions, we must nevertheless choose a particular function; the form above, and the normalization of the interval to [0,1], are adopted for convenience.

Let us first draw the graph of f for $\mu = 0.7$ in Figure VIII.3: it is a parabola that vanishes at $x = 0$ and $x = 1$ and has a maximum equal to μ at $x = 0.5$. Using the graph, we study the iteration of the mapping starting from an arbitrary initial condition x_0, using the method described in Chapter IV. We see on Figure VIII.3 that the iteration converges to x^*, the intersection point of the graph of f with the diagonal, independent of the initial point x_0, with two exceptions: 0 and 1. Choosing $x_0 = 0$ or 1, we find the contrary, a stable fixed point, i.e. an attractor. To designate the fact that at each step of the iteration we find ourselves at the same fixed point, we say that the attractor is of amplified during the iteration; we leave the origin, which is thus an unstable fixed point. The fixed point x^*, towards which any iteration starting from $]0, 1[$ converges, is, on the contrary, a stable fixed point, i.e. an attractor. To designate the fact that, at each step of the iteration, we find ourselves at the same fixed point, we say that the attractor is of period one, where the unit of time is one step of the iteration. Or we could say that we have taken the unit of time to be the period of the limit cycle whose Poincaré section is the fixed point x^*.

VIII.2.2 PERIOD-DOUBLING CASCADE

The situation described by Figure VIII.3 is not the only possibility. The curve $f(x)$ depends on the value of the parameter μ which is, as we have seen, the maximum value of f. By varying μ, we modify the curve, which can have decisive consequences on the sequence of iterates. To convince ourselves of this, let us examine Figure VIII.4, for which $\mu = 0.8$. Now the fixed point x^* is unstable, for the slope of the tangent at this point is greater than one in absolute value. The graphical construction shows that the mapping has two special points x_1^* and x_2^* such that:

$$x_2^* = f(x_1^*) \quad \text{and} \quad x_1^* = f(x_2^*)$$

---

81. The only conditions which must be satisfied by f on $[0, 1]$ are that it:
 . be continuous and differentiable,
 . have an extremum,
 . have a negative Schwartzian derivative.

The second condition requires that f be a nonlinear function. None of the phenomena to be described in this chapter can be generated by a linear function nor, more generally, by any function which is monotonic on the interval. The third condition, which we will not use explicitly, can be seen as a requirement of concavity for f. It insures that all of the bifurcations of the mapping $x \longmapsto f(x)$ be supercritical. The quadratic mapping satisfies this last condition.

VIII.2 THE SUBHARMONIC CASCADE

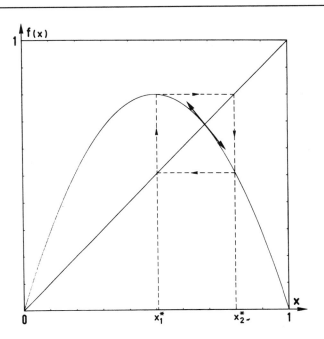

Figure VIII.4 Graph of the mapping $f(x)$ for $\mu = 0.8$.
Both fixed points of f are unstable at this value of μ. We perceive that any initial condition in $]0, 1[$ has as its asymptotic limit the pair of points x_1^* and x_2^* visited in turn.

In other words, the iteration alternates between one point and the other; starting from one of these points, we must iterate twice to return to it. The two points constitute an attractor of period two, also called a 2-cycle. Given that:

$$x_2^* = f(x_1^*) = f(f(x_2^*))$$
$$x_1^* = f(x_2^*) = f(f(x_1^*))$$

these two points — which are not fixed points of f — are fixed points of the function:

$$g(x) = f(f(x)) = f^2(x)$$

as can be verified on Figure VIII.5. More detailed study shows that we pass continuously from the first situation (fig. VIII.3) to the second (fig. VIII.4) by increasing the value of μ. Transition occurs at the threshold value $\mu_1 = 0.75$. At this value, the stable fixed point of f becomes unstable, and, correspondingly, there appear two stable fixed points of f^2. An attractor of period two takes the place of the attractor of period one. The period has indeed doubled, as Floquet theory predicts when the unit circle is crossed at -1.

What happens when we continue to increase μ? The graphs of f and f^2 gradually change, in such a way that the fixed points of f^2 also end up losing their stability. Another simple graphical construction, helps to foretell and to explain the sequence of events. Consider the square around the fixed point x_2^* in Figure VIII.5[82]. Inside the square, we observe a locally parabolic curve containing a stable fixed point — i.e. a situation just like that of Figure VIII.3. Therefore when the fixed point becomes unstable by deformation of the curve, we can expect the same phenomenon as before: the fixed point of g will be replaced by two points which will be the fixed points of the function:

$$h(x) = g(g(x)) = f^4(x).$$

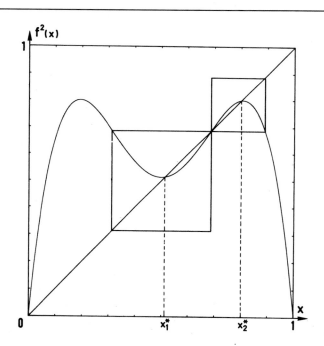

Figure VIII.5 Graph of the mapping $g(x) = f(f(x)) = f^2(x)$ for $\mu = 0.8$.
This mapping has four fixed points, of which two, x_1^* and x_2^*, are stable. f maps each one onto the other.
The two squares drawn around the fixed points serve to emphasize the structural similarity with the graph of $f(x)$ in Figure VIII.3.

82. The same reasoning applies to the fixed point x_1^*, which is also an extremum, but a minimum.

VIII.2 THE SUBHARMONIC CASCADE

This conclusion applies equally to the fixed point x_1^*, which is also an extremum, but a minimum. Both x_1^* and x_2^* simultaneously become unstable at:

$$\mu_2 = \frac{1 + \sqrt{6}}{4} = 0.862\,37...$$

above which g has no stable fixed point. However h, graphed in Figure VIII.6 for $\mu = 0.875$, now has four fixed points. These form an attractor for the iteration of (VIII.1). Starting from any one of these points, four iterations are required to return to it: we now have a 4-cycle. Again, the period has doubled via a subharmonic bifurcation.

By continuing to increase μ in (VIII.1), the same phenomenon will be repeated *ad infinitum*. We will see a cascade of bifurcation, each accompanied by the period doubling associated with a subharmonic instability. The square drawn around one of the fixed points of the function $h(x)$ on Figure VIII.6 emphasizes the preservation of

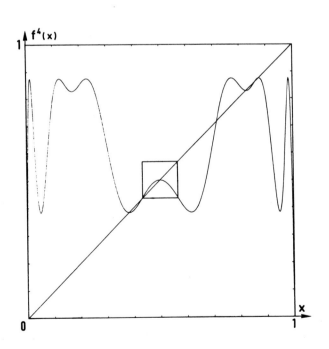

Figure VIII.6 Graph of the mapping $h(x) = g(g(x)) = f^4(x)$ for $\mu = 0.875$. This mapping has eight fixed points, of which four are stable. The square around one of them shows that we again reproduce the situation of Figure VIII.3. This foreshadows the repetition *ad infinitum* of the same process.

the structure exhibited by f in Figure VIII.3. We recognize here the analogy between structures at all scales, already mentioned in connection with the Cantor set. As μ is increased we observe a succession of attractors of period 2^ℓ, or 2^ℓ-cycles, ℓ an integer varying between 0 (for $\mu \leqslant 0.75$) up till infinity. The values of μ at which the bifurcations in the cascade occur have a remarkable property : they form an increasing series converging rapidly towards an accumulation point μ_∞, whose value can only be obtained numerically:

$$\mu_\infty = 0{,}892\,486\,418...$$

Table VIII.1 gives the values of μ corresponding to the first few bifurcations of the subharmonic cascade.

VIII.2.3 SCALING LAWS

Attentive examination of the numbers collected in Table VIII.1 reveals that the convergence towards the accumulation point obeys a simple and rigorous law: the difference between values of μ associated with two consecutive bifurcations is reduced each time by an almost constant factor:

$$\lim_{i \to \infty} \frac{\mu_i - \mu_{i-1}}{\mu_{i+1} - \mu_i} = \delta.$$

An essential result, which cannot be overemphasized, is that the scale reduction factor δ is a universal constant, independent of the details of the function f considered:

$$\delta = 4.669\,201\,609\,102\,990\,9...$$

Table VIII.1.

Periodicity of the attractor	μ value at the bifurcation point
$1 \cdot 2^0 = 1$	
	$\mu_1 = 0.75$
$1 \cdot 2^1 = 2$	
	$\mu_2 = 0.862\,37...$
$1 \cdot 2^2 = 4$	
	$\mu_3 = 0.886\,02...$
$1 \cdot 2^3 = 8$	
	$\mu_4 = 0.892\,18...$
$1 \cdot 2^4 = 16$	
	$\mu_5 = 0.892\,472\,8...$
$1 \cdot 2^5 = 32$	
	$\mu_6 = 0.892\,483\,5...$
....	....
$1 \cdot 2^\infty = \infty$	$\mu_\infty = 0.892\,486\,418...$

VIII.2 THE SUBHARMONIC CASCADE

Distances on the x axis between points on the attractor are reduced by a scale reduction factor given by a second universal constant[83]:

$$\alpha = 2.502\,907\,875\,095\,892\,84...$$

More precisely, in iterating *any* mapping which has a *quadratic extremum* we *always* find the same period-doubling cascade, with the *same* scaling laws as above. The theory is indeed extremely general, which justifies in retrospect the attention we have devoted to this particular function f. What is remarkable is that quantitative predictions can be made provided that a simple qualitative condition is satisfied[84].

A graph of the x values of the points on each attractor, as a function of μ, aids in visualizing the subharmonic cascade just described (fig. VIII.7). The first bifurcations,

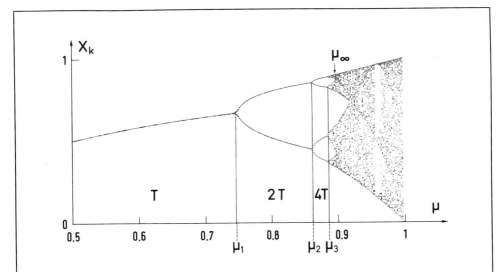

Figure VIII.7 Asymptotic iterates of the mapping $f(x)$ for $\mu \in [0.5, 1]$.
As a function of the parameter μ, we have plotted the value or values of x_k obtained by iteration of $f(x)$ as k tends to infinity. From left to right we see:
— a sequence of periodic attractors, separated by subharmonic bifurcations each of which doubles the number of points on the attractor, as well as its period. The cascade of subharmonic bifurcations has an accumulation point at $\mu_\infty = 0.892\,48...$
— beyond μ_∞, a region where aperiodic and periodic attractors alternate.
A few of the periodic attractors exist over a sufficiently large interval of μ for them to be seen as lighter zones. In particular, the period-three attractor located at $\mu \sim 0.96$ is clearly visible.

83. Like μ_∞, the constants δ and α can only be determined numerically.
84. The analogy between this theory and that of phase transitions is well known, in particular with the procedure and vocabulary of the renormalization group. The constants δ and α can be considered as the "critical exponents" at the accumulation point.

each doubling the number of points of the attractor, appear very clearly. But the bifurcations rapidly become so close to one another that they can no longer be distinguished if μ is represented on a linear scale. On Figure VIII.7, the attractor of period eight is the last that can be discerned without difficulty. This results from the geometric law of convergence mentioned above: the values listed in Table VIII.1 demonstrate why the attractors of period 16, 32, etc., are, on the scale of Figure VIII.7, practically merged.

VIII.3 Characteristics of chaos

VIII.3.1 BEYOND THE SUBHARMONIC CASCADE

As μ increases and tends towards μ_∞, we encounter attractors of increasing period 2^ℓ. However, detecting the periodicity of a phenomenon necessarily requires its observation during a time span of at least several periods. Therefore, an increasingly long time is required, even becoming infinite at the accumulation point. And what happens beyond the accumulation point?

Numerical simulation shows that traversal of the critical value $\mu = 0.892\,486\,418...$ marks the beginning of a very complex domain. On the graph of Figure VIII.7 different zones appear, some lighter and others more shaded. Detailed analysis reveals that in this region, periodic attractors alternate with what is nowadays called chaos.

In the latter case, iteration of f yields a sequence of values of x that:
— never repeat themselves
— depend on the initial condition x_0.

In particular, two arbitrarily close initial conditions give rise to two sequences of iterates — or trajectories — that always eventually diverge from one another. This is certainly unexpected behavior for such a simple transformation! We will now discuss two essential properties of the mapping, which are the key to understanding the chaotic behavior.

VIII.3.2 TWO FUNDAMENTAL PROPERTIES OF THE MAPPING

To understand these two properties, which are the noninvertibility of the transformation and its sensitivity to initial conditions (SIC), it suffices to consider the graph of $f(x)$ for μ close to one. We draw the graph of f (fig. VIII.8 a) and, below the x axis, a horizontal line representing the images $f(x)$. We now examine what happens as we vary x along the interval [0, 1]. The image of $x = 0$ is 0, while that of $x = 1/2$ is 1. The first half [0, 1/2] of the interval is therefore "stretched" by the mapping onto the entire interval [0, 1]. The image of $x = 1$ is clearly 0, and the transformation stretches the second half [1/2, 1] onto the whole interval as well, but reverses its direction. In other words, when x makes a one-way trip along the unit interval, its image executes a

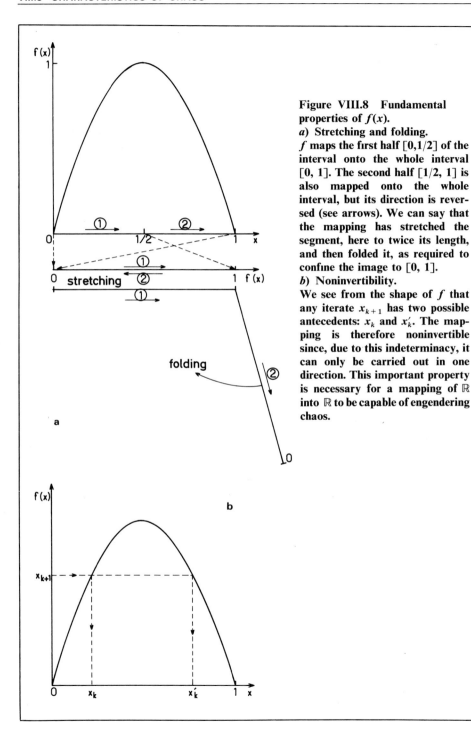

Figure VIII.8 Fundamental properties of $f(x)$.
a) Stretching and folding.
f maps the first half $[0,1/2]$ of the interval onto the whole interval $[0, 1]$. The second half $[1/2, 1]$ is also mapped onto the whole interval, but its direction is reversed (see arrows). We can say that the mapping has stretched the segment, here to twice its length, and then folded it, as required to confine the image to $[0, 1]$.
b) Noninvertibility.
We see from the shape of f that any iterate x_{k+1} has two possible antecedents: x_k and x'_k. The mapping is therefore noninvertible since, due to this indeterminacy, it can only be carried out in one direction. This important property is necessary for a mapping of $\mathbb{R}$ into $\mathbb{R}$ to be capable of engendering chaos.

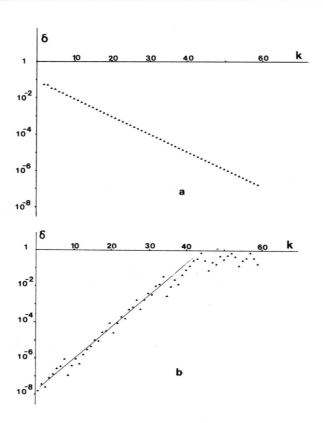

Figure VIII.9 Evolution of the distance δ between two iterations of $f(x) = 4\mu x(1 - x)$ with different initial values.

The mapping is first iterated a hundred times to eliminate transient effects, producing x_0. We then take as initial values x_0 and $x_0 + \delta_0$ to calculate two series of iterates. The (absolute) value of the difference between the two k^{th} iterates varies like:

$$\delta = \delta_0 \exp(\lambda k)$$

where λ is the Lyapunov exponent. The Lyapunov exponent is therefore the average slope of δ vs. k plotted on a semi-logarithmic scale as above.

a) When $\mu = 0.7$, we have $\lambda \simeq -0.2$. The regime is periodic and the distance δ decreases: hence a negative Lyapunov exponent.

b) When $\mu = 0.95$, we have $\lambda \simeq +0.4$. The regime is chaotic and the distance δ increases: hence a positive Lyapunov exponent.

Saturation of δ occurs when the distance attains the size of the interval $[0, 1]$.

complete round trip of the unit interval. This can be constructed as the result of a twofold operation: stretching (by an average factor[85] of two in this example), followed by folding (as required for the image to be confined to the interval [0, 1]). Each of the two operations has an important consequence.

* Stretching causes the image of a small initial displacement δx_0 to be (on the average) multiplied by the stretching factor at each iteration. As soon as the stretching factor is greater than one, this geometric process leads to an exponential increase of the displacement. This occurs above the critical value μ_∞ whenever there exists no periodic attractor: it is the cause of the sensitivity to initial conditions discussed in Chapter VI. The average speed with which a given displacement is amplified is measured by the parameter λ of the exponential law:

$$\delta x_k = \delta x_0 \exp(\lambda k)$$

which is the Lyapunov exponent (see Appendix B) of the one-dimensional mapping. This exponential increase appears clearly in Figure VIII.9.

* Folding, indispensable for the stretching to be compatible with a bounded domain — here the unit interval — causes "mixing" of the images of different points. This too is an important characteristic of chaos. Due to folding, it is impossible to return from a point to its antecedent: every point x_{k+1} has two antecedents x_k at the previous iteration (see fig. VIII.8 b), four at the iteration before that, and so on. In consequence, while the transformation is easily carried out for increasing values of k, it cannot be iterated backwards. Such a transformation is called *non invertible*[86]. The mapping forbids knowledge of the past and, because of SIC, also prediction of future values of x after a period of time. To predict the long term future of x would require that the initial condition be known with infinite precision, since the slightest uncertainty in x_0, even infinitesimal, prevents us from choosing between trajectories which inevitably diverge in the long run. And of course, infinite precision is not possible in practice. Therefore, despite being governed by the deterministic equation (VIII.1), the evolution observed remains fundamentally unpredictable; it appears to us as discodered and chaotic. Contrary to a widespread idea, determinism and chaos are not antithetical concepts.

VIII.3.3 THE INVERSE CASCADE

We have seen that the approach to chaos via period doubling is a highly structured process. We can therefore imagine that the chaotic behavior observed for $\mu > \mu_\infty$ would not be entirely bereft of order, undetectable at first sight. One element of order,

---

85. Only an average stretching can be defined, given the parabolic form of $f(x)$. The regions close to $x = 0$, $x = 1$ are the most stretched, while the region near the maximum $x = 1/2$ is actually contracted.

86. More generally, any continuous mapping of $\mathbb{R}$ into $\mathbb{R}$ must be non invertible in order to lead to chaos.

to be discussed in the next section, is the existence of "windows of periodicity": periodic attractors in the interval $[\mu_\infty, 1]$. Some of these can be seen as lighter zones[87] in Figure VIII.7. The periodic and aperiodic attractors are closely interwoven. Another kind of order is found in the aperiodic attractors: detailed analysis shows them to be "noisy limit cycles" of period 2^ℓ, ℓ an integer tending to infinity as μ approaches μ_∞ from above. More precisely, the iteration visits sequentially a set of 2^ℓ disjoint segments of $]0, 1[$. After 2^ℓ iterations, we again find ourselves in the same segment: this is why it is called a cycle. However the behavior inside each segment is completely chaotic: hence the qualifier "noisy". The iterates of order 2^ℓ are entirely contained in a small segment, but within the segment, they are completely disordered.

As μ is increased, we see that at certain parameter values the segments fuse, two by two. The noisy cycle of period 2^ℓ is replaced by another of half the period, or $2^{\ell-1}$. As μ increases, the process is repeated, until we attain the "period" $2^0 = 1$. The diagram of Figure VIII.10 summarizes this schematically and brings out the fact that facing the subharmonic cascade, there is another cascade, of similar structure, but in the opposite

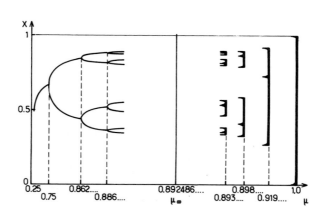

Figure VIII.10 Direct and inverse cascades.
The purpose of this figure is to bring out a fact not visible in Figure VIII.7: it shows the inverse cascade, a kind of fuzzy reflection above μ_∞ of the direct cascade. A nonlinear scale, defined by $\sqrt[5]{\tanh(4\mu - 4\mu_\infty)}$, is used for the abscissa; this graphic artifice should not occult the fact that the "natural" representation of the solutions is that of Figure VIII.7.

From S. Grossmann, S. Thomae.

---

87. In the next section we will return to what are called windows of periodicity and we will see that they form a strictly ordered set.

direction along the μ axis. We call the first a direct cascade and the second, an inverse cascade. One result — remarkable, to say the least — is that the parameter values μ at which the bifurcations of the inverse cascade occur also converge towards μ_∞ with the same scale factor $\delta = 4.669...$ as for the direct cascade. Here is yet more proof, were it still necessary, that there is sometimes order in chaos!

VIII.3.4 WINDOWS OF PERIODICITY

As we have said, the interval $[\mu_\infty, 1]$ does not contain only chaos. In it, we also encounter narrow ranges of μ within which the trajectories are strictly periodic. A p-cycle ($p > 1$) first appears, then engenders a sequence of attractors of period $p \cdot 2^\ell$ ($\ell = 1, 2, ..., \infty$) by a subharmonic cascade obeying the same scaling laws as described above. The accumulation point ($\ell \longrightarrow \infty$) marks the end of what we call a *period-p window*. Each window occupies a segment of the μ axis, the length of which is

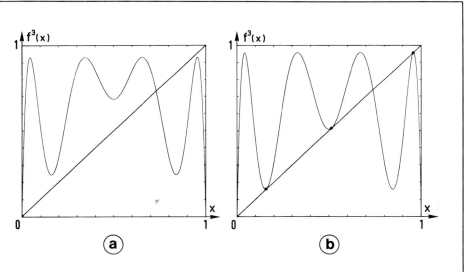

Figure VIII.11 Birth of a 3-cycle.
The largest window of periodicity (0.957 1... $< \mu <$ 0.962 4...) contains cycles of period 3.2^ℓ ($\ell = 0, ..., \infty$). To explain how a 3-cycle can arise we graph the third return map:

$$x_{k+3} = f^3(x_k).$$

a) Slightly below ($\mu = 0.9$) the threshold value, the mapping has no stable fixed point.
b) At the threshold ($\mu = 0.975$ 1...), the graph of f^3 becomes tangent to the diagonal at three points. The three points, at first marginally stable, become stable just above the threshold. A tangent bifurcation has produced the 3-cycle.

highly variable but, most of the time, extremely small. This is why it is generally not possible to detect them on Figure VIII.7, where almost the whole interval $[\mu_\infty, 1]$ seems to be chaotic. The largest of the windows (0.957 1... $< \mu <$ 0.962 4...) contains cycles with a basic period of three. We will use the period-3 window as an illustration.

To say of a cycle that its period is p means that after p iterations we return to the point of departure. To identify a p- cycle it is convenient to plot a graph of the pth return map, that is, the function $x_{k+p} = f^p(x_k)$. This is shown on Figure VIII.11 for $p = 3$ and for two values of μ, one just under and the other exactly at the threshold of appearance of the cycle. Below the threshold, the fixed points — the points of intersection of the identity map with the curve — are unstable (slope greater than one in absolute value). As μ is increased the curve becomes tangent to the identity map at three points. These three fixed points of f^3, marginally stable at the threshold and asymptotically stable above the threshold, form a 3-cycle for f. This type of bifurcation is called, for obvious reasons, a *tangent bifurcation*. If we continue to increase μ, we expect a subharmonic instability to destabilize this periodic regime as before. This expectation is borne out, and the cycle of period three has a destiny similar to that of the cycle of period one, leading to chaos[88] via a subharmonic cascade, still governed by the same scaling laws.

VIII.3.5 THE UNIVERSAL SEQUENCE

Another extraordinary property concerns the relative location of the windows of periodicity along the μ axis. While the absolute location and the length of a window depend on the particular mapping f considered, the order in which we encounter these windows as μ is increased is immutable. In other words, for any function f (provided that it has an extremum, quadratic or not), the periodic attractors form a sequence, always in the same order, which is therefore called the *universal sequence*.

To characterize a periodic attractor, we note that two attractors of the same period can differ from one another by the order in which their points are visited. Using the simplest example of attractors of period four, we represent the four points schematically along the x axis (fig. VIII.12). We see that there exist two 4-cycles[89], corresponding to the circuits 1-2-3-4 and 1-3-2-4.

It is useful to identify a periodic attractor by the location along the x axis of its constituents. There are two ways of characterizing different attractors of the same period. One way consists of, first, numbering the points of the attractor according to the

---

88. When the sequence of numerically computed iterates seems to be chaotic we are naturally led to ask: is the sequence truly chaotic, or is it periodic with a very large period, accompanied by very long-lived transients? This question has as yet received no definitive answer, although some facts support the idea of genuine chaos.

89. Only two different attractors can exist given the form of the function f considered here. Its shape (fig. VIII.3) imposes that the image of the rightmost point be the leftmost point: hence the necessity for any cycle beginning at point 1 to end at point 4.

VIII.3 CHARACTERISTICS OF CHAOS

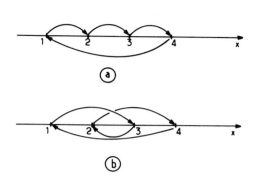

Figure VIII.12 The two 4-cycles attractors.
Each periodic attractor is characterized by the order in which its constituent points are visited. Two different 4-cycles can then be distinguished:
a) 1-2-3-4 *b)* 1-3-2-4.
There can be no other, given the form of the mapping f under consideration.

Table VIII.2.

μ	Period p	Sequence of points	RL sequence
	6	2-0-4-3-5-1	RLRRR
	5	2-0-4-3-1	RLRR
	3	2-0-1	RL
	6	2-5-3-0-4-1	RLLRL
	5	2-3-0-4-1	RLLR
	6	2-3-0-4-5-1	RLLRR
	4	2-3-0-1	RLL
	6	2-3-4-0-5-1	RLLLR
	5	2-3-4-0-1	RLLL
	6	2-3-4-5-0-1	RLLLR

order in which they are visited, and then, writing down the order in which they are located on the x axis. In the second method, we assign to each point a letter, R or L, according to whether it is to the *right* or to the *left* of the maximum of f. Listing the letters in the order in which the points are visited leads to a characteristic sequence of letters RL. It can be shown that in any window of periodicity, there exists a value of μ such that the maximum is itself part of the attractor. Therefore it suffices to begin iterating from the maximum, which we therefore designate by 0 (not to be confused with the leftmost point of the interval $x = 0$). Since 0 is located neither to the right nor to the left of the maximum, the RL sequence contains one letter less than the periodic attractor has points.

Table VIII.2 gathers this data for the different attractors of the universal sequence having a basic period less than or equal to six.

Due to the form of $f(x)$, the image of the maximum (point 0) is always the rightmost point (point 1), whose image is in turn the leftmost point: this is why the sequence of points always begins with 2 and finishes with 1, while the sequence of letters always begins with RL, for any periodic attractor.

VIII.4 Experimental illustrations

VIII.4.1 NATURE OF THE OBSERVATIONS

The experimental identification of some of the phenomena that we have described regarding the quadratic mapping of the interval should not be very problematic. If a periodic regime of frequency f undergoes a subharmonic instability, resulting in another periodic regime of twice the period, this will entail:

— multiplication by two of the number of points of a Poincaré section,

— the appearance in the Fourier spectrum of the frequency $f/2$ and its odd harmonics ($3f/2$, $5f/2$, etc.).

In a subharmonic cascade this process occurs repeatedly and is therefore easily observed, as long as we can regulate the control parameter with sufficient accuracy.

Similarly in the case of an inverse cascade we expect to see:

— a lengthening of the segments which constitute the Poincaré section as they merge in pairs (see fig. VIII.10),

— the noise associated with chaotic behavior successively destroy the subharmonics $...f/16$, $f/8$, $f/4$, $f/2$ in the opposite order of their appearance in the direct cascade, before totally invading the spectrum.

Study of the universal sequence presupposes precise identification of the type of periodic attractor. This task, inaccessible to spectral analysis, requires the use of Poincaré sections.

Many experiments, of which a large proportion were performed on the three systems described in Chapter V, have established the relevance of the theoretical

conclusions we have arrived at here in Chapter VIII. The observations leave no doubt that:
- the subharmonic cascade is indeed one of the modes of transition to chaos,
- if the control parameter is increased beyond the accumulation point μ_∞ of the subharmonic cascade, we really do observe an inverse cascade. The inverse cascade is accompanied by a gradual broadening of the lines of the Fourier spectrum, and hence the growth of chaos,
- beyond the direct cascade, we encounter periodic attractors belonging to the universal sequence.

When the measurements are sufficiently precise, we can even verify the scaling laws: agreement between theory and experiment is no longer merely qualitative, but quantitative.

VIII.4.2 SUBHARMONIC CASCADE : R.B. CONVECTION

We shall first describe an experiment carried out on liquid mercury at very low Prandtl number (Pr $\sim$ 0.03). To stabilize the convective structure, we use cells of small aspect ratio ($\Gamma = 4$ or $\Gamma = 6$) which, moreover, we place in a magnetic field. Two purposes are served in doing this. First, given the high electrical conductivity of mercury, the convection rolls have a strong tendency to align themselves in the direction parallel to the magnetic field. This fixes the spatial order and prevents the structure from "melting" as Ra is increased (see Section V.3.2). In addition, the magnetic field damps certain modes causing oscillation of the rolls; it intensifies dissipation, which we have seen to be favorable to use of the first return map.

The mercury is placed between two thick copper plates. The convective motions are measured using bolometers, since optical methods cannot be used in an opaque medium. In a first phase of the experiment, we fix the magnetic field strength at zero and increase the temperature difference until the onset of convection at a value Ra_c of the Rayleigh number. Continuing to increase Ra, we notice, at a value close to $2Ra_c$, the onset of a new instability. The signal recorded by the bolometer begins to oscillate in time with a frequency f_1. This oscillatory instability can be attributed to a wave propagating along the roll axes. As Ra is further increased, the periodic regime in turn becomes unstable, and there appears in the power spectrum of the signal a second frequency[90] f_2 close to, but nonetheless distinct from $f_1/2$. The physical oscillation associated with this second frequency is not yet identified with certainty. For a slightly larger value of Ra, the frequencies of the two oscillators lock when the condition of subharmonic resonance $f_2 = f_1/2$ is satisfied.

This frequency locking marks the beginning of the second phase of the experiment. A constant and uniform magnetic field is applied, whose intensity is such that the

---

90. A transition has occurred from a limit cycle to a torus T^2 via the kind of bifurcation described in Chapter VII.

amplitudes of the two oscillators become comparable. By gradually increasing the Rayleigh number, we encounter a succession of well-defined values of Ra at which one periodic regime bifurcates to another of twice the period. Figure VIII.13 a shows recordings from several consecutive periodic regimes. The emergence in the Fourier spectrum of the subharmonics $f_1/4$, then $f_1/8$, $f_1/16$, $f_1/32$ (and their odd harmonics) is the signature of period doubling (see fig. VIII.13 b). From these results we can attempt to evaluate the convergence ratio of the successive bifurcations. We find a value of 4.4, extremely close to the universal asymptotic limit 4.669... predicted by theory.

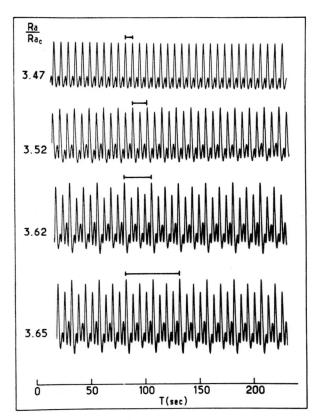

Figure VIII.13 a Subharmonic cascade in a thermoconvection experiment. The changing shape of the signal (temperature of the fluid at one point as a function of time) clearly shows the period-doubling process that takes place as the control parameter Ra/Ra_c is increased. The line segments indicate the length of one period, defined by a basic pattern which is repeated indefinitely.

From A. Libchaber, S. Fauve, C. Laroche.

VIII.4 EXPERIMENTAL ILLUSTRATIONS

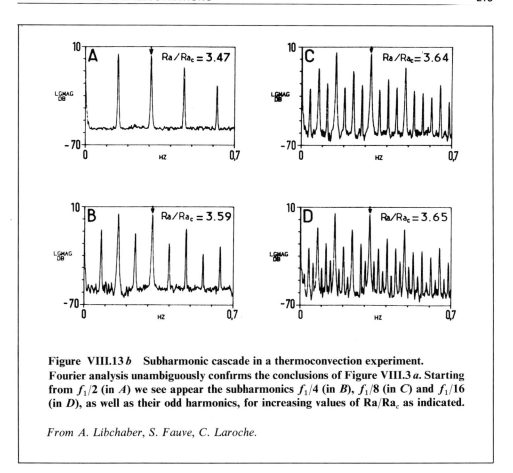

Figure VIII.13 *b*   Subharmonic cascade in a thermoconvection experiment. Fourier analysis unambiguously confirms the conclusions of Figure VIII.3 *a*. Starting from $f_1/2$ (in *A*) we see appear the subharmonics $f_1/4$ (in *B*), $f_1/8$ (in *C*) and $f_1/16$ (in *D*), as well as their odd harmonics, for increasing values of Ra/Ra_c as indicated.

From A. Libchaber, S. Fauve, C. Laroche.

This R.B. convection experiment offers indisputable confirmation of the existence of a subharmonic cascade in a physical system. The confirmation is even stronger in that measurements of quantities such as the ratio between amplitudes of the consecutive subharmonics support theoretical predictions. Let us also mention that thermoconvection in other fluids (liquid helium, water, etc.) also gives rise to a period-doubling cascade.

VIII.4.3   INVERSE CASCADE: THE COMPASS

The subharmonic cascade is also easily observed in the compass, further demonstrating the universality of the theory. The axis of the magnet (cf. Chapter V) is submerged in oil of high viscosity so as to produce strong energy dissipation. Having fixed the amplitude and frequency of the rotating magnetic field, we then gradually vary the amplitude B_1 of the stationary magnetic field which modifies the control parameter M of Equation (12) of Chapter V. The experiment uses the following parameter values:

$$\alpha = 0{,}174; \quad P = 0{,}335; \quad M \in [0.160\,0,\ 0.232\,1].$$

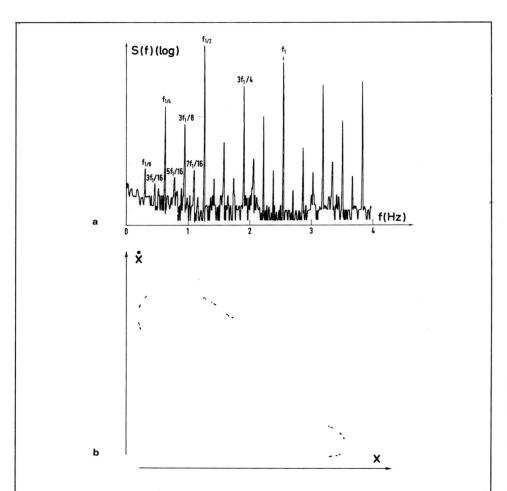

Figure VIII.14 The $f_1/16$ regime of the compass.
The compass described in Chapter V also undergoes a subharmonic cascade. The Fourier spectrum (*a*) and the Poincaré section (*b*) in an $(x, \dot{x})$ plane are taken from the limit cycle obtained after four subharmonic bifurcations. The form of the spectrum, as well as the division of the Poincaré section into sixteen groups of points are as predicted by theory.

From V. Croquette and C. Poitou.

VIII.4 EXPERIMENTAL ILLUSTRATIONS

By varying M, we generate a sequence of periodic regimes characterized by a doubling of the period at each stage. As an example, Figure VIII.14 shows results obtained after the fourth bifurcation. On the Fourier spectrum we distinguish the subharmonics $f_1/2$, $f_1/4$, $f_1/8$, and $f_1/16$ of the initial frequency f_1 as well as their odd harmonics. Similarly, the Poincaré section contains sixteen points subdivided into two

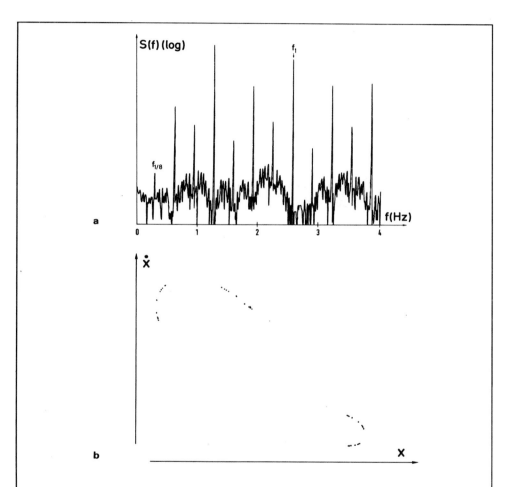

Figure VIII.15 Disappearance of $f_1/16$ in the inverse cascade.
a) The peaks corresponding to the odd harmonics $f_1/16$ are now swamped in noise and are no longer visible in the Fourier spectrum. However the $f_1/8$ frequencies are still clearly identifiable.
b) The points of the Poincaré section are divided into eight groups within which they no longer join in pairs.

From V. Croquette and C. Poitou.

groups. Each group is in turn subdivided into four pair of points, making a total of eight pairs. This is exactly the kind of distribution that we would obtain by tracing a vertical line on Figure VIII.7 in the domain of existence of the attractor of period $16T$. If we continue to increase M above a value close to 0.228 5 (the estimated accumulation point) we see a tangible change in the evolution of the Fourier spectrum and Poincaré

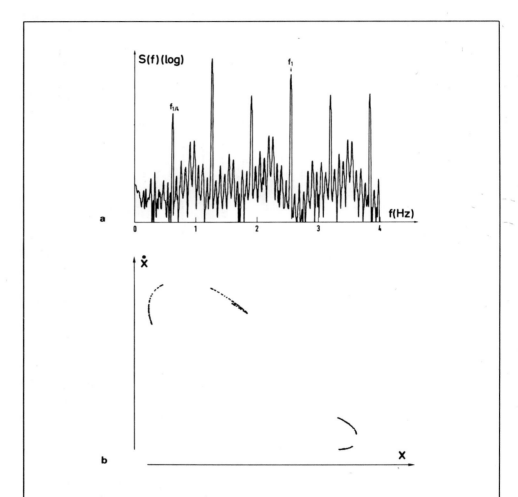

Figure VIII.16 Disappearance of $f_1/8$ in the inverse cascade.
a) Same as Figure VIII.15, but this time peaks that are multiples of $f_1/8$ are no longer visible.
b) The Poincaré section now consists of only four arcs of curves, noticeably lengthened.

From V. Croquette and C. Poitou.

VIII.4 EXPERIMENTAL ILLUSTRATIONS

section. No additional subharmonic higher-order peaks appear in the spectrum. Instead, noise begins to widen the base of certain peaks before absorbing them altogether. This process occurs first on the odd harmonics of $f_1/16$ (fig. VIII.15 a), then those of $f_1/8$ (fig. VIII.16 a), and so on. Similarly, in the Poincaré section, the formation of new pairs of points by "doubling" of existing points ceases. Instead, the additional points disperse, little by little, along short segments which tend to lenghten as M is increased before merging two by two. This is illustrated by Figure VIII.15 b, on which we can still distinguish eight groups, and by Figure VIII.16 b, on which we can only distinguish four. This is the behavior we expect if, on Figure VIII.10, we move a vertical line to the right, beyond μ_∞. As predicted by theory, an inverse cascade follows the direct cascade.

We note that, up to measurement accuracy, the points of the Poincaré section (Figure VIII.16 b) do not cover a surface but are located along two arcs of curves. This is because the energy dissipation is sufficient to mask the fine structure of the aperiodic attractor. This justifies analyzing the behavior in terms of a mapping of an interval onto itself.

VIII.4.4 UNIVERSAL SEQUENCE : THE B.Z. REACTION

The B.Z. reaction in an open reactor has a sequence of periodic regimes with period doubling at each step of a cascade of bifurcations, exactly as in R.B. convection and in the compass. Only the first few bifurcations of the cascade, the regimes of period T, $2T$, $4T$, $8T$, are actually observed. The reason is experimental: the control parameter here is the inverse of the mean residence time of the chemical species in the reactor, which is varied via the flow rate through the supply pumps. It is not yet possible to stabilize the flow rate to better than 0.5-1 %, at least not over substantial lengths of time. Given the value of the universal constant δ of the geometric law, we see how far we are from being able to observe regimes with periods greater than $8T$.

Aside from the subharmonic cascade, better illustrated by other experiments, the B.Z. reaction does nonetheless offer a very interesting illustration of the theory. A whole set of periodic regimes is observed, recapitulated in Table VIII.3 using the notation previously introduced. In comparing Tables VIII.2 and 3, we identify the first few elements of the universal sequence. Even regimes whose basic period is greater than six are observed. It is altogether likely that other periodic regimes are present but not detectable since their domains of existence on the control-parameter axis is too one might say that these regimes are unstable with respect to experimental perturbations. The absence of certain elements of Table VIII.2 in Table VIII.3 can be interpreted in the same way and should not concern us.

Let us now examine Table VIII.3 in more detail. During each experiment, we record the variation with time of the potential of a Br^- specific electrode. This enables us to follow the progress of the reaction as measured by the logarithm of the ionic concentration. When we vary the mean residence time of the chemical species while

Table VIII.3.

	Period	Sequence of the points	RL sequence
Cascade	1	0	—
	2	0-1	R
	4	2-0-3-1	RLR
	8	2-6-0-4-3-7-5-1	RLRRRLR
Universal sequence	10	2-8-6-0-4-3-9-5-7-1	RLRRRLRLR
	6	2-0-4-3-5-1	RLRRR
	5	2-0-4-3-1	RLRR
	3	2-0-1	RL
	6	2-5-3-0-4-1	RLLRL
	9	2-8-5-3-0-6-4-7-1	RLLRLRRL
	5	2-3-0-4-1	RLLR
	4	2-3-0-1	RLL
	8	2-6-3-7-4-0-5-1	RLLLRLL

holding all other parameters (e.g. chemical concentrations in the supply, reactor temperature) constant, we find various regimes — some periodic, others not. Figure VIII.17 shows recordings taken from several of the periodic regimes observed[91]. How can we use this information to identify the type of attractor with which we are dealing?

To answer this question, the reader must recall that the experimentalist travels the same path as the theorist, but in the opposite direction! The theorist begins with the function defining the successive iterates. The experimentalist first identifies the iterates from measurements, and hopes to then reconstruct the function being iterated — or at least its graph — which is the unknown of the problem.

Having measured only one of the dynamical variables, the attractor is not accessible in the original phase space. However, using the technique described at the end of Chapter IV, we construct a projection of the attractor in a three-dimensional phase space by taking as coordinates the value of the signal at a time t, and at two later times $t + \tau$ and $t + 2\tau$. We then look for the points of intersection of the trajectory with a plane, for example the plane perpendicular to Figure VIII.18 indicated by the dashed line. We find that the intersection points, far from being distributed at random, are located along a single curve (almost a straight line). This is the sign of strong dissipation. Under these circumstances, we can plot the abscissa of a point along this curve as a function of that of its antecedent, obtaining the first return map: the graph of x_{k+1} as a function of x_k. If our hypotheses are satisfied, we would expect to find the set of points distributed along a curve. This must be the graph of the function being

---

91. Note the resemblance between these signals and those of Figure VIII.13 for R.B. convection.

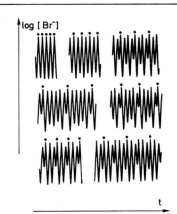

Figure VIII.17 Periodic regimes in the B.Z. reaction.
The signal from an electrode specific to Br^- ions is used to identify several periodic regimes, corresponding to different values of the reactant flux. These signals, a sampling of those obtained after a subharmonic cascade, describe motion on attractors belonging to the universal cascade of Table VIII.3. The dots marked above the signal serve to indicate the length of the period.
From R. Simoyi, A. Wolf, H. Swinney.

iterated. And this is indeed what happens, as proves Figure VIII.19 *a*: we have a curve, meaning that the dispersion in the (x_k, x_{k+1}) plane is very weak. Note in particular that this curve is not monotonic: it has a maximum, a "bump", like the quadratic function used to present the theory. At this stage, comparison with theory is almost self-evident. By slowly varying the control parameter we can induce a gradual modification of the

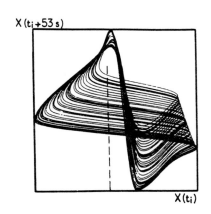

Figure VIII.18 An aperiodic attractor.
An image of the attractor in the space $X(t)$, $X(t + \tau)$, $X(t + 2\tau)$ with $\tau = 53$ sec. is reconstructed from the signal $X(t)$. We see its projection on the $X(t)$, $X(t + \tau)$ plane. We then take a Poincaré section of this image with a plane perpendicular to the figure whose intersection with it is indicated by the dashed line.

From R. Simoyi, A. Wolf, H. Swinney.

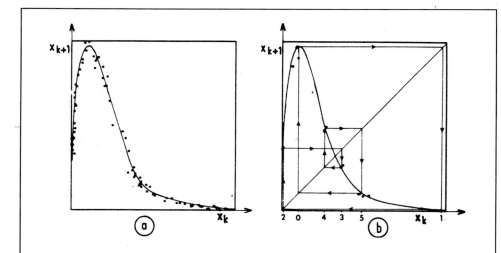

Figure VIII.19 First return map.
a) Once the Poincaré section of the attractor has been obtained, the dynamics are studied by graphing the coordinate of a point x_{k+1} as a function of that of its antecedent x_k. The experimentalist aims towards reconstructing the graph of the first return map from these observations. The figure shows that we in fact obtain a curve rather than dispersed points, and that this curve has an extremum.
b) By gradual deformation of the curve obtained in (*a*), we see how transition takes place from an aperiodic attractor to a neighboring periodic attractor with a period of six.

From R. Simoyi, A. Wolf, H. Swinney.

shape of the curve and hence a succession of regimes, some chaotic, others periodic. Figure VIII.19 *b* shows how the iteration curve of Figure VIII.19 *a* yields the first period-6 attractor of the universal sequence, identified as 2-0-4-3-5-1, or *RLRRR*.

References for Chapter VIII

P. Collet, J. P. Eckmann, *Iterated Maps on the Interval as Dynamical Systems*, Birkhaüser, Boston (1980).

V. Croquette, "Déterminisme et chaos", *Pour la Science*, **62**, p. 62 (1982).

M. J. Feigenbaum, "Quantitative universality for a class of non-linear transformations", *Journal of Statistical Physics*, **19**, p. 25 (1978).

S. Grossmann, S. Thomae, "Invariant distributions and stationary correlation functions of one-dimensional discrete processes", *Zeitschrift für Naturforschung*, **A32**, p. 1353 (1977).

A. Libchaber, S. Fauve, C. Laroche, "Two-parameter study of the routes to chaos", *Physica*, **7D**, p. 73 (1983).

R. M. May, "Simple mathematical models with very complicated dynamics", *Nature*, **261**, p. 459 (1976).

M. Metropolis, M. L. Stein, P. R. Stein, "On finite limit sets for transformations of the unit interval", *Journal of Combinatorial Theory*, **A15**, p. 25 (1973).

R. H. Simoyi, A. Wolf, H. L. Swinney, "One-dimensional dynamics in a multi-component chemical reaction", *Physical Review Letters*, **49**, p. 245 (1982).

C. Tresser, P. Coullet, "Itérations d'endomorphisme et groupe de renormalisation", *Compte-rendus de l'Académie des Sciences de Paris*, **A287**, p. 577 (1978).

B. Derrida, A. Gervois, Y. Pomeau, "Iteration of endomorphisms on the real axis and representation of numbers", *Annales de l'Institut Henri Poincaré*, **29**, p. 305 (1978).

B. Derrida, A. Gervois, Y. Pomeau, "Universal metric properties of bifurcations of endomorphisms", *Journal of Physics*, **A12**, p. 269 (1979).

CHAPTER IX

Intermittency

IX.1 Introduction

In the introduction to the second part of this book, we have already signaled intermittency as being one of the typical routes of transition from a periodic state to chaos. In this chapter, we will justify this assertion both by theoretical arguments and by the analysis of typical and well chosen experiments. In this introduction we will first explain what is covered by the word "*intermittency*". We will then put the sections which follow into perspective.

Intermittency has, strictly speaking, no canonical definition. We call a signal intermittent if it is subject to infrequent variations of large amplitude. In this somewhat vague framework we can place the many natural random phenomena which are not easily amenable to a traditional statistical description in terms of probability distributions, mean values, variances, etc. To give a theoretical example of such a situation, suppose that we wanted to sample a real random variable x with probability distribution $P(x) = (1 + x^2)^{-1/2}$. This distribution is not admissible in elementary probability theory because it is not normalizable: the integral $\int_{-\infty}^{+\infty} P(x)\,dx$, which measures the probability of drawing some value, and should therefore be equal to one, is in fact logarithmically divergent for large $|x|$. This means that a histogram of the values of x taken from a finite sampling of size $N(x_1, x_2, ..., x_N)$ will be very spread out and, once normalized, will tend to zero for any value of x. In contrast, the histogram of a normalizable probability distribution $P_0(x)$ such that $\int_{-\infty}^{+\infty} P_0(x)\,dx$ is a convergent integral, tends to $P_0(x)$ for each value of x when the number N of samples tends to infinity. For a normalizable distribution, the non-normalized histogram increases like N, on the average, for N large (there are approximately $P_0(x) \cdot \Delta x \cdot N$ values of x drawn in the interval $(x - (\Delta x/2), x + (\Delta x/2))$. For the "distribution" $(1 + x^2)^{-1/2}$, the non-normalized histogram grows only like $N/\ln N$. The reason for this is that, by increasing N, we sample larger and larger values of x, with sufficient weight to continually decrease the relative weight of any finite value of x. From this comes one possible definition of intermittency: the variable x is intermittent if the relative weight of the large fluctuations grows continually for increasingly long statistical samples. However we will not use this formal mathematical definition in practice.

Nature and physics offer numerous examples of intermittent processes. Particularly in hydrodynamics, intermittency appears in several classes of experimental phenomena. If we perturb a rapid (high Reynolds number) flow by inserting a plate parallel to the flow's average direction, a boundary layer forms near the plate. Averaged over time, this turbulent boundary layer has a well-defined spatial frontier. But, for a reason which is not yet well understood, this frontier undergoes very large instantaneous displacements at infrequent time intervals. The corresponding fluctuations in the velocity field become more and more intermittent as we take measurements at increasing distances from the boundary layer. We can also cite the small-scale intermittency in fully developed turbulence, a phenomenon whose explanation remains controversial. Transition flows in pipes — i.e. flows whose Reynolds numbers are just at the experimental transition point between the laminar and turbulent regimes — also have an intermittent structure: the turbulence tends to be concentrated in spatially well-defined zones, as Reynolds showed a little more than a century ago.

It is probable — but not certain — that all of the intermittency phenomena we have mentioned in hydrodynamics originate in the spatial structure of the phenomena: the "large fluctuations" in question are localized both in space and in time. However the purpose of this book is primarily the description of dynamical phenomena in which spatial structure does not play an important role: we have confined the description of spatial structure to the amplitudes of a few well-defined modes. It has been shown — theoretically as well as experimentally — that *temporal intermittency* can exist in dynamical systems with a small number of modes during transition between periodic[92] and chaotic (or quasiperiodic) regimes. Even in this limited framework there exist several types — three, in fact — of transition through intermittency, each with its own characteristics.

Let us briefly describe the phenomenology shared by all three types. For a value r of a control parameter less than some critical value r_i, the dynamical system in question (which can be experimental like the R.B. or B.Z. systems, or theoretical like the Lorenz model, or a mapping on the circle, or on a higher dimensional space) has a limit cycle (or a stable discrete cycle for mappings). The system oscillates in a regular fashion and is stable against small perturbations. When r slightly exceeds r_i (the intermittency threshold), we have an *intermittent* dynamical regime. The time signal consists of oscillations which appear regular and which resemble the stable oscillatory behavior for $r \lesssim r_i$. But now, the oscillations are interrupted from time to time by "abnormal" fluctuations, whose amplitude and direction are approximately the same from one fluctuation to another, and which depend little on r. We call this transition *intermittent*

---

92. We can also imagine a transition via intermittency directly from a stationary state to a turbulent regime. This kind of transition can occur if, for example, the stable and unstable manifolds of the fixed point of a flow are both two-dimensional. For the Lorenz model, this cannot happen since the dimension of the phase space is only three. This type of intermittency, which has not been extensively studied either experimentally or theoretically, will not be considered here.

because, as r approaches r_i in the intermittent region ($r \gtrsim r_i$), the fluctuations become increasingly rare, disappearing altogether for $r \lesssim r_i$. We emphasize that at the transition it is *neither the maplitude, nor the duration* of the exceptional fluctuations which tend to zero, but only their *average frequency*.

The theory of intermittent transitions has two parts. The first, Floquet theory, deals with the linear instability of a limit cycle, explaining the "spontaneous" growth of fluctuations starting from a regime close to the periodic regime. The second is the process of "reinjection" or "*relaminarization*" via which the intermittent fluctuation ceases, to be replaced by another phase of regular oscillations. Classification of intermittency into types I, II, and III is based on the three types of linear instabilities of periodic trajectories: crossing of the unit circle by the Floquet multiplier at $+1$ (type I), -1 (type III), or at two complex conjugate eigenvalues (type II). In the three sections which follow we will give a more complete theoretical description of each of the three types, complemented by experimental illustrations of types I and III.

The paths of the Floquet multipliers in the complex plane depend only on the linearization of the dynamical equations about the limit cycle. This approximation which is linear in the amplitude of the fluctuations, would lead to unchecked growth of the fluctuations. Yet, in our qualitative outline of the phenomenon of intermittency, we have described the fluctuations as attaining a finite level as time passes. The theory therefore requires an analysis of effects which are at least weakly nonlinear in the amplitude of the fluctuations. For types II and III, intermittency is possible only if the bifurcation is *subcritical*, i.e. if nonlinear effects tend to *augment* the instability. But for type I intermittency, we cannot ask that the bifurcation be supercritical or subcritical, for there exists only one type of bifurcation which is always — in a certain sense — subcritical (see Appendix A).

IX.2 Type I intermittency

IX.2.1 GENERAL THEORETICAL CONSIDERATIONS

In this case the periodic trajectory which is stable for $r < r_i$ is destabilized (and in fact disappears) at $r = r_i$ because an eigenvalue of the Floquet matrix exits from the complex unit cercle at $+1$. Our analysis will follow the two steps described above. We will first consider phenomena in the neighborhood of the destabilized periodic trajectory. We will accomplish this by a local Landau-type analysis, including the lowest order nonlinear effects in a Taylor series in the fluctuation amplitude. Despite its relatively rudimentary nature, this approach leads to a certain number of quantitative predictions which will be later compared to a numerical simulation of the Lorenz model, and to data from a B.Z. experiment.

In the second stage of our analysis, we will discuss the process of reinjection, or relaminarization, defined above. This process depends on the global structure of the flow in the phase space, in the sense that once it has left the neighborhood of the destabilized limit cycle, the trajectory must explore parts of phase space at a finite distance from the limit cycle, and thus becomes sensitive to the global structure of the flow. In particular we will see that relaminarization can be explained very simply for flows on the torus T^2. We know that the only regimes possible on T^2 are periodic or quasiperiodic. Therefore this implies that type I intermittency (but not the other types) can occur for some transitions between two regular (non-chaotic) regimes, such as periodic to quasiperiodic. Our discussion of the Lorenz model will show that a type I transition can also take place between a periodic and a chaotic regime. All of this will be discussed in more detail in the section devoted to relaminarization.

IX.2.2 LOCAL ANALYSIS

Generically the eigenvalue of the Floquet matrix that crosses the unit circle has a unique eigenvector. This is the case for the Lorenz model, for example, as we will see in Section IX.2.5. Let us then define a coordinate u in the plane of the Poincaré section which points in the direction of the eigenvector. Multiplication by the Floquet matrix is then reduced to simple multiplication of u by the eigenvalue $\lambda(r)$:

$$u' = \lambda(r)u \tag{IX.1}$$

Crossing at $+1$ for $r = r_i$ means that $\lambda(r_i) = 1$ and $(d\lambda/dr)|_{r_i} \neq 0$. To have an idea of what happens to the Poincaré mapping when r varies near r_i, we will make the approximation (to be justified further on) that Equation (IX.1) for $r = r_i$ is the beginning of the Taylor series of a function $u'(u, r)$ in the neighborhood of $u = 0$ and $r = r_i$. Taking higher order terms into account will suffice to explain type I intermittency.

A limit cycle for the dynamical system shows up as a fixed point of the Poincaré transformation. Near the fixed points for $r < r_i$, the graph of $u'(u)$ is almost tangent to the identity map, since $(du'/du)|_{r_i} = \lambda(r_i) = 1$. We therefore find ourselves in the situation represented schematically in Figure IX.1 a. At the bifurcation point $(r = r_i)$, the curve $u'(u)$ is exactly tangent to the identity map, since $(du'/du) = 1$ at the fixed point (fig. IX.1 b)[93]. Passage from Figure IX.1 a to IX.1 b is accomplished by translating the curve $u'(u)$ upwards. If r continues to increase, the curve continues to move up, forming a small channel between itself and the identity map (fig. IX.1 c).

The one-parameter family of curves near the point of contact is represented by the following generic form for u':

$$u' = u + \varepsilon + u^2 \tag{IX.2}$$

---

93. This bifurcation by coalescence of two fixed points, one stable, the other unstable is called a saddle-node bifurcation. Whether it is supercritical or subcritical has no meaning in that it is accompanied by the disappearance of all solutions, be they stable or unstable (cf. Appendix A).

IX.2 TYPE I INTERMITTENCY

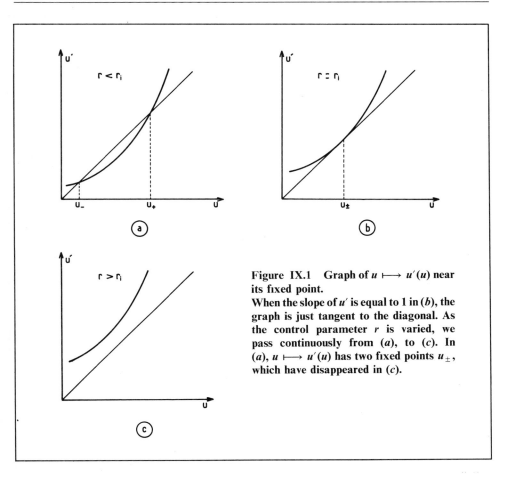

Figure IX.1 Graph of $u \longmapsto u'(u)$ near its fixed point.
When the slope of u' is equal to 1 in (b), the graph is just tangent to the diagonal. As the control parameter r is varied, we pass continuously from (a), to (c). In (a), $u \longmapsto u'(u)$ has two fixed points $u_\pm$, which have disappeared in (c).

The coefficient of u^2 (the second term of the Taylor series of u' about $u = 0$ and $r = r_i$) has been taken to be 1 by an appropriate choice of scale for u. The parameter ε, proportional to $r - r_i$, is the control parameter which is varied to produce the transition from Figure IX.1 a to Figure IX.1 c.

For $\varepsilon = 0$, we have the situation of Figure IX.1 b: the equation $u'(u) = u$ has a double root at $u = 0$, with slope $(du'/du)|_{u=0} = +1$. For $\varepsilon \neq 0$, fixed points of (IX.2) are solutions of the equation $u'(u) = u$, i.e. $u_\pm = \pm(-\varepsilon)^{1/2}$, and exist only if $\varepsilon < 0$. Their Floquet eigenvalues are $(du'/du)|_{u_\pm} = 1 \pm 2(-\varepsilon)^{1/2}$, so that u_- is stable, and u_+ is unstable. This is easily illustrated by using the graphical construction for iterating the transformation $u \longrightarrow u'(u)$ (fig. IX.2).

The stable fixed point u_- is an attractor for initial conditions $u < u_+$, and the fixed point u_+ is unstable. Iterations beginning from $u > u_+$ diverge rapidly like $(u^{(0)})^{2^n}$, for n large, with $u^{(0)}$ close to the initial condition. The graphical representation of the transition from the $\varepsilon < 0$ regime to the $\varepsilon > 0$ regime also shows that truncating the

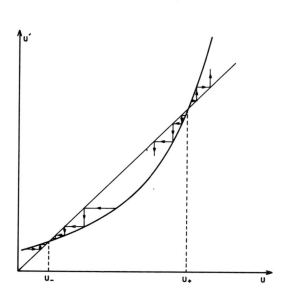

Figure IX.2 Iteration of $u \longmapsto u'(u)$ when two fixed points $u_\pm$ exist. The point u_- is locally stable: iterations starting near it converge towards it. However u_+ is unstable: iterations diverge from it.

Taylor series of $u'(u, r)$ at the first order in $(r - r_i)$ and at the second order in u gives a correct description of the transition neighborhood: the curvature of $u'(u)$, the slope close to $+1$ in the region of contact, and the rise of the curve are all well represented.

We have seen that when ε is small and positive, a narrow channel is created between the graphs of the identity map and of $u'(u)$. This has two important consequences:

i) First, for $\varepsilon > 0$ ($r \geqslant r_i$) there no longer exists a fixed point of the first return map in the region considered. This is clear since the fixed points are $u_\pm = \pm(-\varepsilon)^{1/2}$.

ii) An iteration that starts from a negative value of u will systematically drift towards the $u > 0$ region (see fig. IX.3): because ε and u^2 are positive, we have $u' = u + \varepsilon + u^2 > u$.

But if ε is very small, the successive iterates accumulate in the narrowest part of the channel. This brings us back to the question of intermittency. The region of the channel — more specifically, the u values that we will actually observe — is very close to the fixed points that existed for $\varepsilon < 0$. Therefore if we go back to the continuous time signal

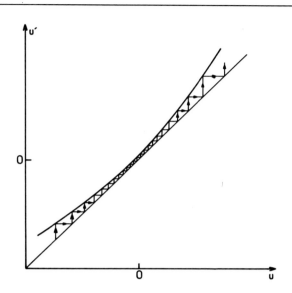

Figure IX.3 Iteration of $u \longrightarrow u' = u + \varepsilon + u^2$ for $\varepsilon > 0$.
Unlike that of Figure IX.2 ($\varepsilon < 0$), this mapping has no fixed point. For ε small and positive, iterations beginning at negative values of u spend a long time in the narrow channel separating the graphs of u' and the identity map.

of the flow that corresponds to passage through the channel for the Poincaré mapping, we find a time dependence very similar to that of the stable oscillations taking place for $\varepsilon < 0$. But the oscillations are no longer stable since the first return map $u \longmapsto u'(u, \varepsilon)$ no longer has a fixed point near 0, and so there is no periodic solution, stable or unstable, to the equations of motion near the limit cycle.

This first theoretical description of the intermittent transition leaves several open questions.

i) Does there exist a simple dynamical model exhibiting an intermittent transition of the kind we have just described? We will see in Section IX.2.5 that the Lorenz model exhibits just such a transition.

ii) Does the theory give quantitative predictions that could be compared with experiment?

iii) Once the iteration has passed through the channel and the trajectory has left the almost regular oscillatory regime, how can the channel be re-entered from the left, to start another phase of oscillation?

We now examine the last two points.

IX.2.3 QUANTITATIVE PREDICTIONS OF THE MODEL

In our theoretical model, the physical parameters have been eliminated, leading to the universal form of the transformation $u \longrightarrow u'(u)$:

$$u' = u + \varepsilon + u^2$$

a form which is valid in a neighborhood of $u = 0$. The quantitative predictions deduced from this description will therefore be of the form of *scaling laws* governing, for example, the average duration of laminar phases near $\varepsilon = 0$. To find these laws, we notice that, if u' is close to u, we can replace $u' - u$ by du/dk. That is, the index k of iteration necessary to go through the channel is of order $\varepsilon^{-1/2}$, as is the interval of k is small. We then replace the *difference* equation by the *differential* equation:

$$\frac{du}{dk} = \varepsilon + u^2 \qquad (IX.3)$$

This equation has as its general solution:

$$u(k) = \varepsilon^{1/2} \, \text{tg} \, (\varepsilon^{1/2}(k - k_0)) \qquad (IX.4)$$

k_0 being approximately the step at which the iteration goes through the narrowest part of the channel. For simplicity we take $k_0 = 0$.

The function $u(k)$ diverges for $k = \pm (\pi/2)\varepsilon^{-1/2}$ (in fact for $k = (n\pi/2)\varepsilon^{-1/2}$, for any odd integer n). The meaning of this divergence is that when k has attained values of the order of $\varepsilon^{-1/2}$, $u' - u$ is no longer small, so that the difference Equation (IX.2) can no longer be replaced by the differential Equation (IX.3). This tells us that the number of iteration necessary to go through the channel is of order $\varepsilon^{-1/2}$, as does the interval of k values between two consecutive divergence of Equation (IX.4). We have found a scaling law: the average duration of the laminar phases diverges like $\varepsilon^{-1/2} \sim |r - r_i|^{-1/2}$ as the threshold is approached.

When an intermittent transition occurs between periodic and turbulent regimes, the Lyapunov number $\gamma(\varepsilon)$ characterizing the level of instability of the turbulent trajectories is of the order of the inverse of the correlation time of the signal with itself. If we assume that the correlation disappears almost completely during the turbulent bursts, we deduce that the correlation time is of order $\varepsilon^{-1/2}$, like the mean time for channel traversal, and that its inverse, the Lyapunov number, obeys the law $\gamma(\varepsilon)_{\varepsilon \to 0+} \sim \varepsilon^{1/2}$. This law is indeed verified by the Lorenz model (see Section IX.2.5).

Another interesting quantity is the statistical distribution of the duration τ of the laminar phases, that is $P(\tau, \varepsilon)$. In a sample of N laminar phases (N integer), there are $N \int_0^\tau d\tau' \, P(\tau', \varepsilon)$ phases which last less than τ. We already know that the average duration of a laminar phase is $\int_0^\infty d\tau \, \tau P(\tau, \varepsilon) \sim \varepsilon^{-1/2}$. For ε fixed, the duration of laminar phases is bounded from above by a quantity of order $\varepsilon^{-1/2}$, and the closer we

IX.2 TYPE I INTERMITTENCY

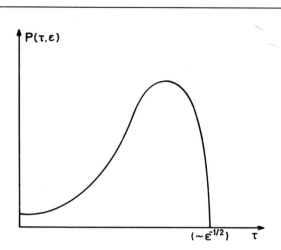

Figure IX.4 Distribution of the lengths of laminar phases in type I intermittency. The figure has only qualitative significance, for the distribution depends on the details of the relaminarization process. The upper bound of order $\varepsilon^{-1/2}$ close to $\varepsilon = 0_+$ is specific to type I intermittency, and corresponds to the maximum time for traversal of the channel of Figure IX.3.

get to the threshold, the less the duration fluctuates. These considerations allow us to draw a qualitative shape for $P(\tau, \varepsilon)$ in an intermittent regime (fig. IX.4).

The exact form of $P(\tau, \varepsilon)$ depends upon the details of the problem, as the fluctuations of τ reflect the fluctuations of the process of reinjection into the channel. The distribution $P(\tau, \varepsilon)$ is more easily measured than the preceding scaling laws, which all require pinpointing the threshold $\varepsilon = 0$ and thus a series of experiments at very closely spaced values of the control parameter. We note also that, for the other intermittency types (types II and III), the distribution of laminar phase duration follows a law very different from that of type I. This other distribution law has a maximum for short times and decreases algebraically at large times.

IX.2.4 RELAMINARIZATION

As yet we have not considered the behavior of the intermittent solutions outside of the channel region. This question is related to relaminarization, the process allowing reentry into the channel. We will restrict ourselves here to qualitative considerations, since the reinjection process leads to exploration of regions of phase space which are of finite size. Thus only a topological approach allows us to reach conclusions of any generality.

CHAPTER IX INTERMITTENCY

In what follows we describe three possible means of reinjection for: i) a flow on a torus, ii) the baker's transformation (which, with a few modifications, applies to the Lorenz model), and iii) the Smale attractor.

Relaminarization on the torus

Recall that when we consider a flow on the torus T^2, the Poincaré transformation is simply an invertible mapping f of the circle into itself. When represented as a graph in rectangular coordinates (fig. IX.5), fictitious discontinuities are created, which disappear when 0 and 1 are identified. The function $\theta \mapsto f(\theta)$ is strictly monotonic (e.g. increasing). Figure IX.6 shows how an intermittent transition of type I takes place for a diffeomorphism of the circle. The graph of f becomes tangent to and then leaves the identity map. In this case, the reinjection process is very simple: the iterates, once out of the channel go around the circle to enter the channel from the other side. This is not a transition to turbulence. Indeed, we already know that a flow on T^2 is either periodic or quasiperiodic, and never chaotic or turbulent. The arrival of "turbulent" bursts is therefore approximately periodic in this kind of intermittency, and its periodicity is either commensurate or incommensurate with the periodicity of the rapid motion on the torus, according to whether the intermittent regime is periodic or quasiperiodic.

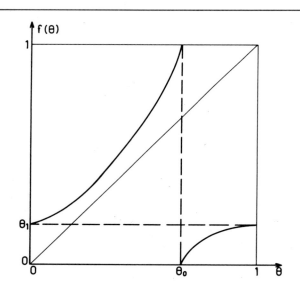

Figure IX.5 Qualitative form of the first return map of a flow on T^2.
The Poincaré section is a circle, parametrized here by [0, 1] with the endpoints identified. The discontinuities of the graph at 0, θ_0, and 1 are only apparent, resulting from the mode of representation.

IX.2 TYPE I INTERMITTENCY

When the channel closes off, as in Figure IX.6 a, a pair of fixed points of f appears: there is frequency locking (see Chapter VII and Appendix C). In this case, after having gone around the large circle of the torus n times, the trajectory returns to the same point on the small circle. Knowledge of the Poincaré map does not determine the value of n; the winding number is defined only modulo 1. In the intermittent situation (fig. IX.6 c) preceding frequency locking, the time signal shows "phase slippage" between two "quasiregular" periods of oscillation. We can explain this in the following way. While passing through the channel, the rotations along the large and small circles of the torus have a well-defined mutual phase. In the time signal, this gives the appearance of regular oscillation, with only one period. Between two channel crossings there is an extra rotation around the small circle. This means that one of the oscillators

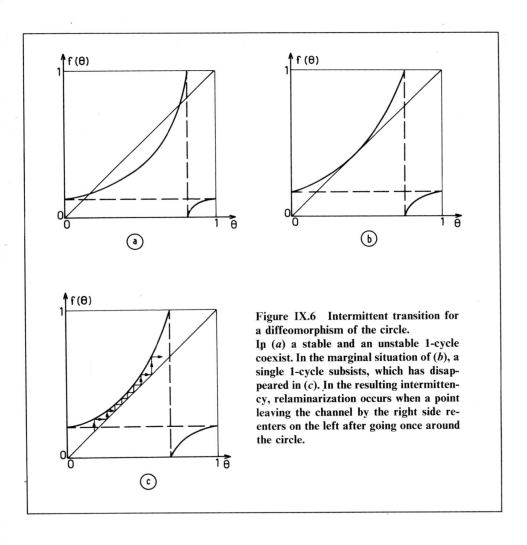

Figure IX.6 Intermittent transition for a diffeomorphism of the circle.
In (a) a stable and an unstable 1-cycle coexist. In the marginal situation of (b), a single 1-cycle subsists, which has disappeared in (c). In the resulting intermittency, relaminarization occurs when a point leaving the channel by the right side reenters on the left after going once around the circle.

has gained a period with respect to the other: hence the appearance of phase slippage in the time signal. The phase slippage is the signature of the abnormal "fluctuation" of the intermittent regime. It can even happen that for some dynamical variable, phase slippage is the only manifestation of the abnormal fluctuation, without there being any noticeable variation in the oscillation amplitude. This is true for example in a relaxation oscillator.

Transition by intermittency and the baker's transformation

In what follows we will explain a relaminarization process which appears when the baker's transformation is involved. With a few modifications, this will apply to the intermittent transition in the Lorenz model as described later. We will first give several results about the shift map and the closely related baker's transformation. We will show that, despite apparent discontinuities, the baker's transformation can be the first return map of a continuous flow. Finally we will explain how an intermittent transition from a periodic to a chaotic regime can occur and, in particular, how relaminarization takes place.

A transformation of the interval $[0, 1[$ into itself which typically exhibits S.I.C. is the shift map $x \longmapsto \{2x\}$, where $\{2x\}$ is defined to be the fractional part of $2x$, for example $\{2 \cdot (3/5)\} = 1/5$. With each iteration of the transformation all distances are multiplied by two: thus even an initially infinitesimal distance grows geometrically as iteration proceeds. The shift map cannot be a first return map for two reasons: it is not uniquely invertible (for example $\{2 \cdot (1/4)\} = \{2 \cdot (3/4)\} = 1/2$) and it is discontinuous ($\{2x\}_{x \to (1/2 + 0)} \longrightarrow 0$ and $\{2x\}_{x \to (1/2 - 0)} \longrightarrow 1$). At first glance the presence of *discontinuity* seems incompatible with the *continuity* of the equations of motion such as the Lorenz equations. In fact this is not so. Without entering into detail we can say that since there always exists an unstable fixed point for the Lorenz equation (for example the conductive state $(X, Y, Z) = (0, 0, 0)$), there also exists a set of initial conditions leading to the unstable fixed point at long times.

This set, called the stable manifold[94] of the fixed point, forms a surface of zero measure (and therefore initial conditions belonging to it are highly improbable, unless chosen deliberately). If the unstable fixed point is not contained in the plane of the Poincaré section, then the intersection, if it exists, of the stable manifold with the Poincaré section contains points which have no image under the first return map. It is along this curve of intersection that the Poincaré mapping can be discontinuous. An example of this is to be found in the Lorenz model for the plane of section $Z = r - 1$.

---

94. This surface is called the *stable* manifold of the fixed point. The *unstable* fixed point also has an *unstable* manifold (consisting of the initial conditions which reach the point at *negative* infinite times). In contrast, a stable fixed point has only stable manifolds.

IX.2 TYPE I INTERMITTENCY

We thenefore wish to construct from the shift map an invertible but discontinuous transformation. In the shift map there are exactly two antecedents for each point, i.e. two numbers x_0 and x_1 such that:

$$0 \leqslant x_0 \leqslant 1/2, \quad x_1 = x_0 + 1/2$$

and: $\{2x_0\} = \{2x_1\}$.

To make the transformation invertible, we must introduce another coordinate y which will distinguish between the two antecedents. We define a transformation of the square $[0, 1] \times [0, 1]$ into itself, called the baker's transformation, by:

$$(x, y) \longmapsto (\{2x\}, \; y'(y, x))$$

where: $y'(y, x) = y/2 \quad$ if $\; 0 \leqslant x \leqslant 1/2$
$y'(y, x) = (y + 1)/2 \quad$ if $\; 1/2 < x \leqslant 1$.

This invertible transformation conserves areas since the Jacobian:

$$\begin{vmatrix} \dfrac{\partial x'}{\partial x} & \dfrac{\partial y'}{\partial x} \\ \dfrac{\partial x'}{\partial y} & \dfrac{\partial y'}{\partial y} \end{vmatrix} = \begin{vmatrix} 2 & 0 \\ 0 & \dfrac{1}{2} \end{vmatrix} = 1$$

We observe that we have preserved the shift map of the x coordinate. To illustrate the transformation we draw the head of a cat ("Arnold's cat") in the unit square of the (x, y) plane, and its image, after is has been cut in two and stretched by the baker's transformation (fig. IX.7). The stretching along the abscissa reflects the S.I.C. of the shift map.

A simple generalization of the preceding construction allows us to construct an invertible and discontinuous mapping (in two dimensions) from any discontinuous mapping $x \longrightarrow f(x)$ which maps the interval $[0, 1]$ into itself twice (i.e. every point has two antecedents) and which is everywhere increasing. Another generalization gives a mapping which contracts areas, reflecting the dissipative nature of the underlying dynamical system. When iterated, the mapping has an attractor with a layered structure, whereas the baker's transformation of Figure IX.7 does not, since it conserves areas. For example, instead of mapping y according to $y \longmapsto y/2$ and $y \longmapsto (y + 1)/2$, we could use $y \longmapsto \beta(y/2)$ and $y \longmapsto \beta(y + 1)/2$, where $0 < \beta < 1$. The area is contracted by a factor of $\beta (< 1)$ at each iteration. Any mapping f satisfying the conditions set forth above (f is monotonically increasing and each point has two antecedents) can exhibit an intermittent transition analogous to that presented for a circle, and so this is true for the mapping $(x, y) \longmapsto (f(x), y'(y, x))$ as well. Figure IX.8 shows the way in which f changes as the intermittency threshold is crossed.

By iterating the mapping of Figure IX.8c we see that the reinjection (or relaminarization) process occurs by crossing the discontinuity of f. Once out of the channel, the iterates increase until they go beyond the discontinuity (denoted by x_0 on

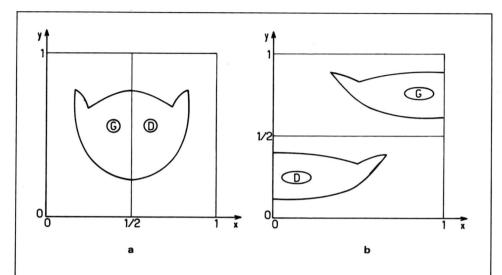

Figure IX.7 The baker's transformation.
The effect of the transformation is shown by considering a pattern drawn on the unit square, here a cat's head (this example is due to Arnol'd). The transformation takes its name from the following description:
A square of dough (*a*) is flattened along *y* and lengthened along *x*. Cut vertically in two, the halves are placed one on top of the other to form (*b*). The volume of the dough is not changed by the operation. This discontinuous transformation conserves areas, contraction by a factor of 1/2 along *y* being exactly compensated for by dilation by a factor of 2 along *x*.

Figure IX.8 c). Ordinarily no iterate is located exactly at x_0, so that the next iteration places the point at a small positive value of x. Since the origin is an unstable fixed point, the iterates start to grow again, eventually re-entering the channel, and so on. Note that, unlike in relaminarization on T^2 considered earlier, here there always remains an (unstable) fixed point of the mapping, which was taken to be the origin in Figure IX.8. Finally, we note that between two "laminar phases", the trajectory explores unstable regions of phase space. In such regions, for example for $x_0 < x < 1$ on Figure IX.8, the derivative of f is greater than one. The fluctuation of the corresponding time signal will therefore be essentially unpredictable, although its average characteristics will be about the same from one turbulent burst to another, since it explores the same part of phase space. We can then speak of a true "turbulent burst". This is manifested in, for example, the Lorenz model to be discussed later.

IX.2 TYPE I INTERMITTENCY

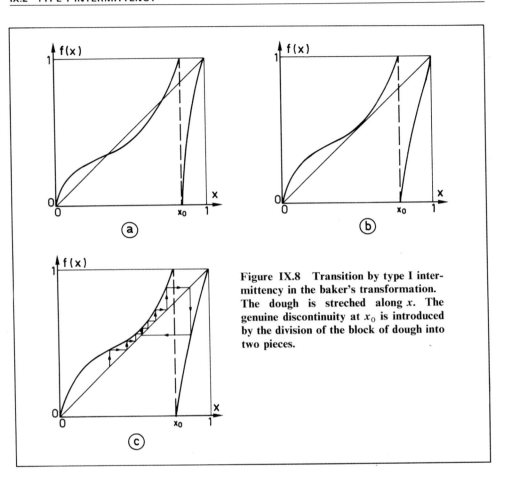

Figure IX.8 Transition by type I intermittency in the baker's transformation. The dough is streched along x. The genuine discontinuity at x_0 is introduced by the division of the block of dough into two pieces.

Relaminarization for the Smale attractor

In what follows, we will explain how to make the shift map, and analogous mappings invertible, by adding a mapping on two other real coordinates which contracts areas. This mapping (already known to topologists in the 1920's and 30's) allowed Smale to show that a continuous dynamical system could exhibit S.I.C. in a robust fashion (that is, for the system under consideration and all of its close neighbors). In this model, the first return map is the same for one of the coordinates, as that derived previously from the dyadic mapping. Since it is after all the shift map which causes the intermittency, there is nothing fundamentally novel with respect to the model based on the baker's transformation. However, the interpretation of the relaminarization process differs from the previous case, because there is no longer any discontinuity in the first return map. We will briefly discuss this last point after having examined the Smale attractor.

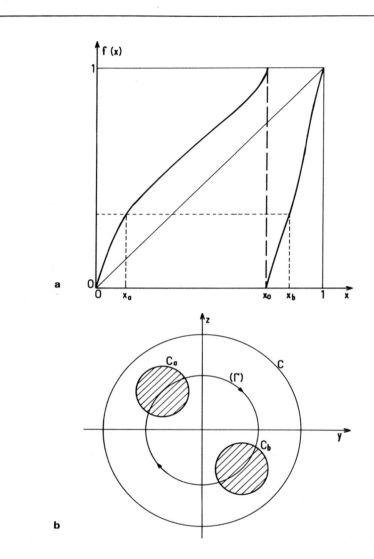

Figure IX.9 Effect of Smale's transformation on the coordinates.
a) **First return map on the angular variable x.**
The mapping is not invertible: when x varies once over the circle (represented here by $[0, 1]$) its image $f(x)$ goes twice around the circle.
b) **Transformation of the y and z coordinates.**
The disk C is mapped onto either C_a or C_b, according to whether x is x_a or x_b (see (a)). The global transformation is thus invertible. As x varies over the circle, the disks C_a and C_b make a complete circuit of Γ, remaining always diametrically opposed.

IX.2 TYPE I INTERMITTENCY

We again consider a mapping $x \longrightarrow f(x)$, mapping the interval $[0, 1]$ twice onto itself and exhibiting an intermittent transition via disappearance of a pair of fixed points, shown schematically in Figure IX.9. But contrary to the preceding case, we eliminate the discontinuity by letting x be an angular variable so that 0 and 1 are identified in Figure IX.9 a. The discontinuity disappears since then:

$$f(x_0 - 0) = +1 \quad \text{and} \quad f(x_0 + 0) = 0.$$

To make the mapping invertible and continuous we add two more variables y and z to the phase space, instead of the one variable added previously. For the coordinates y and z, the domain of the mapping is the interior of a circle C. Consider the image of the circle for two values of x, x_a and x_b, such that $f(x_a) = f(x_b)$. The image is constructed to be two small non intersecting circles, diametrically opposed and located inside of C (see fig. IX.9 b). One of the small circles, C_a, is the image of C with $x - x_a$, and the other,

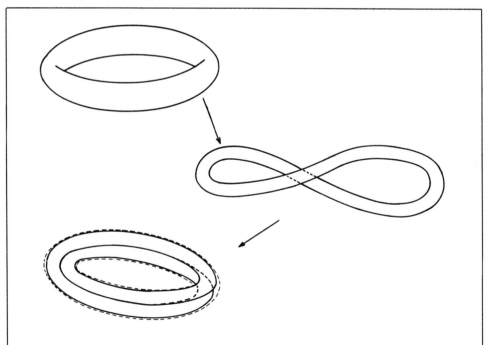

Figure IX.10 Diagram of the mapping of a "tire" into itself by Smale's transformation.
The tire is stretched along its large circle and folded into a figure eight. The two loops of the eight are folded over one another and reinserted into the initial tire. The reinsertion is made possible by the fact that the section and volume of the tire have been diminished by dissipation.

C_b, is the image of C with $x = x_b$. As x_a varies from 0 to x_0 and x_b from x_0 to 1, C_a and C_b each rotate by an angle π along the curve Γ shown in Figure IX.9 b. Therefore, as x goes from 0 to 1, completing an entire rotation on its domain [0, 1], the small circles will also have completed one rotation, insuring the continuity of the mapping for y and z.

We can also decompose this mapping by starting with a solid torus. The solid torus is the phase space: it is generated by the disk inside the circle C of the (y, z) plane when it is rotated about the circle on which x varies. We first decrease the volume contained in the torus while doubling its circumference. Having formed a figure eight and then having folded the two loops of the figure eight onto one another, we reinsert it into the initial solid torus (see fig. IX.10). This construction illustrates a possible Poincaré mapping for a *four-dimensional* flow whose Poincaré section gives *three-dimensional* objects. It also shows that this mapping of the solid torus into itself can define a first return map, since it results from a sequence of continuous deformations. Indeed, the question of knowing if a given mapping of a more or less high dimensional phase space into itself is a possible first return map can become highly nontrivial. Let us say as well that, undoubtedly because of its complexity, such a mapping has never been observed in a physical system.

The qualitative description of the intermittent transition in phase space and of the relaminarization process is, for the present continuous case, different from the one we

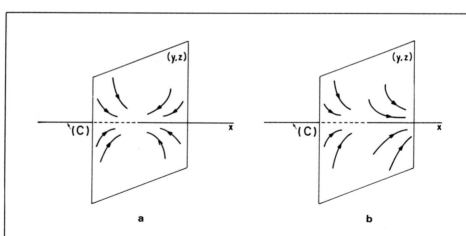

Figure IX.11 Neighborhood of the fixed point of Smale's transformation. The iterates are strongly attracted towards the x axis, along the y and z directions, while drifting slowly along the x axis itself. After (*a*) and before (*b*) the fixed point disappears.

IX.2 TYPE I INTERMITTENCY 241

have outlined for the (discontinuous) baker's transformation. We must therefore find a new explanation of the relaminarization process which does not rely on discontinuity. Consider the neighborhood of the fixed point disappearing at the transition to intermittency. This fixed point is weakly stable in the x direction and has a finite stability in the y and z directions (this stability corresponds to the contraction of areas in the (y, z) plane — see fig. IX.9). When the fixed point disappears to form the channel (in the mapping of the x coordinate), the finite attraction in the y and z directions persists. The stable manifold in the x direction now becomes a line (c) which attracts nearby points situated in a neighborhood of it, and along which the first return map is the mapping $x \longrightarrow f(x)$ with the channel (see fig. IX.11). If during its erratic motion (i.e. the turbulent burst), the trajectory visits the neighborhood of the "ghost" of the fixed point, it will first be attracted towards (c). The process of attraction towards (c) takes place at a finite speed, while the motion parallel to (c) is very slow close to the intermittency threshold. Once the point is almost at (c), it begins to feel the (slow), dynamics of the one-dimensional mapping along x and thereby to drift slowly in the channel. Therefore the laminar phases last a long time, much longer than the "turbulent" exploration of phase space between two visits to the "ghost" of the fixed point. The laminar phases will thus comprise the major part of the temporal behavior.

IX.2.5. INTERMITTENCY IN THE LORENZ MODEL

As the Lorenz model was introduced in Chapter VI and is the subject of Appendix D, we will only recall here that it consists of three ordinary differential equations meant to model Rayleigh-Bénard convection. The equations are:

$$\dot{X} = \Pr Y - \Pr X \qquad \text{(IX.5.1)}$$
$$\dot{Y} = -XZ + rX - Y \qquad \text{(IX.5.2)}$$
$$\dot{Z} = XY - bZ \qquad \text{(IX.5.3)}$$

With the usual parameter values $\Pr = 10$, $b = 8/3$, the behavior of the model has been extensively studied as a function of r, and also sometimes of initial conditions. These studies have revealed results of great diversity. In particular two scenarios of transition towards turbulence have been observed: transition via a subharmonic cascade (Chapter VIII) and transition via intermittency. In this section we will outline how this transition occurs concretely for the Lorenz model.

One of the possible types of dynamical behavior the Lorenz model can have is a limit cycle. For $r = r_0 = 166$, initial conditions taken at random lead, after transients have disappeared, to perfectly regular oscillation as can be seen from the graphs of $X(t)$, $Y(t)$, $Z(t)$, or any function of these quantities (fig. IX.12).

For values of r slightly greater than r_0, we observe stable oscillations like those of Figure IX.12 over the whole interval $[r_0, r_i]$ where $r_i = 166.07$. This "robustness" of the limit cycle regime expresses the profound property of *structural stability* of stable oscillation. However, when r becomes slightly greater than r_i, typical solutions to the

Figure IX.12 Regular oscillations of the Lorenz model for $r = 166$.

From Y. Pomeau and P. Manneville

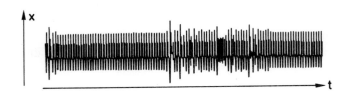

Figure IX.13 Intermittent regime of the Lorenz model for $r = 166.1$. Between turbulent bursts the oscillations are nearly the same as those of the stable limit cycle of Figure IX.12.

From Y. Pomeau and P. Manneville.

Lorenz model become "turbulent" but in a very characteristic way. For $r = 166.1$ (recall that $r_i = 166.07$) the time signal $X(t)$ (fig. IX.13) shows oscillations that are almost as regular as those of Figure IX.12, but that are interrupted from time to time by bursts of turbulence. These erratic fluctuations have a well-defined average duration and terminate by what we have called relaminarization, i.e. a return to the laminar regime. Although this description certainly evokes transition via intermittency, we cannot immediately conclude that it is type I intermittency: type II or type III intermittency might also be responsible for the phenomenon we have described.

Type II intermittency is ruled out for the Lorenz model (with positive parameter values), as it would imply *local* dilation of volumes in phase space whereas the Lorenz system contracts areas *everywhere*[95]. We can also exclude type III intermittency by examination of Figure IX.13, although somewhat less firmly. In this recording we do

---

95. It is for the same reason that the Lorenz model cannot have an invariant torus T^2: if it did, the volume contained inside the torus would be globally conserved by the dynamics. This is excluded by the exponential contraction of all volumes.

not observe the growth of a subharmonic oscillation during the laminar phase, a characteristic feature of type III intermittency (see Section IX.3). The proof that type I intermittency is at work here relies mainly on the numerical construction of a first return map showing the "channel opening" phenomenon typical of type I. Figure IX.14 allows us to conclude with certainty that it is type I intermittency that we observe in the Lorenz model close to $r \sim r_i$. Since a numerical model all quantities are accessible, other theoretical predictions can be tested. For example, Figure IX.15 shows that the Lyapunov number $\gamma(\varepsilon)$ increases like $\varepsilon^{1/2}$ near the intermittency threshold. This also confirms that the turbulent bursts are chaotic, in the sense that they destroy the temporal correlation of the signal with itself. A transition does indeed take place from a periodic regime to a chaotic regime with a positive Lyapunov number. Let us mention finally that the relaminarization process is analogous to that of the baker's transformation. The first return map has discontinuities resulting from the intersection of the stable manifold of the fixed point (0, 0, 0) with the plane of the Poincaré section. These are consequences of the complexity of the (stange) attractor of the trajectories in the intermittent regime.

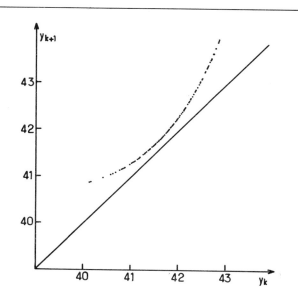

Figure IX.14 First return map in the intermittent regime ($r = 166.2$). This demonstrates both the presence of the channel and the complexity of the mapping, which contains several discontinuities.

From Y. Pomeau and P. Manneville.

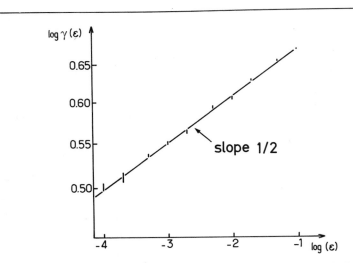

Figure IX.15 Variation of the Lyapunov number in the vicinity of the intermittency threshold.
The growth observed is, in accordance with theory, proportional to the square root of ε, the distance from threshold, as we can verify via the logarithmic scale used for the abscissa.

From Y. Pomeau and P. Manneville.

IX.2.6 INTERMITTENCY IN THE B.Z. REACTION

Introduced in Chapter V, the B.Z. chemical reaction displays a rich variety of spatio-temporal behavior. We will now focus our attention on a transition via type I intermittency which has been observed experimentally in an open reactor of homogeneous phase. The control parameter is the mean residence time T_r of the reactants. The physico-chemical conditions are stabilized as precisely as possible to the following values:

    Reaction volume: 28 ml
    Temperature   : 39.6°C

Concentration of the reactants in moles/liter:

    $NaBrO_3$ 1.8 10^{-3}
    $CH_2(COOH)_2$ 5.6 10^{-3}
    $Ce_2(SO_4)_3$ 5.8 10^{-4}
    H_2SO_4 1.5

The concentration of Ce^{4+} ions, determined by measurement of the optical density of the reacting medium at 340 nm, is the variable used to follow the progress of the reaction.

For a mean residence time T_r of 100 minutes, the reaction oscillates with perfect regularity (fig. IX.16 a) with a period on the order of a minute. When T_r becomes less than a critical value $T_{r,i} \sim 76$ minutes, the oscillations subsist most of the time, but are interrupted from time to time by a large fluctuation (fig. IX.16 b). And when T_r goes further below $T_{r,i}$, the time between two large fluctuations decreases (fig. IX.16 c).

Naturally, this brings to mind transition via intermittency. This hypothesis is confirmed by the existence of a channel in the first return map. The first return map is graphed experimentally (fig. IX.17) by plotting the amplitude of one oscillation as a function of that of the previous one. We mention that this transition does seem to lead to a chaotic regime, notably because of the spread in the probability distribution of intervals between large fluctuations (fig. IX.18). Note that this distribution has the shape predicted by theory (cf. fig. IX.4).

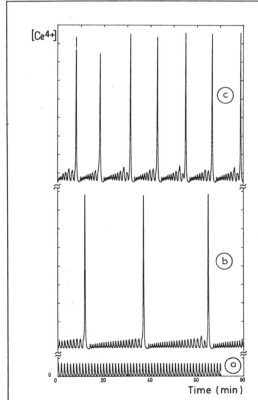

Figure IX.16 Oscillations in the concentration of Ce^{4+} ions in the vicinity of the intermittency threshold in the B.Z. reaction.
a) Below threshold, the reaction oscillates regularly; the regime is a limit cycle (mean residence time of 100 min).
b) Just above threshold (residence time 76 min), the amplitude of the small oscillations grows slowly with time until the sudden appearance of a large amplitude peak.
c) For a considerably smaller residence time (35 min), the small oscillations are fewer and less regular, while the large amplitude peaks arrive more frequently.
From Y. Pomeau, J. C. Roux, A. Rossi, S. Bachelart, and C. Vidal.

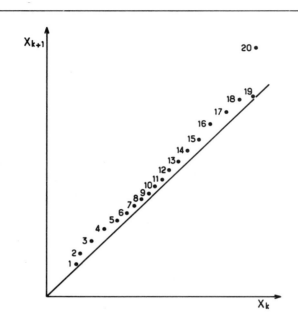

Figure IX.17 First return map $x_{k+1} = f(x_k)$.
Here the coordinate x used for this graph is the amplitude of the small oscillations recorded in the signal of Figure IX.16 b. The experimental points have been numbered in chronological order to demonstrate the slow passage along the diagonal while the channel is traversed.

From Y. Pomeau, J. C. Roux, A. Rossi, S. Bachelart and C. Vidal.

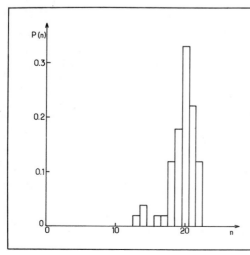

Figure IX.18 Histogram of the length of regular oscillation phases near the intermittency threshold.
Plotted is the probability $P(n)$ of observing n regular oscillations between two large amplitude peaks. The data is taken from an experimental signal, part of which is shown in Figure IX.16 b. The result definitely corroborates the theoretical prediction illustrated by Figure IX.4.
From Y. Pomeau, J. C. Roux, A. Rossi, S. Bachelart and C. Vidal.

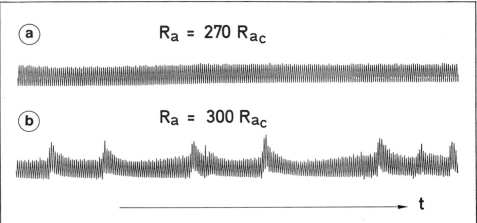

Figure IX.19 Behavior of the velocity near the threshold of type I intermittency in R.B. convection.
The convective structure in this experiment is composed of two rolls along each of the two horizontal directions of the cell.
The only thermoconvective oscillator is of the "warm droplet" type.
a) Periodic regime for $Ra < Ra_i$, where $Ra_i = 295 Ra_c$ is the threshold for onset of intermittency.
b) Intermittent regime for $Ra > Ra_i$. Notice the long sequences (of unequal length) of barely perturbed oscillations (passage through the channel) interrupted by sudden brief events. The resemblance between these experimental signals and the behavior of the Lorenz model shown on Figures IX.12 and IX.13 is striking.
From P. Bergé, M. Dubois, P. Manneville and Y. Pomeau.

Other results, also in accord with theoretical predictions, have been obtained in thermoconvection experiments. Figure IX.19 shows two time signals below and above the intermittency threshold, whose resemblance to numerical simulations of the Lorenz model (fig. IX.12 and fig.IX.13) is compelling.

IX.3 Type III intermittency

In this case an eigenvalue of the Floquet matrix crosses the unit circle of the complex plane at (-1) and the corresponding bifurcation is subcritical, so that the lowest order nonlinear effects do not stabilize the trajectories (for a flow) or the successive iterates (for a mapping) in the neighborhood of the former attractor[96]. We will now develop the theory of type III intermittency and illustrate its occurrence experimentally in the R.B. instability.

---

96. In the case of period doubling (possibly a cascade), the nonlinear effects on the contrary stabilize the linearly unstable subharmonic fluctuations (see Chapter VIII).

IX.3.1 THEORY OF TYPE III INTERMITTENCY

Like that of type I intermittency, the theory of type III intermittency comprises several parts:

1) A description of local phenomena in the weakly unstable region (corresponding to laminar phases in the dynamics).

2) Quantitative consequences for the statistics of the laminar phases close to the threshold.

3) A description of the relaminarization process: how can a laminar phase begin again after a turbulent burst?

Description of local phenomena

As in type I intermittency, from a simple model we can make quantitative predictions about the laminar phases of the dynamics since these phases essentially explore the neighborhood of the former attractor. We have already indicated, in the introductions to this chapter and to the second part of the book, that the phenomenology of intermittency was relatively independent of its type. For control parameter values that are less than a critical value, we have a stable limit cycle. When the control parameter slightly exceeds the critical value, we still find periodic behavior most of the time. But it is interrupted from time to time by turbulent bursts. The bursts are more infrequent as we approach the critical value of the control parameter, where the Floquet multiplier crosses (-1). As can be seen on experimentally recorded signals (observed for the R.B. instability, Figure IX.20), the laminar phase can be seen as an interval during which the subharmonic is amplified until a kind of final catastrophe marks the beginning of the turbulent burst. Following an approach close to that used to analyze type I intermittency, we will study the dynamics of laminar phases using a one-dimensional mapping. The mapping is meant to describe the first return map along the weakly unstable manifold of the closed trajectory.

In the *linear* approximation, this mapping takes the form:

$$x \longmapsto x' = -(1 + \varepsilon)x \qquad (\text{IX.6})$$

The fixed point $x = 0$ attains marginal stability at $\varepsilon = 0$. For ε slightly positive (negative), this fixed point is unstable (stable). The Floquet multiplier $(\partial x'/\partial x)|_{x=0}$ is equal to $-(1 + \varepsilon)$, crossing the unit circle at (-1) for $\varepsilon = 0$. The fixed point has coordinate $x = 0$ along the weakly stable (for $\varepsilon < 0$) or weakly unstable (for $\varepsilon > 0$) manifold. As in type I intermittency, we will complement the linear analysis by adding to Equation (IX.6) terms that are nonlinear in x. As will be shown later, to be consistent we must add terms of second and third order in x. The new mapping is then:

$$x \longmapsto x' = -(1 + \varepsilon)x + \alpha x^2 + \beta' x^3 \qquad (\text{IX.7})$$

For differential equations the coefficients α and β' can be determined by explicit formulas. Here we will just assume that they are given. Note that their values imply a

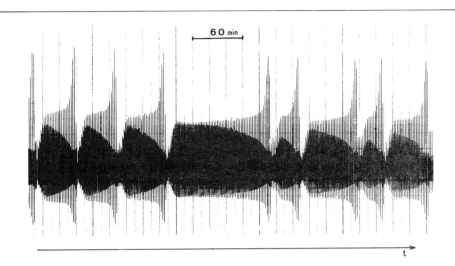

Figure IX.20 Time dependence of the horizontal temperature gradient near the threshold of type III intermittency in R.B. convection.
This R.B. experiment is carried out with a ratio $Ra/Ra_c \simeq 416.5$. Note the continuous growth and then the abrupt increase of the amplitude of the subharmonic, and the concomitant decrease of the amplitude of the fundamental.

From M. Dubois, M. A. Rubio and P. Bergé.

choice of scale for x since they appear in a third degree polynomial which is not homogeneous in x.

To analyze the mapping defined by Equation (IX.7) it is preferable to iterate twice, for a reason which will soon appear. Limiting ourselves to terms which are cubic in x and linear in ε (since we are interested only in x and ε in a neighborhood of 0), this second return map becomes:

$$x \longmapsto x'' = x'[x'(x)] \simeq (1 + 2\varepsilon)x + \beta x^3 \qquad \text{(IX.8)}$$

where:

$$\beta = -2(\beta' + \alpha^2).$$

In the expression for β, we have included only terms independent[97] of ε, since we wish to examine what happens in the limit $\varepsilon \longrightarrow 0$.

---

97. We also assume that β has a well-defined sign near $\varepsilon = 0$. Indeed β, the coefficient of the cubic term in the Taylor expansion of $x'(x)$, is a function of ε. We will not examine the case $\beta(\varepsilon = 0) = 0$, termed of codimension two.

We remark two features of Equation (IX.8):

i) The coefficient of the linear term is close to $+1$ (instead of -1 in Equation (IX.7)). This simply comes from the fact that the Floquet multiplier is multiplied by itself at each iteration. Hence, at the second iteration, the initial (-1) becomes $(-1)^2 = +1$.

ii) The quadratic term in x has disappeared in Equation (IX.8) (at least to lowest order in ε). This is in marked contrast to the model of type I intermittency developed earlier and results from algebraic cancellation (and in fact of all terms in $x'(x)$ of even order) after two iterations, to lowest order in ε.

We can rewrite Equation (IX.8) in the form:

$$x'' \doteq x((1 + 2\varepsilon) + \beta x^2).$$

The nonlinear term βx^2 has two totally opposite effects according to the sign of β. If β is negative, we have a supercritical bifurcation with period doubling. In the weakly nonlinear domain ($\varepsilon \gtrsim 0$), the mapping defined by Equation (IX.8) has a stable fixed point (the point $x = 0$ is unstable). Therefore the mapping x' has a stable two-cycle: there exist values x_1 and x_2 such that:

$$x''(x_{1,2}) = x_{1,2}; \quad x'(x_1) = x_2; \quad x'(x_2) = x_1$$

where
$$x_{1,2} \underset{\varepsilon \to 0^+}{\sim} \pm(-2\varepsilon/\beta)^{1/2}$$

and
$$|(dx'/dx)|_{x_1} \cdot (dx'/dx)|_{x_2}| < 1 \quad \text{(stability condition)}.$$

If, on the contrary, β is positive, the nonlinearity βx^2 has the same effect as a positive ε: it increases the multiplicative coefficient of x with each iteration. Then the nonlinearity *reinforces* the instability, and we have a subcritical bifurcation. It is this latter case which leads to type III intermittency and which interests us now.

As in type I intermittency, we will analyze the subcritical bifurcation ($\beta > 0$) by replacing the difference $x'' - x$ by dx/dk near the fixed point, and the difference Equation (IX.8) by the differential equation:

$$\frac{dx}{dk} = x(2\varepsilon + \beta x^2) \tag{IX.9}$$

Scaling x by $(2\varepsilon/\beta)^{1/2}$ and k by ε^{-1}, it takes the universal form:

$$\frac{dx}{dk} = x(1 + x^2) \tag{IX.10}$$

This scaling is by itself already useful in making predictions from the theory. The index k, which counts the number of oscillations at the fundamental frequency, is a measure of time. The natural scale for k being ε^{-1}, we deduce that the average duration of laminar periods diverges like ε^{-1} close to the threshold of intermittency. Close to this same threshold, the average correlation time of the signal is also on the order of the average duration time. This time is also the inverse of the Lyapunov exponent characterizing the average rate of divergence of trajectories in phase space. We deduce that the Lyapunov exponent, characterizing turbulence, increases like ε in the intermittent phase. In the same way, we can say that since the scale of x is $(2\varepsilon/\beta)^{1/2}$, the average amplitude of the subharmonic fluctuations measured during the laminar phases is of order $\varepsilon^{1/2}$ close to the threshold.

Instead of integrating Equation (IX.10) we will now explain how to deduce from it the statistical distribution of the lengths of the laminar periods.

Statistics of the laminar periods

Equation (IX.9) shows the existence of two regimes for a given small value of ε. If x is sufficiently larger than $(2\varepsilon/\beta)^{1/2}$, it is the term in βx^2 which dominates. Under these conditions, Equation (IX.9) reduces to:

$$\frac{dx}{dk} = \beta x^3 \qquad (IX.11)$$

which has the solution:

$$x = \frac{x_0}{(1 - 2\beta k x_0^2)^{1/2}}$$

with the initial condition $x(k = 0) = x_0$. This solution diverges when the denominator vanishes, i.e. when:

$$k = \frac{1}{2\beta x_0^2} \qquad (IX.12)$$

The time at the end of which the solution to Equation (IX.11) diverges can be identified as the length of a laminar phase measured in basic oscillation periods.

This provides us with a first estimate of the statistical distribution of the laminar phases. Suppose that the initial value x_0 is taken at random, in other words from a uniform distribution in the neighborhood of the unstable fixed point (a hypothesis to be justified later in studying the relaminarization process). The relationship between x_0 and k (length of the laminar phase) having been determined by Equation (IX.12), we have the following equation relating differentials:

$$dk = \frac{1}{\beta x_0^3} dx_0$$

which can also be written:

$$dx_0 = \frac{k^{-3/2}}{2^{3/2} \beta^{1/2}} dk.$$

This equation allows us to relate the distribution of k to that of x_0. If x_0 is arbitrary, then in a sufficiently large statistical sample the fraction of x_0 values contained between x_0 and $x_0 + dx_0$ will be proportional to dx_0. The preceding relation between differentials shows that the relative number of laminar phases contained between k and $k + dk$ is of order $P(k)\, dk \sim \beta^{-1/2} k^{-3/2}\, dk$.

This distribution law is of course not valid for small values of k. By definition, small values of k mean that the length of the laminar phase is on the order of several fundamental periods. For such short phases, the successive iterates diverge too rapidly for us to be able to consider $x'' - x$ as small and replace Equation (IX.8) by Equation (IX.9)[98]. Therefore the distribution law $P(k) \sim k^{-3/2}$ is valid only for sufficiently large values of k. But these values must not be too large, either. If k becomes very large, then Equation (IX.12) shows that the initial point of the iteration approaches 0, contradicting the hypothesis that $x \gg (\varepsilon/\beta)^{1/2}$, used to derive Equation (IX.11) from Equation (IX.9).

If this hypothesis is not satisfied (i.e. if the point of departure is too close to the unstable fixed point), then the laminar phase is divided into two parts. At first, the linear term in x is dominant in Equation (IX.9) and so $x(k) \sim x_0 \exp(2k\varepsilon)$. This part ends approximately when $x \sim (\varepsilon/\beta)^{1/2}$ so that $k \sim (1/4\varepsilon) \ln(\varepsilon/\beta x_0^2)$. Reasoning analogous to that developed above shows that afterwards, for $k \gg \varepsilon^{-1}$ (realized when $x_0 \ll (\varepsilon/\beta)^{1/2}$), the probability distribution $P(k)$ decreases like $\exp(-2\varepsilon k)$.

We summarize by saying that $P(k)$ has two asymptotic limits: for $1 \ll k \ll \varepsilon^{-1}$, we have $P(k) \sim k^{-3/2}$, and for $k \gg \varepsilon^{-1}$, $P(k) \sim \exp(-2\varepsilon k)$.

Relaminarization

As was already the case for type I intermittency, the study of relaminarization cannot be as quantitative as that of the laminar phase: relaminarization depends on the global structure of the flow and its study can therefore only be of a topological or qualitative nature.

We begin by remarking that the Floquet multiplier cannot cross the unit circle at (-1) for a flow defined on T^2 or on the plane. Indeed, by drawing diagrams one realizes that the weakly unstable trajectory detaching itself from the closed trajectory must be drawn on a Moebius strip centered on the closed trajectory. A Moebius strip is nonorientable[99] and so cannot be part of the plane or of T^2, which are orientable

---

98. Strictly speaking, the difference Equation (IX.8) predicts that the iterates will diverge like:
$$x(k) \simeq \beta^k (x_0)^{3^k}.$$
Yet this extremely rapid divergence does not lead to a singularity after a finite time, unlike what happens in the differential Equation (IX.10). In any case, the approximation we are using implies that x is small enough for us to truncate the Taylor expansion of $x'(x)$ after a few terms. The rapid divergence of the iteration, or the singularity in the differential equation, causes x to leave the domain of validity of the equations.

99. A surface is orientable if we can uniquely define a direction of rotation on closed curves. A plane, the surface of a sphere, a torus T^2 are orientable. Drawing a small circle on one of these surfaces and moving it about continuously, we uniquely define a direction of rotation (clockwise, for example) at every point on the surface. In contrast, a Moebius strip is not orientable, nor is a Klein bottle. Displacing a small (oriented) circle on a Moebius strip, we can reach the back of the sheet which forms the Moebius strip. The circles drawn in the front and on the back of the sheet have opposite orientations once the two sides of the sheet are identified. This shows the non orientability of the Moebius strip.

surfaces. This excludes the relaminarization process for type III intermittency from taking place on a torus, as may occur for type I. However relaminarization by a baker's transformation or on the Smale attractor are possibilities for type III intermittency, just as they are for type I.

To go from type I to type III in the construction of Figure IX.9, it suffices to add to the end of the process (or equivalently to the beginning, since the system is iterative) a 180° rotation of the torus about a horizontal axis. This reverses the sign of the Floquet multiplier, moving it from a neighborhood of $(+1)$ (type I intermittency) to a neighborhood of (-1) (type III intermittency). In this case, the mapping of Figure IX.9 changes from $x \longmapsto f(x)$ to $x \longmapsto \hat{f}(x) = f(1-x)$ (fig. IX.21). These topological constructions do not relate type I to type III intermittency in a very detailed way. In particular, the fixed points are in general not preserved by these global constructions.

These considerations apply to the problem of relaminarization. In type I intermittency, relaminarization results from the required passage through the channel which has appeared between the graphs of $f(x)$ and of the identity map.

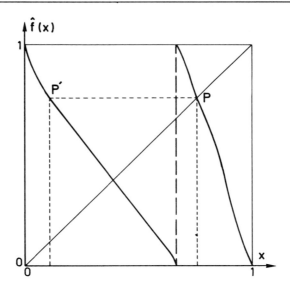

Figure IX.21 Mapping of the circle in type III intermittency.
If the slope at P becomes less than -1 and if the accompanying bifurcation is subcritical, one obtains type III intermittency. The (unstable) fixed point subsists and has two antecedents: itself and another point P'. When the iteration arrives in the vicinity of P', the mapping then sends it to the vicinity of P: this is the relaminarization process.

In type III intermittency, we find nothing of the kind, since crossing of the unit circle by the Floquet multiplier is not accompanied by disappearance of the fixed point of the mapping. For the mapping shown schematically in Figure IX.21, type III intermittency can appear when the slope of $\hat{f}$ at the fixed point P becomes less than -1, and when a condition on the first few terms of the Taylor series of $\hat{f}$ near P insuring the subcriticality of the accompanying bifurcation is met. We see that these conditions do not imply the disappearance of the fixed point P. Rather, relaminarization will result from the fact that the fixed point has two antecedents: itself and the point marked P' on Figure IX.21. Therefore if, while wandering during a turbulent burst, the trajectory visits the neighborhood of P', the following iterations will bring the trajectory back to the neighborhood of P, the weakly unstable fixed point. This will mark the beginning of a new laminar phase. One might think that this process is excluded for real systems, governed by deterministic equations which allow us to go back in time[100] in a unique way. This would exclude, *a priori*, a non-invertible Poincaré mapping. But we have already seen that, when a Poincaré mapping is sufficiently contracting along certain directions of the phase space, the resulting first return map could be exempted from the constraint of invertibility. For intermittency, even a weak contraction along the stable manifold of the closed trajectory suffices. If the initial point falls close to the stable manifold of the trajectory, it will yend exponentially fast towards the periodic trajectory (or fixed point of the Poincaré mapping) along this manifold, whereas its divergence along the unstable manifold will be slow, since the instability is only marginal near the intermittency threshold. By using the mapping of Figure IX.21 as a model, we have reduced the approach to the unstable fixed point along its stable manifold to a single iteration: the one taking P' into P.

Under these conditions we can justify taking the initial point of iteration during a laminar phase at random in the neighborhood of P. Indeed, at least in Figure IX.21, the mapping $\hat{f}$ is strongly dilating outside a region close to P. This dilation, or SIC, is, as we have seen, synonymous with loss of memory; hence the fact that the points of reinjection are effectively taken at random. In the section which follows, we will present an R.B. experiment where such a transition by type III intermittency has been observed.

---

100. The possibility of going back in time is clear when we consider ordinary differential equations of the form $(d\vec{x}/dt) = F(\vec{x})$. On the other hand, for partial differential equations with dissipation, such as the equations of hydrodynamics, it is in general impossible to go backwards in time, even over infinitesimal intervals. The Fourier, or heat, equation $T_t = kT_{xx}$ ($k > 0$) is an *ill-posed* problem (in the sense of Hadamard) if we try to go back in time, i.e. to find $T(t, x)$ from $T(t + \tau, x')$ for $\tau > 0$. It is therefore not completely clear that the condition of uniqueness of trajectories under time reversal is as much of a constraint as it seems to be. This non invertibility comes from the existence of damped fluctuations with arbitrarily small time constants as time passes. When the direction of time is reversed, these same fluctuations are amplified arbitrarily quickly, and therefore become uncontrollable. Where hydrodynamics is concerned, we can adopt the point of view that in a *reduced description* (like that leading to ordinary differential equations such as the Lorenz model) these fluctuations are neglected no matter what the direction of time. This makes (formal) invertibility of the reduced description possible. A more detailed justification would lead us too far astray from the subject of this book.

IX.3.2 TYPE III INTERMITTENCY IN R.B. CONVECTION

The experimental conditions are as follows: the convective liquid is silicone oil with Prandtl number Pr $\sim$ 38, the height of the cell is $d \sim 1.5$ cm, its length is $L_x = 2d$ and its width $L_y = 1.2d$. The Rayleigh number Ra is on the order of 300 to 400 times the critical value Ra_c, and the quantity measured is the deviation of a narrow light beam due to the temperature gradients in the convecting fluid. The raw observations are as follows. For Ra/Ra_c between about 333 and 337, the thermoconvective currents oscillate[101] with almost perfect regularity, the oscillation frequency being on the order of 1.7×10^{-2} Hz (i.e. a period of about a minute). At $Ra/Ra_c \sim 416.7$, the subharmonic frequency appears in the signal, accompanied by intermittency. We return to Figure IX.20 to show a typical signal. Between two turbulent bursts, the increase in the subharmonic is clearly visible as well as an apparent decrease in the fundamental frequency. When the amplitude of the subharmonic attains a value of the same order of magnitude as the fundamental, a fluctuation or turbulent burst appears, shattering the signal's regularity. Immediately afterwards, regular behavior reppears with a small random amplitude for the subharmonic. This initial amplitude determines the length of the laminar phase until the next turbulent burst. The smaller the initial amplitude of the subharmonic, the longer the laminar phase lasts. The intermittent behavior is reversible, and this confirms the idea that transition by intermittency to the chaotic state is reversible.

Using the experimental data we can show that during the laminar phases, the second return map has the form predicted by Equation (IX.8). Let I_k be the value of the kth maximum of the physical signal like that recorded on Figure IX.20. The second return map will be the graph of $I_{k+2} = f(I_k)$, for a sequence of values of k. Recall that the theory predicts the following form for the function:

$$I_{k+2} = (1 + 2\varepsilon)I_k + bI_k^3 \tag{IX.13}$$

This form agrees well with the experimental data, as demonstrated by Figure IX.22 b. We note in particular the existence of an inflection point in the neighborhood of the fixed point of $f(I_k)$. This corresponds to the vanishing of the quadratic term in the Taylor series of $f(I_k)$.

Another signature of the intermittency is the statistical distribution of the lengths τ of the laminar phases (fig. IX.23). A calculation along the lines of

---

101. The physical nature of these oscillations is rather complicated and is related to the geometry of the convection currents in the cell. In this experiment, the oscillation is related to the pulsation of a cold plume. A plume is a local fluctuation in the fluid temperature, here of lower temperature than its surroundings, which detaches itself at more or less regular intervals from the boundaries responsible for the temperature gradient in the fluid. Plusmes exist mainly in high Prandtl number fluids, for the following reason. For a fluid to have a high Prandtl number means that its kinematic viscosity is much greater than its thermal diffusivity. The effectiveness of thermal diffusion is therefore low, permitting high gradients of temperature and also of density to exist for a long time before relaxing. This causes large fluctuations in the Archimedean force which drives the plumes.

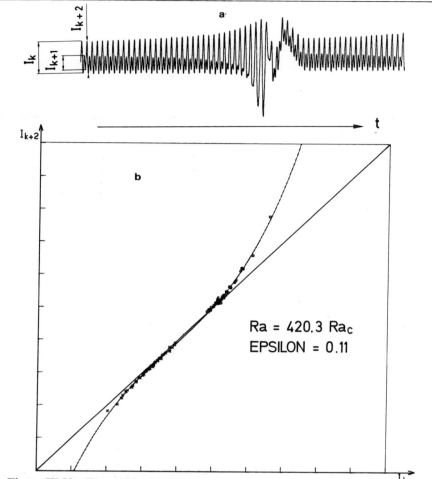

Figure IX.22 Type III intermittency in R.B. convection.
a) Sample of a time signal from the same regime as Figure IX.20. Expanding the time scale provides a better illustration of the growth of the subharmonic and the correlated decay of the fundamental. To the left are defined the quantities I_k used in graphing the second return map.
b) Graph of the second return map $I_{k+2} = f(I_k)$.
Two different symbols are used to construct this graph from the experimental results of Figure IX.20: one of them (□) corresponds to the subharmonic (increasing amplitudes) and the other (×) to the fundamental (decreasing amplitudes). The continuous curve is the graph of the function:

$$f(I) = (1 + 2\varepsilon)I + aI^2 + bI^3$$

(a and b constants with $a \ll b$) predicted by theory to be the functional form of the second return map near the intermittency threshold. We note the excellent agreement with experimental results obtained by adjusting the value of the parameter ε ($\varepsilon = 0.11$).
From M. Dubois, M. A. Rubio and P. Bergé.

IX.3 TYPE III INTERMITTENCY

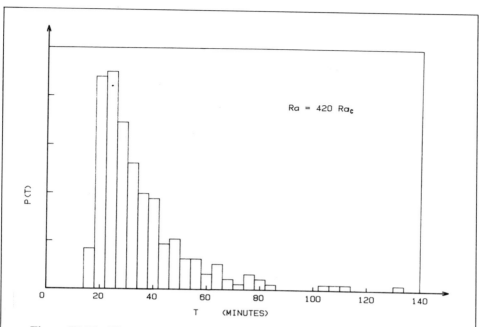

Figure IX.23 Histogram of the lengths of laminar phases.
The lengths observed vary from eighteen minutes to more than two hours. The most significant feature is the long tail for T large, characteristic of type III intermittency. This shape is clearly distinguishable from that occurring in type I intermittency (see figs. IX.4 and IX.18).

After M. Dubois and M. A. Rubio.

Section IX.3.1 shows that if $P(\tau)\,d\tau$ is the fraction of laminar phases which last between τ and $\tau + d\tau$ ($d\tau$ small), then:

$$P(\tau) \sim \frac{\exp(-2\varepsilon\tau)}{[1 - \exp(-4\varepsilon\tau)]^{3/2}}.$$

For $\tau \gg \varepsilon^{-1}$ we have $P(\tau) \sim e^{-2\varepsilon\tau}$, while for $1 \ll \tau \ll \varepsilon^{-1}$, we have $P(\tau) \sim (4\varepsilon\tau)^{-3/2}$, both conforming to the theoretical predictions. The time $\tau = 1$ which bounds the domain of validity of the latter estimate can be understood as the period of oscillation at the fundamental frequency. Indeed, the idea of a laminar phase makes sense only if, while it lasts, we can observe a large number of these oscillations. Theory and experiment are compared in the present case by counting the number of laminar phases $N(\tau > \tau_0)$ lasting longer than τ_0. For $\tau_0 \gg 1$, we should have:

$$N(\tau > \tau_0) \sim \int_{\tau_0}^{\infty} d\tau\, P(\tau) \sim \left(\frac{\exp(-4\varepsilon\tau_0)}{1 - \exp(-4\varepsilon\tau_0)}\right)^{1/2} \tag{IX.14}$$

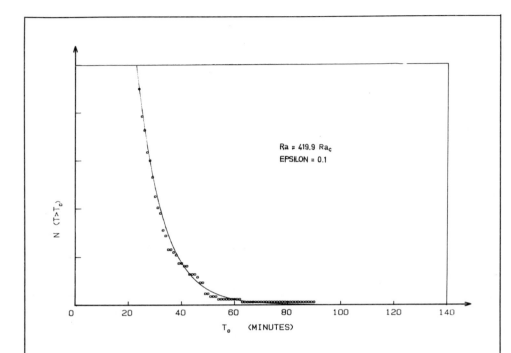

Figure IX.24 Number of laminar phases lasting longer than τ_0.
The experimental points are obtained by summing the histogram of Figure IX.23. The continuous curve represents the functional form predicted by theory and is adjusted to agree with experiment via the parameter ε. It is striking that the value thus calculated ($\varepsilon = 0.1$) is very close to that determined independently from the second return map (fig. IX.22).

From M. Dubois, M. A. Rubio and P. Bergé.

This function of τ_0 is entirely specified by the value of the parameter ε. Figure IX.24 gives a comparison between Equation (IX.14) and the experimental data. The agreement is excellent, since the best ε for Equation (IX.14) to fit the data is $\varepsilon \sim 0.1$, while the experimental data by themselves (figs. IX.22 b and IX.23) lead to $\varepsilon \sim 0.11$. This identifies the observed phenomenon as type III intermittency. We also note that the hypothesis formulated in Section IX.3.1 that the laminar phases start off randomly in the neighborhood of the fixed point, is borne out by the statistics of the measurements leading to Figure IX.24.

IX.4 Theory of type II intermittency

We will limit our analysis to weakly nonlinear, or local, phenomena, i.e. those taking place in the neighborhood of the weakly unstable closed trajectory. We can find a model of a first return map which is continuous, invertible, and exhibits a transition via type II intermittency. This model provides, among other things, a possible pathway for relaminarization. It is based on a rather complicated geometrical construction in a five-dimensional space, the basic idea being the same as that of the Smale attractor.

A periodic trajectory loses its stability as two complex conjugate Floquet multipliers leave the unit circle of the complex plane. The unstable perturbation then grows exponentially in modulus, in the linear approximation, rotating at the same time by a finite angle in its plane at each iteration. It is natural (and traditional) to use complex numbers to describe linear and weakly nonlinear phenomena for these bifurcations. Let (x, y) be the Cartesian coordinates of the small perturbation from the fixed point in the direction of the eigenvectors associated with the two complex conjugate eigenvalues leaving the unit circle. We define the complex number $z = x + iy$, so that, in the linearized approximation for small perturbation amplitudes, the first return map acts by multiplying z by $\lambda\, e^{i\phi}$, λ a positive real, ϕ real. At the instability threshold, λ equals one, while ϕ is the angle rotated at each iteration (cf. Appendix C). In this notation the two complex conjugate eigenvalues of the Floquet matrix are $\lambda\, e^{\pm i\phi}$ and determination of linear stability is reduced to the study of the extremely simple iteration:

$$z \longmapsto z' = \lambda\, e^{i\phi}\, z. \tag{IX.15}$$

As in the two intermittency types studied earlier, we introduce a small control parameter ε, whose passage through 0 coincides with marginal linear stability, i.e. with $\lambda = 1$. We therefore have $\lambda = 1 + \varepsilon$. The angle ϕ as it appears in Equation (IX.15) is also *a priori* a function of ε. But we will assume that $\phi(\varepsilon = 0)$ is not one of certain special angles such as $0, \pi/2, \pi, 2\pi/3$. By excluding these situations we can, as we will see later, neglect the dependence of ϕ on ε in the neighborhood of $\varepsilon = 0$.

As in type I and III intermittencies, analysis of the weak nonlinearities will be based on seeing the right hand side of Equation (IX.15) as the beginning of a Taylor series of a function $z'(z)$. But here there is no reason for the series to be entire in z. This means that we must consider powers of z and of z^* (the complex conjugate of z) as being on equal footing, classifying terms only by their order in the modulus of z. As in type III intermittency, we must keep terms of up to third order to arrive at a theory which is first order in the nonlinearities. We then have:

$$\begin{aligned}z \longmapsto z' = \lambda\, e^{i\phi}\, z &+ a_1 z^2 + a_2(z^*)^2 + a_3 z z^* \\ &+ b_1 z^3 + b_2 z^2 z^* + b_3(z^*)^2 z + b_4(z^*)^3 + \ldots\end{aligned} \tag{IX.16}$$

where the a_i and b_i are complex coefficients, which can be calculated in an analytic example following the general principles of a Landau-type expansion. Like the linear terms, the nonlinear terms have two effects: they rotate z about the origin, and also change its length. How to separate these two effects is not obvious, in general, given that they are closely linked by the nonlinearity of Equation (IX.16). However here we can use the simplifications appearing near $z = 0$ and $\lambda = 1$, the vicinity for which the expansion (IX.16) is in fact valid. The effect of the nonlinearities is very different according to whether we consider the *modulus* or the *angle* of z. The instability primarily affects the behavior of the modulus, which is marginally stable for $\varepsilon = 0$. This means that its evolution is very slow if λ is close to 1 (or ε close to 0), as we will assume. However, the angle by which z rotates at each iteration remains finite at $\varepsilon = 0$. We can therefore make what is called an *adiabatic approximation*[102], by which we assume that the angle of z has rotated considerably while its modulus remains almost unchanged. To illustrate, let us consider the case where the nonlinearities in Equation (IX.16) are limited to the single term $a_1 z^2$:

$$z \longmapsto z' = \lambda\, e^{i\phi}\, z + a_1 z^2. \tag{IX.17.1}$$

The equation for the complex conjugate is:

$$z^* \longmapsto z'^* = \lambda\, e^{-i\phi}\, z^* + a_1^*(z^*)^2 \tag{IX.17.2}$$

and, multiplying the two expressions term by term, we get:

$$|z'|^2 = \lambda^2\, |z|^2 + |a_1|^2\, |z|^4 + \lambda\, |z|^2 (e^{i\phi}\, z^* a_1^* + e^{-i\phi}\, z a_1) \tag{IX.18}$$

We are now interested in the average evolution of $|z'|^2$ during a number of iterations large enough for the angle z to have taken on many values, but small enough for the modulus of z to have varied only slightly. To a first approximation the mapping on the angles is rotation by a constant angle ϕ with initial value ψ_0. If ϕ is an arbitrary irrational angle, we know[103] that the angles $(\psi_0 + k\phi)$ mod 2π are uniformly distributed over $[0, 2\pi]$ when k takes on a large number of different integer values. The average we seek will therefore be a simple average over the angle of z, with a uniform distribution on $[0, 2\pi]$. In averaging Equation (IX.18), we leave unchanged the terms $\lambda^2 |z|^2$ and $|a_1|^2 |z|^4$ which are independent of the angle. In contrast, consider the last term of Equation (IX.18): if ψ is the angle of z, this last term depends on ψ like $\cos(\overline{\phi} - \psi)$ which, averaged over ψ, gives zero. An extension of this argument shows

---

102. An adiabatic expansion is one in which a parameter varies slowly whereas the system undergoes rapid variation (in general, oscillation). This nomenclature comes from thermodynamics: a transformation is adiabatic when it takes place sufficiently slowly to allow us to neglect irreversible phenomena due to viscous friction, proportional to the speed of the transformation. Despite their apparent simplicity, adiabatic theories are subtle and profound: they are related to problems such as the long term stability of the solar system, the microscopic origin of the second principle of thermodynamics, and the classical limit of quantum mechanics.

103. It is here that the condition that $\phi(\varepsilon = 0)$ not be a "special" angle such as $\pi/2$, $2\pi/3$, etc. enters in. Otherwise, we would not be allowed to make the hypothesis that the values of $(\psi_0 + k\phi)$ are uniformly distributed over the circle. Since ϕ depends on $|z|$, the situation is much more complicated in this case: it is then necessary to consider explicitly the coupling between ϕ and the modulus $|z|$. The corresponding theory resembles the famous KAM (Kolmogoroff-Arnol'd-Moser) theory.

IX.4 THEORY OF TYPE II INTERMITTENCY

that the average evolution of $|z|$ (which we will call ρ) is governed, in the adiabatic approximation, by:

$$\rho \longmapsto \rho' = (\lambda^2 \rho^2 + A\rho^4 + ...)^{1/2} \tag{IX.19}$$

The sum in parentheses is an entire series in ρ^2, whose coefficients can be calculated [104] order by order by expanding Equation (IX.16) and averaging over ψ. The coefficient A in Equation (IX.19) depends explicitly on the $a_i's$ and $b_i's$ of Equation (IX.16).

In the approximation of interest to us, ρ is close to 0 and we can consider $A\rho^4$ as small compared to $\lambda^2 \rho^2$. This allows us to expand the square root in (IX.19), to obtain:

$$\rho' = \lambda\rho\left(1 + \frac{A\rho^2}{2\lambda^2} + ...\right) \tag{IX.20}$$

(Recall that λ is positive and close to 1.) Here too the terms in the parentheses are the beginning of an entire series in ρ^2. In what follows we will limit ourselves to considering only the terms written explicitly in Equation (IX.20), that is, to the iteration on the modulus defined by:

$$\rho \longmapsto \rho' = \lambda\rho\left(1 + \frac{A}{2\lambda^2}\rho^2\right). \tag{IX.21}$$

The distinction between a subcritical and supercritical bifurcation depends upon the sign of A. To understand this point, we can consider the effect of the nonlinear terms in $\rho'(\rho)$ to a first approximation, as merely changing the factor λ to $\lambda(1 + (A/2\lambda^2)\rho^2)$. If λ is greater than 1 (i.e. if here is linear instability), the instability will be counteracted by the lowest order nonlinearities if A is negative, and amplified if A is positive. To be more quantitative, consider the neighborhood of $\lambda = 1$, so that $\lambda = 1 + \varepsilon$ with $|\varepsilon| \ll 1$. The mapping defined by Equation (IX.21) has as fixed point(s) the solution(s) to $\rho'(\rho) = \rho$, which are $\rho = (-2\varepsilon/A)^{1/2}$ if ε/A is negative, and $\rho = 0$. (We see here the square root dependence of the Landau theory of bifurcations). If A is negative, there is a supercritical bifurcation and the fixed point $\rho = (-2\varepsilon/A)^{1/2}$ is linearly stable for $\varepsilon > 0$, while $\rho = 0$ is unstable. On the other hand, if A is positive (the case of interest to us here), then the bifurcation is subcritical. This means that for $\varepsilon < 0$ the linearly stable point $\rho = 0$ coexists with the unstable fixed point $\rho = (-2\varepsilon/A)^{1/2}$. At $\varepsilon = 0$, the unstable fixed point (which is in fact an invariant circle on the (x, y) plane of the unstable directions of the Floquet matrix) merges with $\rho = 0$, the projection of the limit cycle onto the (x, y) plane. Therefore for $\varepsilon > 0$, there no longer exists a stable attractor in the region of phase space near to the closed trajectory that interests us. It is in this case that type II intermittency can occur, which we will now study using the mapping defined by Equation (IX.21). Non dimensionalizing ρ by $(2\lambda^2/A)^{1/2}$, we put Equation (IX.21) into standard form:

$$\rho \longmapsto \rho' = (1 + \varepsilon)\rho + \rho^3. \tag{IX.22}$$

---

104. Notice that in these expressions, all explicit references to the angle ϕ have disappeared. This reflects a profound property of the adiabatic approximation: the "fast" variable (here ϕ) is decoupled from the "slow" variable (here ρ) at all algebraic orders of the expansion. Coupling appears only at transcendental orders in ρ such as $\exp(-1/\rho)$.

This is — formally — the same iterative relation as that considered for type III intermittency. Therefore, the relationship between the length of laminar periods (denoted by N), the value of ρ at the beginning of the period (denoted by ρ_0), and ε is also essentially the same as for type III. If $\varepsilon^{1/2} \ll \rho_0 \ll 1$, the iteration of Equation (IX.22) is dominated by the cubic term and we have $N \sim \rho_0^{-2}$. If, on the other hand, $\rho_0 \ll \varepsilon^{1/2}$, the increase in the oscillation amplitude during the laminar phase is controlled by the $\varepsilon\rho$ term in Equation (IX.22) and so we have $N \sim -\varepsilon^{-1} \ln \rho_0$. Therefore, as in type III intermittency, the histogram of the lengths N of the laminar phases is composed of two parts, according to whether $N \gg \varepsilon^{-1}$ or $1 \ll N \ll \varepsilon^{-1}$. This elementary theory of the distribution does nevertheless differ from the analogous theory developed for type III intermittency. In the latter case, the unstable manifold is a line and the laminar phase starts off on the line, at a random point near the unstable fixed point. In other words, if s_0 is the distance between the initial point and the unstable fixed point, the probability that s_0 is located in the interval $[s_0, s_0 + ds_0]$ is simply ds_0 times a normalization factor. But for type II intermittency, the unstable manifold is two-dimensional. The length of the laminar phase is still determined by the distance ρ_0 between the initial point and the unstable fixed point, measured on the surface in the usual way. If the initial point is located at random on the surface, the probability that ρ_0 lies in the interval $[\rho_0, \rho_0 + d\rho_0]$ will be $2\pi\rho_0\,d\rho_0$ multiplied by a constant, since $2\pi\rho_0\,d\rho_0$ is the area of the annulus bounded by the circles of radius ρ_0 and $(\rho_0 + d\rho_0)$ centered on the fixed point. Since $\rho_0 \sim N^{-1/2}$ we have the following relation between differentials:

$$d\rho_0 \sim N^{-3/2}\,dN.$$

Now suppose that the initial point is taken randomly on the unstable manifold, that is, with a measure $d\mu$ proportional to the area of the manifold. We then have $d\mu \sim \rho_0\,d\rho_0$, or $d\mu \sim \rho_0 N^{-3/2}\,dN \sim N^{-2}\,dN$. If we are exactly at the instability threshold ($\varepsilon = 0$), this means that the average length of a laminar phase is divergent since it is given by an integral which, for N large, behaves like $\int N(dN/N^2)$. But for ε finite (and small), the average value $\langle N \rangle$ of N does not diverge. Indeed, the probability distribution $P(N)$ behaves like $1/N^2$ for $\varepsilon^{-1} \ll N \ll 1$ (which corresponds to $\varepsilon^{1/2} \ll \rho_0 \ll 1$) and like $\exp(-2\varepsilon N)$ for $N \gg \varepsilon^{-1}$ (or, equivalently, $\rho_0 \ll \varepsilon^{1/2}$). Thus for ε finite, the integral defining $\langle N \rangle$ converges for large N, since in this domain $P(N)$ decreases exponentially. The characteristic time scale limiting this length of laminar phase varies like ε^{-1} close to $\varepsilon = 0$. Since the integral defining $\langle N \rangle$ at $\varepsilon = 0$ diverges logarithmically for large N, we deduce that $\langle N \rangle$ varies like $\ln \varepsilon^{-1}$ close to $\varepsilon = 0$. Now suppose that the Lyapunov number $\gamma(\varepsilon)$ is of the same order of magnitude as $\langle N \rangle^{-1}$, so that the signal loses its memory after each turbulent burst. We then deduce the law $\gamma(\varepsilon) \sim (1/\ln(1/\varepsilon))$. Let us finish by indicating that careful numerical studies point to subtle statistical effects that alter this elementary description of the statistics.

References for Chapter IX

P. Bergé, M. Dubois, P. Manneville, Y. Pomeau, "Intermittency in Rayleigh-Bénard convection", *Le Journal de Physique-Lettres*, **41**, p. L341 (1980).

M. Dubois, "Approach of the turbulence in hydrodynamic instabilities" in *Symmetries and Broken Symmetries in Condensed Matter Physics*, I.D.S.E.T. Paris (1981).

M. Dubois, M. A. Rubio, P. Bergé, "Experimental evidence of intermittencies associated with a subharmonic bifurcation", *Physical Review Letters*, **51**, p. 1446 (1983).

P. Manneville, Y. Pomeau, "Intermittency and the Lorenz Model" in *Symmetries and Broken Symmetries in Condensed Matter Physics*, I.D.S.E.T. Paris (1981).

P. Manneville, Y. Pomeau, "Different ways to turbulence in dissipative dynamical systems", *Physica*, **1D**, p. 219 (1980).

P. Manneville, "Intermittency, self similarity and 1/f spectrum", *Le Journal de Physique*, **41**, p. 1235 (1980).

Y. Pomeau, *Approche de la turbulence*, Actes de la conférence de la Société française de Physique, Clermont-Ferrand, Les éditions de physique, Orsay (1981).

Y. Pomeau, "Approche de la turbulence" *Actes de la conférence de la Société française de Physique*, Clermont-Ferrand, Les éditions de physique, Orsay (1981).

Y. Pomeau, P. Manneville, "Intermittent transition to turbulence in dissipative dynamical systems", *Communications in Mathematical Physics*, **74**, p. 189 (1980).

CONCLUSION

Debate

Christian Vidal: To begin it seems appropriate to evoke a primordial aspect which has not been developed in this book: the fact that the domain of validity and the applicability of the theory of dynamical systems remains uncertain at this time. We have available to us descriptions of a number of events and of sequences of events that lead to chaotic situations: this is what the second part of the book consists of. Thus when certain conditions are satisfied, we are capable of predicting the way in which aperiodic behavior will appear, i.e. how the transition will occur. But in contrast, what the theory does not at all specify is the set of circumstances which must be united for a given sequence of events ending in chaos to occur. The theory does not define, at least not yet, the *prerequisites* for chaotic behavior and this is undoubtedly its major shortcoming at the present time.

Yves Pomeau: I agree entirely with this point of view. We can say that the analysis which has been made of transitions to chaos and of chaotic phenomena certainly has, in spite of its deficiencies, great predictive power, or it should have, since experimentalists have seen the transitions as well as the phenomena. We can therefore imagine that by analyzing these experiments more closely, by trying to understand in more detail the phenomena that have occurred, we should be able to understand, in a given experiment, why a given scenario of transition towards chaos or why a certain type of turbulence has been seen, whether a small or a large number of degrees of freedom has been seen. What is not clear right now is if, in order to know this, enormous numerical calculations will be necessary, which would make the predictions not all that worthwhile, requiring prohibitively long computation time, and in the final analysis, not very enlightening, or if, on the contrary, we will be able to develop an intermediate theory less general than that of dynamical systems as it stands, but sufficiently detailed for it to be possible to predict what is going to occur in a given situation, without entering into the details of the particular equations governing the system.

Pierre Bergé: I would like to underline, with regard to these viewpoints on prediction, a certain *practical* importance of the concepts introduced in this book. We can, for example, hope that one day we will refine weather predictions, were it only by generalizing the method of first return maps. We have seen that through these mappings the future, although admittedly not the long-term behavior of a chaotic system, can indeed be determined: in short, to find "order within chaos". Can we, by relying on these kinds of methods, make better weather predictions? Will the study of living microorganism populations depending on two parameters, the evolution of populations of animal species subjected to several kinds of cycles, be better understood and, therefore, better controlled? We believe so. We must also note the importance that theories of deterministic dynamical systems have taken recently for understanding hydrodynamics, at least as regards certain kinds of turbulence. No one doubts today that we have, in many experiments in which the effective number of degrees of freedom is small, understood several well-defined routes to chaos. The big question is: will we be able to go much further, towards developed turbulence, where the number of degrees of freedom is very high? We must mention some very recent work in this direction on models that are in some ways intermediate between those with a tiny number of degrees of freedom examined in the framework of this book, and models which have a very large, practically infinite number of degrees of freedom. We are beginning to understand the mechanisms arising in model systems of fifty or so degrees of freedom. It is interesting to note the progress taking place in these intermediate cases, which are preparing us to cross part of the gulf that exists at present between deterministic turbulence and the vast domain of developed turbulence about which, at the moment, there remains a great deal to understand.

Yves Pomeau: It is interesting that models have been proposed to describe relatively concrete phenomena such as chemical turbulence or instabilities in flames. These models are intermediate between systems with a small number of degrees of freedom and fully developed turbulence, with all its attendant problems. In particular, these models have the advantage of helping us to understand the interaction between effects of complicated dynamics described in this book and the equally complex effects of spatial structure, which are certainly the two elements that should allow us to explain developed turbulence. It is important to realize that the studies that are carried out on these models involve enormous numerical calculations, but we must also realize that the overall goal is not to perform enormous numerical calculations, but to analyze the numerical calculations in a certain way so as to make progress. What has been lacking in studies of developed turbulence up until now is not so much the means of performing enormous calculations, but rather the way in which these calculations have been analyzed using somewhat too rudimentary concepts, which has not greatly contributed to understanding the statistical mechanisms that lie at the basis of the theory of turbulence. We are now coming out of this situation of analyzing problems of nonlinear evolution too crudely and we are beginning to use the relatively sophisticated methods of analysis which have been used for systems with a small number of degrees of freedom.

Christian Vidal: If we try to look to the future, as you have just done, we can reasonably expect to progress downstream, that is, to make our way from systems with a small number of degrees of freedom to systems with a larger number of degrees of freedom, before arriving at a more general description, bearing the elements of comprehension of fully developed turbulence. Furthermore, we must also endeavor to progress upstream, that is, to try to find something resembling a first, indeed even a second principle of the theory of dynamical systems to make an analogy with thermodynamics. A highly ambitious task, I grant. These general principles could mean attaining the minimal conditions required to observe the manifestation of this or that behavior, increasing tenfold the theory's power.
Construction is already underway on the first road, as we have just seen: it is the natural continuation of what has been done during the last decade or so. The other plunges much further towards the deepest roots of the analysis: if indeed there ever is any progress, it is highly probable that it will be extremely slow. Having taken stock of the prospects for advancement as we can apprehend them today, we must single out another element of great importance: it is that of the simultaneously conceptual, methodological, and pedagogical roles which this theory of dynamical systems deserves to occupy within what are called the exact sciences.

Yves Pomeau: We have signalled in the book the practical importance of the idea of sensitivity to initial conditions, and will return to it shortly. We must realize that introduction of sensitivity to initial conditions constitutes truly essential progress in the knowledge of the exact sciences. The analysis of many phenomena we have already alluded to (meteorology, biology, turbulence, economics) has been reconsidered using this concept. It can be said that this shows that an apparently exact mathematical formalism can imply the idea of statistical variation without the necessity of introducing fluctuations of external origin, and we know that, in spite of this statistical aspect, we are able to do something, to analyze the probability distributions of the fluctuations or the short-term evolution, to analyze the way in which chaos settles in via Lyapunov exponents or the way in which unpredictability appears. In a way, this is an aspect of the phenomena that we have studied, absent in the mathematics we have learned up until now, but which should appear more and more important for understanding the phenomena to which we have alluded.

This is a good time to underline the novelty of the methods used and put into practice. This novelty is practical only: while their use goes back ten or twenty years, the fundamental ideas had been launched by people like Poincaré and Hadamard in the beginning of the century. It so happens that their practical penetration into physics is very recent and they are effective, as we have been able to see, and also at the same time have a particularly rich educational aspect which it would be desirable to introduce into higher education. The other aspect is that of universality.

Pierre Bergé: I would like to complement what has been said by shedding light on the "virtues" of this domain, some of which at least have not passed undetected by the reader of our book. The first is the art of simplifying, ending up with an extreme simplification which nevertheless has some meaning. I take a specific example: Rayleigh-Bénard convection, even in a small container where only two convective rolls are present, is governed by the equations of hydrodynamics, the Navier-Stokes equations, which are in general perfectly insoluble. Without simplifications and approximations, the most powerful computers are incapable of providing us with numerical solutions. First approach: we try to make a model by drastically truncating the original equations; this has been done, we have seen, with the Lorenz model. This approach is already extremely rich; but we can simplify further by passing from differential equations to finite difference equations, ending up with two — and even one-dimensional mappings. By doing this we arrive at an extremely simple formulation, exploitable by particularly modest computational means, and we succeed in discovering, among other things, the route to turbulence via period-doubling cascades which is verified perfectly by experiment.

The second point I would like to bring out is the power of the idea of universality. We have simplified to the extreme and now the model is universal. What do we mean by this? We mean that is is not *only* the equation $f(x) = 4\mu x(1 - x)$ which describes a period-doubling cascade, but rather that *almost any* nonlinear function of x will do: the scenario is insensitive to the details of the model. Another aspect related to universality is that this period-doubling leading to chaos is found experimentally: in a parametric pendulum, in certain systems of electronis oscillators, in a synchronous motor launched in a particular way, in certain non monotonic chemical reactions, in the motion of a convecting fluid, etc.. There is truly a universality of behavior here that is quite reassuring and gives an enormous scope to the analysis made of it. Finally, among the general qualities which I am pleased to recognize in the study of dynamical systems, there is a very enriching collaboration between theoreticians and experimentalists and — even more remarkable — between mathematicians and physicists. In addition, it is a domain in which, with relatively modest material tools, but a good dose of imagination, one can contribute: it illustrates, as it were, the triumph of "light physics".

Christian Vidal: A closely related idea which cannot help but strike the physicist is the close relationship of methodology, even to the train of thought, which exists between the theory of dynamical systems and the techniques that have been applied for a certain number of years to the study of phase transitions. I am, of course, referring particularly to the renormalization group to which we can turn to determine the famous parameter $\delta = 4.669\,2...$ of the subharmonic cascade. It is clear that one day a global synthesis will have to be made, for there are too many formal similarities between the two for them not to be joined in a single, more vast, approach.

On a larger plane, it is undoubtedly appropriate to underline an element rich in epistemological consequences, to which we have merely alluded in the debate: this is the challenge to the meaning and to the scope of the ideas of determinism and chance, as we are accustomed to practicing them today. Obviously a serious revision of their scientific definitions is imperative, and we must now go beyond the ideas stated precisely for the first time by Laplace close to two centuries ago. Incidentally, it is also remarkable to note that the concept of strange attractor, while incontestably enlarging the recognized domain of determinism, simultaneously establishes on a much more solid basis than before, the necessity of recourse to statistical methods. Indeed, this is no longer justified merely by considerations of a pragmatic nature, but is imposed, as it were, by mathematical logic itself.

Yves Pomeau: Although it is not treated in the book, we can perhaps also mention the case of Hamiltonian systems which are important, on the one hand in celestial mechanics but, in a more immediate way, in the problems of fusion plasmas which are, as we know, among the dominant technological and scientific problems of our era. What different people have shown is that within the framework of Hamiltonian systems as can be seen in, for example, the motion of particles in inhomogeneous magnetic fields, there also exist laws of universality whose principles were discovered by methods related to the renormalization groups for critical phenomena. There too, a kind of conceptual regrouping has taken place; for example, in the case of Hamiltonian systems, a relationship has been demonstrated between one-dimensional crystals, problems of quantum mechanics in quasiperiodic potentials, and this problem of particle motion in the relatively complex magnetic fields of fusion machines. Here too, we can say that nonlinear physics, the physics which we have tried to talk about in this book, is sufficiently powerful to bring out universal phenomena. We can say that the blocks which existed in the conceptual problems related to nonlinearity are beginning to disappear little by little and we see emerging a king of new, unifying concept in nonlinear phenomena, a relatively unexpected development.

Christian Vidal: From the philosophical point of view, there is an extremely powerful idea that will undoubtedly take some time to penetrate and to permeate existing attitudes: it follows from this central concept of sensivity to initial conditions, to which is sometimes given the name of weakly stable dynamics. The fundamental idea is that through this concept, science is once again on the process of recognizing, identifying its own limits. Take a parallel: Heisenberg's uncertainty principle. According to this principle, the most sophisticated equipment that can be imagined will never allow us to measure simultaneously with arbitrary precision, both the position and the momentum of a particle. Here, somewhat analogously, the idea of sensivity to initial conditions forbids us from ever being able to predict the destiny of a dynamical system whose flow is on a strange attractor, no matter what we do. Certainly the parallel must not be carried too far, for there are multiple differences. The most obvious is that the theory of dynamical systems is a phenomenology of the macroscopic, whereas the smallness of $\hbar$, the quantum of action, limits the substantive consequences of the Heisenberg principle to the microscopic level. Nevertheless both cases are concerned with identification of the intrinsic limits of science and, *ipso facto*, of its power. This power exists, of course, but it is bounded in the most mathematical sense of the word. That science, as it progresses, thus reveals itself capable of finding some of the frontiers of its own sphere of activity is after all not so banal, nor unworthy of interest.

Appendices

APPENDIX A

Local bifurcations of codimension one

1 Local bifurcations

In Chapter II, bifurcations appear in a relatively anecdotal fashion, in the context of the forced and parametric oscillators. The idea is much more fundamental: it is important enough for a whole branch of mathematics, called *bifurcation theory*[1], to be devoted to it. We summarize the rudiments of the theory in this appendix.

Consider an autonomous flow in $\mathbb{R}^n$:

$$\dot{\vec{X}} = F_\mu(\vec{X}) \qquad \vec{X} \in \mathbb{R}^n.$$

This flow depends on a set of parameters, represented symbolically by the letter μ in the subscript of F ($\mu \in \mathbb{R}^k$ if there are k parameters). These could be Ra and Pr in a thermoconvection experiment; Pr, b, and r in the Lorenz model; the reactant concentration $[X]_0$, volume flux J, and temperature T in a chemical reaction; etc. The solutions to the system of algebraic equations:

$$F_\mu(\vec{X}) = 0$$

are fixed points of the flow, i.e., the singular points of the associated vector field, or the steady states of the physical system. The flow can of course have other solutions of a different nature, such as periodic solutions. In the parameter space, the existence of a given solution can be followed using a graph describing its dependence on μ: the solution as a function of μ is called a *solution branch*. A point in parameter space from which several branches emerge is called a *bifurcation point*.

Often, for reasons having to do with the methods of analysis available, we are led to limit our investigation to the immediate neighborhood of the bifurcation. The idea is to retain only the lowest order terms in the Taylor series in order to derive essential properties. This should not be considered too serious a handicap, even though the most visible consequence is to restrict, in principle, the domain of validity of the conclusions.

---

1. Bifurcation theory originated in the work carried out by Henri Poincaré in the beginning of this century on systems of differential equations. The word bifurcation itself was coined by Poincaré to designate the emergence of several solutions from a given solution.

This mathematically local description carries the name of *local bifurcations*, as opposed to *global bifurcations*, meaning that we concern ourselves with the vicinity of a given solution, and not with what may happen at a finite distance from it.

II Codimension

Although the dimension k of the parameter space depends on each problem under consideration, we wish to develop a general approach, free from restrictions arising from particular circumstances. This is why we define the *codimension of a bifurcation* to be the smallest dimension of parameter space in which such a bifurcation can occur. Let us illustrate this idea in a three-dimensional parameter space ($k = 3$). The locus of points in parameter space where some condition is satisfied, for instance change of stability of a solution, is generally a surface Ω (see fig. A.1 a).

Almost any line D crosses this surface: by moving along D (dimension one), we eventually intersect Ω and observe the corresponding bifurcation. Thus we say that the bifurcation is of codimension one. If we now set, not one, but two conditions, these will both be verified only along the line of intersection L of the two surfaces Ω_1 and Ω_2 associated with the two conditions. *A priori*, an arbitrary line has no chance of intersecting L: we must move along a surface (dimension two) to cross the line of bifurcations of codimension two (see fig. A.1 b).

What we have just said for $k = 3$ is easily generalized to any other value of k, simply by replacing the words: space, surface, etc., by hyperspace, hypersurface, etc. Thus the essential element, is not the value of the dimension k itself but that of the codimension, i.e. the minimum number of parameters that must be varied in order to observe a certain type of bifurcation. The study of bifurcations is thereby not only simplified, since it uses only the geometrical properties of the appropriate space, but its conclusions are also more generally applicable.

For reasons of simplicity, we will examine only local bifurcations of codimension one, referring the reader interested in bifurcations of codimension two and three to more specialized books and articles.

III Supercritical codimension-one bifurcations of fixed points

Mathematicians have proved that, under fairly general conditions, the local study of codimension-one bifurcations from a fixed point can be reduced to a few archetypes. More precisely, it is proved that there exist series expansions and changes of variable such that, near the fixed point, the behavior is described by a small number of differential equations depending only on one parameter denoted by μ. Intuitively this conclusion seems reasonable enough for bifurcations of codimension one . A more surprising result is that only one variable in phase space is also sufficient. After having

III SUPERCRITICAL CODIMENSION-ONE BIFURCATIONS OF FIXED POINTS

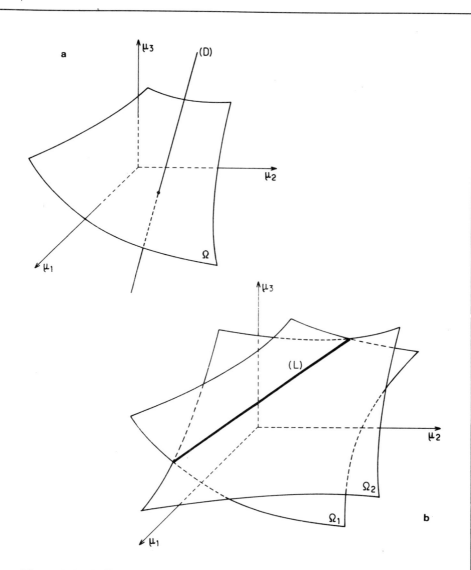

Figure A.1 Codimensions of bifurcations in a three-dimensional parameter space. *a)* A simple condition is generally represented by a surface Ω. An arbitrary line D intersects the surface (except under exceptional circumstances). It therefore suffices to move along D in order to encounter the bifurcation, said to be of codimension one. *b)* The two conditions represented by Ω_1 and Ω_2 are simultaneously satisfied only along the line of intersection L. Only a surface (not drawn in the figure) has a nonzero probability of intersecting L. This is why the bifurcations which occur on L are said to be of codimension two.

Table I Codimension-one bifurcations of fixed points

Normal form (supercritical)	Name of bifurcation
$\dot{x} = \mu - x^2$	Saddle-node (or turning point)
$\dot{x} = \mu x - x^2$	Transcritical (or with exchange of stability)
$\dot{x} = \mu x - x^3$	Pitchfork
$\dot{z} = (\mu + i\gamma)z - z|z|^2$	Hopf

x : real variable
z : complex variable
μ : distance from the bifurcation point
i : imaginary number
γ : arbitrary constant.

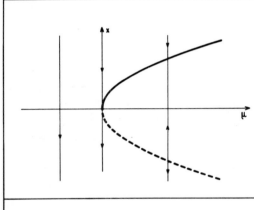

Figure A.2 Saddle-node bifurcation diagram.
The fixed point is located at the origin ($x = 0$, $\mu = 0$). Two branches of steady states emerge from the bifurcation point: one stable (heavy solid line), the other unstable (dashed line). The vertical lines represent the lines of force of the vector field, the arrows indicating the direction.

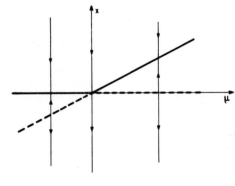

Figure A.3 Diagram of a transcritical bifurcation.

applied transformations that bring the fixed point to the origin, the equations become one of the classical *normal forms*. Limiting ourselves to third order, we will briefly summarize the four forms most often encountered, assembled in Table I. Once the typical cases have been compiled, described, and analyzed, it then suffices to decide which of them fits our problem, and then to avail ourselves of the complete (local) theory.

III.1 SADDLE-NODE BIFURCATION $\dot{x} = \mu - x^2$

The steady solution is $x = \pm\sqrt{\mu}$. It is defined only for $\mu > 0$, first appearing at $\mu = 0$. There exists no solution, stable or unstable, for $\mu < 0$. This leads to the bifurcation diagram of Figure A.2.

III.2 TRANSCRITICAL BIFURCATION: $\dot{x} = \mu x - x^2$

Two steady solutions $x = 0$ and $x = \mu$ coexist. It can be seen that the $x = 0$ solution is stable if $\mu < 0$ and unstable if $\mu > 0$, and vice versa for the $x = \mu$ solution: there is exchange of stability between the two solutions at the bifurcation point (see fig. A.3).

III.3 PITCHFORK BIFURCATION: $\dot{x} = \mu x - x^3$

The steady solutions are in this case $x = 0$ and $x = \pm\sqrt{\mu}$, the latter two defined only for $\mu > 0$. Note that this normal form is invariant under the transformation $x \longrightarrow (-x)$. Each time a problem is insensitive to a reflection symmetry (e.g. formation of R.B. rolls at the first convective instability) we must expect to encounter this type of normal form, whose bifurcation diagram is shown on Figure A.4.

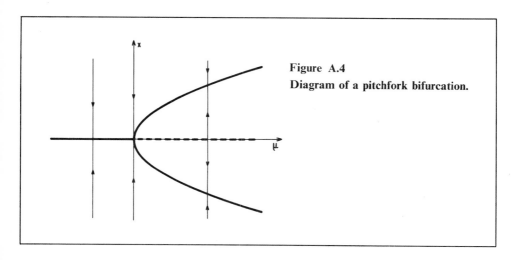

Figure A.4
Diagram of a pitchfork bifurcation.

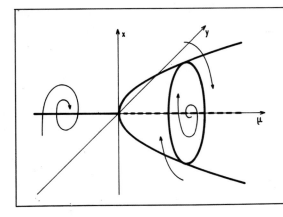

Figure A.5 Diagram of a Hopf bifurcation.
This diagram includes an additional dimension y. The arrows represent lines of force of the vector field.

III.4 HOPF BIFURCATION: $\dot{z} = (\mu + i\gamma)z - z|z|^2$

This time z is a complex variable, while γ is an arbitrary constant which does not play the role of bifurcation parameter. We note that this normal form is the complex equivalent of that of the pitchfork bifurcation. To find the solution it is convenient to transform to real variables using either Cartesian or polar coordinates. Setting $z = x + iy$, the normal form becomes:

$$\dot{x} = [\mu - (x^2 + y^2)]x - \gamma y$$
$$\dot{y} = \gamma x + [\mu - (x^2 + y^2)]y.$$

Apart from the solution $z = 0$ (i.e. $x = y = 0$), there exists another solution such that $|z|^2$ is independent of time:

$$|z|^2 = x^2 + y^2 = \mu.$$

This condition defines a circle in the (x, y) plane of radius $\sqrt{\mu}$ (see fig. A.5).

IV Subcritical bifurcations

The normal forms considered in the previous section are all *supercritical* (also called normal). By this, we mean that the nonlinear terms in x^2 or x^3 have an effect opposite to that of the instability caused by the term of the lower order. Take the form $\dot{x} = \mu x - x^3$. For x very small, we can retain only the linear term. The solution then obtained $x = \exp(\mu t)$ diverges at infinity when μ is positive. But the solution to the full equation does not diverge exponentially, for the linear term is counterbalanced by the term $(-x^3)$ which rapidly grows too large to be neglected. By putting the equation into the form:

$$\dot{x} = \mu x \left(1 - \frac{x^2}{\mu}\right)$$

we see that the nonlinear term "saturates" the effect of the linear instability at $x^2 = \mu$.

IV SUBCRITICAL BIFURCATIONS

However, nothing forbids the lowest order nonlinear term from also having a destabilizing influence on the solution. The bifurcation in this case is called *subcritical* or sometimes *inverse*. All of the normal forms listed above can be made subcritical merely by changing the sign of the nonlinear term. We then obtain the bifurcation diagrams presented in Figure A.6

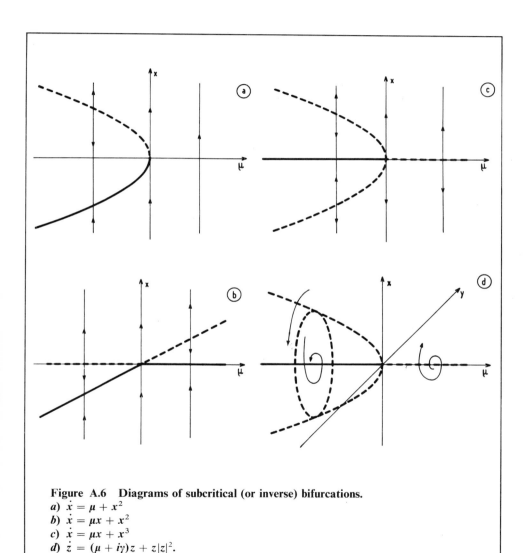

Figure A.6 Diagrams of subcritical (or inverse) bifurcations.
a) $\dot{x} = \mu + x^2$
b) $\dot{x} = \mu x + x^2$
c) $\dot{x} = \mu x + x^3$
d) $\dot{z} = (\mu + i\gamma)z + z|z|^2$.

V Codimension-one bifurcations of periodic orbits

It is natural to ask what kind of bifurcations can take place locally, not from a fixed point, but from a periodic orbit. The temptation to extend the results above is even stronger when we consider that the Poincaré section of a periodic orbit is a fixed point. The question then arises of how far we can push the analogy between a fixed point of a flow and a fixed point of a Poincaré section.

We foresee that the mathematical problem might be more arduous to solve insofar as it is necessary to integrate the equations along the orbit in order to know where a trajectory returns to the plane of section. Fortunately Floquet theory (see Section IV.3.1) circumvents this difficulty, allowing us to study the linear stability of the periodic orbit without having to follow the flow step by step. We can conclude that there exist three possibilities according to the way in which an eigenvalue crosses the unit circle in the complex plane.

If the crossing takes place at $+1$, the theory does not change very much from what has been explained for fixed points of flows. The normal form, topologically equivalent to that of a saddle-node bifurcation, is written:

$$\dot{x} = x + \mu - x^2$$

the term in x being introduced by the eigenvalue $+1$. Similarly there exist normal forms leading to pitchfork or transcritical bifurcations.

When the loss of linear stability results from an eigenvalue equal to -1, the situation has no analogue among the bifurcations of a fixed point of the flow. It is therefore necessary to develop an approach specific to this case, as is done in Chapter VIII. This time the bifurcation leads to the appearance of an orbit with twice the period via a subharmonic instability. The bifurcation diagram is identical to that of a pitchfork bifurcation (see fig. A.4), but the behavior of the solution is radically different. After a subharmonic bifurcation both branches are visited, the periodic solution alternating between them. On the other hand one, and only one, solution becomes established after the choice of one branch *or* the orther at a pitchfork bifurcation.

Finally, if the loss of stability is due to crossing of the unit circle by two complex conjugate eigenvalues, an event occurs which recalls and is in fact often called a Hopf bifurcation, despite several features which distinguish the two bifurcations. In the Poincaré section, the fixed point is replaced by a set of points on a curve, evoking a limit cycle. However, refinements to the analysis are necessary since this curve, which is not an orbit, is traversed in a very particular way. It is here that the winding number and its rational or irrational character become relevant. These questions are treated in Appendix C, devoted to quasiperiodicity and to frequency locking.

APPENDIX B

Lyapunov exponents

I Description

It becomes increasingly difficult to follow the evolution of a chaotic flow as the divergence of the trajectories on the attractor becomes more rapid. This is why we try to estimate or to measure the rate of divergence. The quantity used to characterize the divergence is the *Lyapunov exponent*, sometimes also called the *Lyapunov number*.

To demonstrate the significance of this quantity, let us examine the behavior of a trajectory initially close to a solution $\vec{\phi}(t)$ of a flow:

$$\frac{d\vec{\phi}}{dt} = F(\vec{\phi}).$$

Without loss of generality, we will limit ourselves to a three-dimensional flow $\vec{\phi}(t) = (X(t), Y(t), Z(t))$ and use the Lorenz model (see Appendix D) as an example. Linearizing the flow about $\phi(t)$, we obtain an equation for the evolution of the difference, denoted by $\vec{\delta\phi}(t)$:

$$\vec{\delta\phi} = \frac{\partial F}{\partial \phi}\bigg|_{\phi(t)} \vec{\delta\phi} \qquad (B.1)$$

$\partial F/\partial \phi$ is a matrix which depends both on the flow and on the particular solution considered. For a solution $X(t)$, $Y(t)$, $Z(t)$ of the Lorenz model, we have:

$$\frac{\partial F}{\partial \phi}\bigg|_{\phi(t)} = \begin{vmatrix} -\Pr & \Pr & 0 \\ -Z(t)+1 & -1 & -X(t) \\ Y(t) & X(t) & -b \end{vmatrix}.$$

Analytic integration of (B.1) is, except in special cases, impossible. However, it can always be integrated numerically, yielding a matrix $L(t)$ such that:

$$\vec{\delta\phi}(t) = L(t)\,\vec{\delta\phi}(0)$$

where $\vec{\delta\phi}(t)$ is the solution of (B.1) associated with an initial displacement $\vec{\delta\phi}(0)$. By virtue of the preceding relation, $L(0) = 1$. The matrix L is a square $m \times m$ matrix for a flow of dimension m, and has m eigenvalues.

This is very similar to the Floquet theory for the linear stability of a periodic trajectory. The procedure has the same inspiration: linear analysis of the behavior in the vicinity of a trajectory. But whereas in the Floquet theory we only look at what happens after one period of a closed orbit, here this restriction is lifted. While the eigenvalues of the Floquet matrix give information about the linear stability of a limit cycle, the eigenvalues of L give information about the evolution near a trajectory not constrained to close upon itself.

Analytic integration is nevertheless easy when $\partial F/\partial \phi$ does not depend on time[1] and its eigenvalues are the (possibly complex) numbers $\lambda_1, \lambda_2, \lambda_3$. We arrive at a matrix $L(t)$ which is diagonal in the coordinate system of its eigenvectors:

$$L(t) = \begin{vmatrix} \Lambda_1 & & 0 \\ & \Lambda_2 & \\ 0 & & \Lambda_3 \end{vmatrix} = \begin{vmatrix} e^{\lambda_1 t} & 0 & 0 \\ 0 & e^{\lambda_2 t} & 0 \\ 0 & 0 & e^{\lambda_3 t} \end{vmatrix}$$

Let us designate by L^+ the Hermitian conjugate matrix of L. The trace of L^+L is equal to:

$$\text{Tr } (L^+(t)L(t)) = e^{(\lambda_1 + \lambda_1^*)t} + e^{(\lambda_2 + \lambda_2^*)t} + e^{(\lambda_3 + \lambda_3^*)t}.$$

When t increases, the exponent with the largest real part, $\bar{\lambda}$, eventually dominates the other two terms, so that:

$$\bar{\lambda} = \lim_{t \to \infty} \frac{1}{2t} \ln [\text{Tr } (L^+(t)L(t))].$$

Using this expression, we are able to find the largest real part $\bar{\lambda}$ without having to use the usual methods for determination of eigenvalues. This is a non-negligible practical advantage, as $\bar{\lambda}$ dominates the behavior for long times. We therefore seek to extend it to more general situations.

Most of the time, the matrix $\partial F/\partial \phi$ depends on time and $L(t)$ cannot be put into the simple form above. There are two reasons for this: first, the eigenvalues of $\partial F/\partial \phi$ are not constant, and second, L is not diagonalizable in a fixed reference frame. We can, however, still define the quantity:

$$\lambda_{[\phi]} = \lim_{t \to \infty} \frac{1}{2t} \ln [\text{Tr } (L^+(t)L(t))]$$

because what are called multiplicative ergodic theorems establish the existence of this limit for a large category of situations. This limit is called the Lyapunov exponent (or

---

1. In a nonlinear flow, this will be the case only if the solution $\phi(t)$ itself is stationary, as we can see immediately from the equations of the Lorenz model. Although this does not correspond to the problem at hand of the evolution near a trajectory, examination of this hypothesis is nevertheless useful for what will follow.

Lyapunov number) associated with the solution $\vec{\phi}(t)$. It expresses the fact that, at long times, the difference $\partial \phi$ grows or decays exponentially on the average, according to whether the sign of $\lambda_{[\phi]}$ is positive or negative.

II Analysis of simple cases

Among the many possibilities, let us focus on three cases which are easier to understand in that $\partial F/\partial \phi$ — and thus L — is diagonalizable in a fixed coordinate frame.

When $\partial F/\partial \phi$ has constant eigenvalues, the eigenvalues of $L(t)$ are of the form $e^{\lambda_i t}$ with λ_i independent of t for any i, so $\lambda_{[\phi]}$ is indeed the largest real part[2] of the $\lambda_i's$.

Consider as a second case a matrix $\partial F/\partial \phi$ corresponding to multiplication by a scalar α fluctuating randomly in time (a "multiplicative noise" effect). Then:

$$\dot{\overrightarrow{\delta\phi}}(t) = \alpha(t)\, \overrightarrow{\delta\phi}(t)$$

$$\overrightarrow{\delta\phi}(t) = \exp\left[\int_0^t dt'\alpha(t')\right] \overrightarrow{\delta\phi}(0)$$

we see that the matrix $L(t)$ is reduced to a scalar:

$$\ln L(t) = \int_0^t dt'\alpha(t').$$

Let us consider the average value of this quantity:

$$\overline{\ln L(t)} = \int_0^t dt'\overline{\alpha(t')}.$$

If the statistical ensemble over which we calculate this average is a stationary (time-independent) ensemble, the ergodic theorem[3] allows us to replace $\alpha(t')$ by the (time-independent) ensemble average $\bar{\alpha}$. Therefore:

$$\overline{\ln L(t)} = \bar{\alpha} t$$

---

2. The matrix L has m eigenvalues or exponents, each one referring to behavior in one eigendirection. It is only the method of calculation which singles out the eigenvalue with the largest real part.

3. The ergodic theorem states essentially that if $x(t)$ is a time-dependent random variable with probability distribution $P(x)$, then the mean value of a function $f(x)$ over time, for a realization of $x(t)$, is independent of time and equal to:

$$\langle f \rangle = \int f(x)P(x)\, dx = \bar{f}.$$

and so[4]:
$$\lambda_{[\phi]} = \bar{\alpha}.$$

More generally, when $\partial F/\partial \phi$ and L are diagonalizable in a fixed reference frame, (B.1) takes the form:

$$\delta \dot{X}(t) = A[\phi(t)] \, \delta X$$
$$\delta \dot{Y}(t) = B[\phi(t)] \, \delta Y$$
$$\delta \dot{Z}(t) = C[\phi(t)] \, \delta Z$$

where A, B and C are the eigenvalues, (which we will assume to be real) of $\partial F/\partial \phi$ in the fixed reference frame. The solutions are:

$$\delta X(t) = \delta X(0) \exp\left[\int_0^t dt' A[\phi(t')]\right]$$

where we have taken X to be the direction of the largest average eigenvalue. Since we are interested in the behavior of δX at long times and not in its instantaneous values, it is the quantity:

$$\frac{1}{t} \ln\left|\frac{\delta X(t)}{\delta X(0)}\right| = \frac{1}{t} \int_0^t dt' A[\phi(t')]$$

which draws our attention. The ergodic theorem already mentioned tells us that if $\langle A \rangle$ is the average of $A[\phi]$ over the set of values of ϕ, then $\langle A \rangle$ is also the time average. Hence:

$$\lim_{t \to \infty} \frac{1}{t} \int_0^t dt' A[\phi(t')] = \langle A \rangle.$$

and therefore:

$$\langle A \rangle = \lim_{t \to \infty} \frac{1}{t} \ln\left|\frac{\delta X(t)}{\delta X(0)}\right| \qquad (B.2)$$

We are interested in the average value of the logarithm of $|\delta W(t)/\delta X(0)|$ rather than in the average value of the ratio itself, because the logarithm fluctuates much less, implying that its average value has greater physical significance. To illustrate this, we

---

4. More detailed calculation shows that if $\alpha(t)$ has no "memory", then:

$$(\ln L(t))^2 = \bar{\alpha}^2 t^2 + ct$$

for large times, where the coefficient c depends on the double correlation of $\alpha(t)$ for small times, i.e. on the quantity $\overline{(\alpha(t_1) - \bar{\alpha})(\alpha(t_2) - \bar{\alpha})}$. The correlation in turn depends only on the difference $|t_1 - t_2|$ if α is a stationary random process as indicated above. It tends to zero as $|t_1 - t_2|$ becomes large, because $\alpha(t_1)$ and $\alpha(t_2)$ are then independent variables and $\overline{(\alpha(t) - \bar{\alpha})} = 0$. Therefore at large times the average $\bar{\alpha}t$ of the function $\ln L(t)$ is statistically defined up to fluctuations on the order of $t^{-1/2}$.

II ANALYSIS OF SIMPLE CASES

divide the integration interval into many intervals of length τ, τ representing the typical correlation range of $A[\phi(t')]$. Let N be the number of these intervals. Then:

$$\int_0^t dt' A[\phi(t')] \simeq \tau \sum_{i=1}^N A_i$$

where A_i is the average value of A over one of the small intervals. The A_i's of different indices are statistically independent since the length of the interval is on the order of the correlation range. To this decomposition of the exponent into a sum corresponds the decomposition of the exponential into a product:

$$E(t) = \exp \int_0^t dt' A[\phi(t')] \simeq \prod_{i=1}^N \beta_i \qquad (B.3)$$

where the $\beta_i (= \exp(\tau A_i))$ form a sequence of N independent factors.

Suppose now that the β_i's are equal to 1/2 or to 1 with equal probability. Let us calculate the mean value and the mean square value of $E(t)$:

$$\langle E(t) \rangle = \frac{1}{2^N} \left(\frac{1}{2} + 1 \right)^N = \left(\frac{3}{4} \right)^N$$

$$\langle E^2(t) \rangle = \frac{1}{2^N} \left(\frac{1}{4} + 1 \right)^N = \left(\frac{5}{8} \right)^N$$

The ratio of the mean square value to the square of the mean value is:

$$\frac{\langle E^2(t) \rangle}{\langle E(t) \rangle^2} = \left(\frac{10}{9} \right)^N.$$

When N becomes very large, this ratio also becomes very large. Therefore a few very large values of E suffice for the mean square $\langle E^2 \rangle$ to be much larger[5] than the squared

---

5. To show that this is plausible, we present a simple numerical example. Consider a random variable X equal to 1 with probability 99.9 % and to 1 000 with probability 0.1 %. The mean value of X is $0.999 \times 1 + 0.001 \times 1\,000 \sim 2$, and its mean square is equal to $0.999 \times 1 + 0.001 \times 10^6 \sim 10^3$. This indeed shows that the mean square value of X is much more sensitive to the exceptional presence of a large fluctuation than is the mean value itself. The example also shows that for a highly fluctuating variable, the mean value lacks significance, since to say that the average of X is equal to 2 says little about the distribution of X. It would be much more meaningful to say that the most probable value of X is 1. We can see that this most probable value can be obtained from the mean value of the logarithm:

$$\langle \log_{10} X \rangle = 0.999 \cdot 0 + 0.001 \cdot 3 \sim 0.003$$

Hence:

$$10^{\langle \log_{10} X \rangle} \cong 1$$

which is the result we wanted. This is explained qualitatively by the fact that the logarithm is a very slowly increasing function of its argument. Therefore, large fluctuations of X distort the statistics of $\ln |X|$ much less than those of $|X|$, and *a fortiori* those of X^2.

mean value $\langle E \rangle^2$. This means that the probability distribution of $E(t)$ becomes very dispersed as N increases: we have shown that $E(t)$ defined in (B.3) fluctuates a great deal for N large. As in the example given in the footnote, we can thus expect the mean value of the logarithm of E to be much more "reasonable". And this is indeed what happens: the exponential of the average logarithm of $E(t)$ does give the most probable value of $E(t)$. But we must nonetheless keep in mind that as time elapses, the ratio $|\delta X(t)/\delta X(0)|$ fluctuates enormously about this most probable value. In particular it can always be multiplied by a factor β_i which will divide (or multiply) it by 10 when the time has increased by τ. If we have:

$$\langle \ln E(t) \rangle \underset{t \to \infty}{\simeq} t\lambda$$

we cannot then infer that:

$$\langle E(t) \rangle \underset{t \to \infty}{\simeq} e^{\lambda t}$$

since the ratio $\langle E(t) \rangle / e^{\lambda t}$ has at least finite fluctuations and therefore does not tend to a constant limit, as this last relation would imply.

III Methods of determination

The defining relation (B.2) has great practical importance since it is the basis of certain algorithms for determining $\lambda_{[\phi]}$ from experimented measurements of $\delta X(t)$. The principle is as follows: imagine that we have reduced the continuous experimental dynamics to a first return map $x \longmapsto f(x)$, also experimental, in a finite dimensional phase space ($\mathbb{R}$ or $\mathbb{R}^2$ in practice). From this experimental mapping, we can, by some numerical procedure[6], go to an analytic representation of f. This done, it remains only to calculate the Lyapunov number of the attractor produced by the approximate analytic mapping f. To do this we use the definition directly. We choose a sequence of N iterates $(x_1, x_2, ..., x_N)$ and calculate the product of the Jacobian matrices:

$$J_N = \prod_{i=1}^{N} \frac{\partial f}{\partial \vec{x}}\bigg|_{\vec{x}_i}.$$

---

6. We know (Figures VI.24 and VI.25) that the Poincaré sections of some attractors resemble the Hénon attractor. As the Hénon attractor comes from a transformation depending explicitly on the two real numbers α and β, we seek to determine their values in the experimental situation. This can be done, for example, by a method of least squares, i.e. by minimizing the sum $\Sigma |\vec{X}_{i+1} - F_{\alpha,\beta}(\vec{X}_i)|^2$ over α and β, where $F_{\alpha,\beta}$ is the Hénon transformation and $\vec{X}_i$ the ordered sequence of experimental points.

III METHODS OF DETERMINATION

Since with each iteration, the vector of fluctuations is multiplied (in the linear approximation) by the Jacobian matrix $\left.\frac{\partial \vec{f}}{\partial \vec{x}}\right|_{\bar{x}_i}$, after N iterations we have:

$$\vec{\delta x}_N = \left.\frac{\partial \vec{f}}{\partial \vec{x}}\right|_{\bar{x}_N} \cdots \left.\frac{\partial \vec{f}}{\partial \vec{x}}\right|_{\bar{x}_1} \vec{\delta x}_1 = J_N \vec{\delta x}_1.$$

The temporal average (corresponding to a continuous time) has been replaced by an average over the number N of iterations and the sought-after Lyapunov exponent is then:

$$\lambda = \lim_{N \to \infty} \frac{1}{2N} \ln [\mathrm{Tr}\, (J_N^+ J_N)]$$

J_N replacing $L(t)$. This method clearly works best when f can be reduced to a mapping from the line into itself. Note also λ is dimensionless, while the quantity $\langle A \rangle$ considered earlier has the dimension of frequency (the inverse of time). To go from $\langle A \rangle$ to λ it suffices to multiply $\langle A \rangle$ by the average time between two intersections of the trajectory with the surface of section: λ is the rate of divergence of the trajectory per "average period" (and therefore dimensionless), while $\langle A \rangle$ is the rate or divergence per unit time.

In the numerical simulation of nonlinear flows, determination of the largest Lyapunov exponent is straightforward, given access to adequate computational means. Starting from a previously calculated solution $\phi(t)$, numerical integration of (B.1) leads to the matrix $L(t)$ from which we deduce $\lambda_{[\phi]}$ by applying the definition. We emphasize that according to ergodic theorems, this requires no assumption on the diagonalization of the matrix $\partial F/\partial \phi$. This has been done for the Lorenz model with $\mathrm{Pr} = 10$, $b = 8/3$, and r close to the intermittency threshold r_i (see Chapter IX). We see in this case (fig. IX.15) that the largest Lyapunov exponent increases like the square root of $r - r_i$ (this confirms one of the predictions of the theory of transition via intermittency). Another illustration is provided by the Rössler model, introduced in Chapter IV (Equation 11). For fixed values of the parameters a and b ($a = b = 0.2$), we vary the third parameter c and integrate the flow numerically. We obtain a number of regimes, some periodic and some chaotic. On Figure B.1, a graph of the largest nonzero exponent as a function of c, we see that $\bar{\lambda}$ takes on positive values, implying divergence of trajectories on the attractor.

When, on the other hand, we are confronted with a solution $\vec{\phi}(t)$ that is obtained experimentally, calculating the Lyapunov exponent is a much more delicate task, due to uncertainty in the measurements and inevitable experimental noise. If we assume that $\vec{\phi}(t)$ fluctuates over a certain statistical ensemble, it is theoretically possible to find an interval of time $[t_1, t_2]$ on which the realization (i.e. the particular sequence of values of $\vec{\phi}$) of $\vec{\phi}(t)$ is arbitrarily close to a reference realization given on a time interval $[t'_1, t'_2]$ of the same length. We do this to satisfy the condition of initial proximity of the

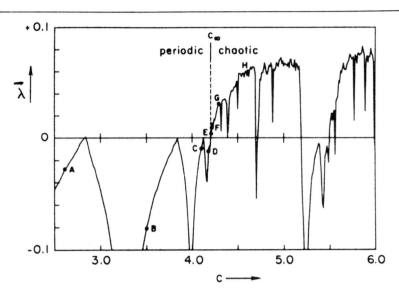

Figure B.1 The largest nonzero characteristic exponent λ of the Rössler model as a function of the parameter c (Equation 11 of the first part of the book).
Positive values of $\bar{\lambda}$ are the signature of divergence of initially neighboring trajectories on the attractor (S.I.C.). The continuous curve passes through three hundred individually calculated points ($a = b = 0.2$, c varies over the interval $[2.5,6]$).
From J. Crutchfield, D. Farmer, N. Packard, R. Shaw, G. Jones, R. J. Donnelly, Physics Letters 76A, p. 1 (1980).

trajectories. By comparing the behavior beyond t_2 and t'_2 we can form conclusions on the evolution of the difference between neighborhing trajectories. The concrete difficulty of this approach comes from the fact that the Lyapunov exponent reflects an average effect: it is therefore necessary to make numerous comparisons to deduce a significant average[7].

IV Characterization of an attractor

For a chaotic solution $\vec{\phi}(t)$ associated with a strange attractor, S.I.C. implies the existence of a positive Lyapunov exponent $\lambda_{[\phi]}$. This is a direct consequence of the average divergence of neighboring trajectories. Therefore, finding a positive Lyapunov

---

7. Several algorithms have been proposed for determining the largest Lyapunov exponent from an experimental time series. In view of the lack of progress, it would seem that a satisfactory result has not yet been formulated. This is why we do not attempt to present these efforts.

IV CHARACTERIZATION OF AN ATTRACTOR

Table

Type of attractor	Sign of exponents
Fixed point	$(-, -, -)$
Limit cycle	$(0, -, -)$
Torus T^2	$(0, 0, -)$
Strange attractor	$(+, 0, -)$

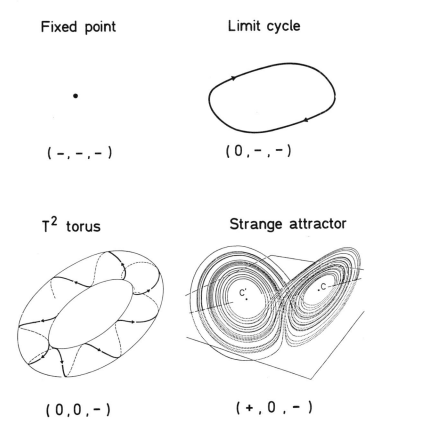

Figure B.2 Signs of Lyapunov exponents of different attractor types in a three-dimensional phase space.
We recognize attractors described elsewhere in the book, notably the Van der Pol limit cycle ($\varepsilon = 0.4$) and the Lorenz attractor ($r = 28$).

exponent is an unambiguous signature of a chaotic regime. Calculating the Lyapunov exponent of a periodic or of a quasiperiodic solution from the definition, we find $\lambda_{[\phi]} = 0$. As soon as the solution is stable, any displacement "perpendicular" to the trajectory decreases as time elapses and the corresponding eigenvalues of the matrix L are therefore always less than one. On the other hand, a displacement along the trajectory itself is neither damped nor amplified: it is equivalent merely to a shift in the initial point x_0. The invariance of such a displacement over time is expressed by the existence of a constant eigenvalue Λ equal to one, which is in this case greater than all of the others. It follows that the largest of the exponents λ_i is then equal to zero.

Extending this simple reasoning, it seems natural to consider identifying an attractor by the sign of the exponents λ_i. For a three-dimensional flow, this leads to the results of the accompanying table, illustrated by Figure B.2.

APPENDIX C

Synchronization of oscillators

I Generalities

The problem of synchronization of coupled nonlinear oscillators requires difficult and profound mathematical theories and has attracted the attention of great minds since Poincaré. It can now be considered as solved, at least from the physicist's point of view. It is out of the question here to make an exhaustive analysis of this immense body of knowledge, which requires difficult concepts from analysis, topology, and number theory. We therefore content ourselves with presenting the most important practical results of the theory.

Broadly speaking, we can say that there are two components essential to the synchronization of oscillators.

a) First, we must have two more or less independent oscillators described by, for example, two Van der Pol equations with different values for the parameter ε, and which can be coupled (later we will give concrete examples of what we mean by coupling).

b) Second, we must have dissipation. Otherwise we would have a Hamiltonian system, for which the coupling of oscillators is described, at least in part, by what is called KAM[1] theory.

The most well-known example of synchronization is the motion of the earth and moon. The frequency of rotation of the moon about its axis is synchronized with its period of revolution about the earth, so that the same face of the moon is always presented to us. Mathematically, this is the simplest kind of frequency locking, 1:1, since the two oscillators have the same frequency. The physical interpretation of the two oscillators is very simple for lunar motion: one is solid body rotation at fixed frequency about an axis of inertia, and the other is revolution on a Keplerian orbit. The coupling mechanism between these two oscillators is much less obvious, and has two origins. First, the distribution of mass in the moon has a certain asymmetry (the moon is not spherically symmetric). Second, this asymmetry is coupled, by the Newtonian gravitational force, to an analogous asymmetry produced on earth by the propagation

---

1. The theory is named after Kolmogoroff, Arnol'd and Moser. But it cannot be said that the whole problem is presently well understood.

of tides. Despite being of weak intensity, the coupling has very important consequences, for its effects accumulate over a considerable length of time (a situation frequently encountered in astronomy). Damping is provided by the friction exerted by the earth on the tides. This frictional force can be estimated by the delay between the tide and the lunar attraction giving rise to it. Nevertheless, we must remark that a quantitative theory of tides remains to be formulated.

Many other examples of synchronization can be cited. We can understand that, between oscillators whose frequencies are very close, a very weak interaction will suffice to synchronize them. For example, it is said that simply by placing two mechanical watches of the same model next to one another, the mechanical transmission of the tick-tock from one to the other ensures their synchronization. Biological examples also come to mind — circadian rhythms synchronized with the alternation between night and day — or other more curious ones, like the synchronization of the menstrual cycles of women within a community.

Numerous examples are provided by the many mechanical and electrical oscillating devices, which surround us. Thus, the subharmonic instability of a parametric oscillator implies an arithmetical relation between its intrinsic frequency and that of the external excitation (see Chapter II). A synchronization process of one kind or another is often the basis of the most precise methods for measurement or detection: heterodyne in radio, vernier in mechanics, moiré in optics. The word "synchronization" itself has passed into the public domain, with the meaning of simultaneity of distant events rather than accordance of close frequencies.

II Analysis of the problem

The theoretical approach to the phenomenon of synchronization that we are going to develop rests mainly on the study of a Poincaré mapping on a circle. Let us briefly recall the connection between synchronization and diffeomorphisms on the circle.

a) The Floquet theory of stability of periodic trajectories shows that a Hopf bifurcation occurs when two complex conjugate eigenvalues of the stability matrix cross the unit circle in the complex plane as the control parameter is varied. The Hopf bifurcation is supercritical when weakly nonlinear effects stabilize the trajectory in the vicinity of the former periodic trajectory, which remains, but is linearly unstable. In general, it r is the diameter of the small circle of the torus and ε the control parameter ($\varepsilon = 0$ at the bifurcation; when $\varepsilon < 0$, the limit cycle along the large circle is linearly stable; when $\varepsilon > 0$, it is linearly unstable and therefore no longer a limit cycle), we have the power law $r \simeq \varepsilon^{1/2}$, if the parametrization by ε is generic[2].

---

2. This does nonetheless exclude some special cases studied by Arnol'd, which occur when for $\varepsilon = 0$, the eigenvalues of the Floquet matrix are simple roots of unity, such as $e^{\pm 2\pi i/3}$ and $\pm i$.

II ANALYSIS OF THE PROBLEM

The conclusion from bifurcation theory which we will use is that, at least in the neighborhood of a supercritical Hopf bifurcation, flows exist for which the attractor is a torus T^2.

b) The flow on T^2 (fig. C.1) defines a Poincaré mapping (or diffeomorphism) with the following properties:

i) It is continuous and differentiable.

ii) It maps a circle T^1 (or a one-dimensional sphere S^1) onto itself.

It is convenient to parametrize the curvilinear coordinate along the circle by "unfolding" it on the unit interval [0, 1], with the understanding that 0 and 1 represent the same point of the circle. Now the diffeomorphism can be represented by a mapping $x \longmapsto f(x)$ of [0, 1] onto itself, such that $f(x) \in [0, 1]$ $f(0) = f(1)$ and the inverse $f^{-1}(x)$ is unique. The last property results from the fact that the flows considered are uniquely reversible. The graph of f in the Cartesian plane has an apparent discontinuity (fig. C.2) resulting from the impossibility of representing a circular coordinate on a planar graph. The condition for continuity of the diffeomosphism on the circle is that, if $f(x_0 + 0) = 1$, then $f(x_0 - 0) = 0$ and $f(0) = f(1) = x_1$ as indicates Figure C.2a. Inversibility is satisfied if $(df/dx) > 0$ everywhere (or possibly $(df/dx) < 0$, but by considering $f \circ f$ instead of f, we are brought back to the case $(df/dx) > 0$).

With this as a starting point we will now study the properties of diffeomorphisms on the circle, like the one in Figure C.2, representing the first return map of a flow on a torus (fig. C.1). One essential element of the analysis is the winding number of f, designated by $\rho_{[f]}$, another idea due to Poincaré.

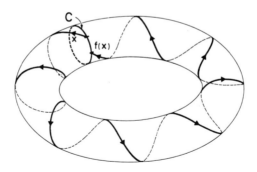

Figure C.1 Poincaré section of a flow on a torus T^2.
This shows how to construct from a flow on a torus the corresponding Poincaré mapping for a circle C on the torus. The trajectory originating from x on C reintersects C at $f(x)$.

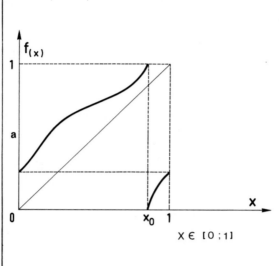
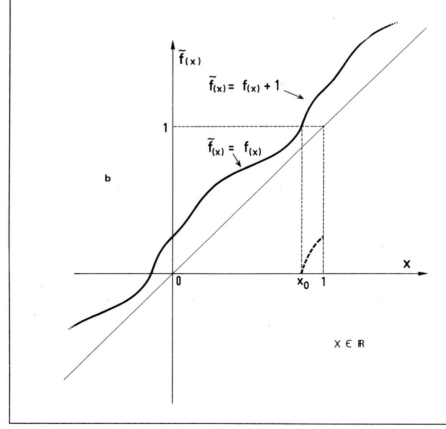

Figure C.2 Cartesian representation of two continuous, invertible mappings.
a) The mapping f of the circle onto itself describes the flow on the torus by means of a Poincaré section (see fig. C.1). The apparent discontinuities at 0,1, and x_0 disappear when we recall that, on both the abscissa and the ordinate, 0 and 1 represent the same point on the cut an unfolded circle.
b) A second mapping $\tilde{f}(x)$ on $\mathbb{R}$ is constructed from f, which can then be used to define the winding number $\rho_{[f]}$ on the torus.

III Winding number

To understand what the winding number means, consider first the simplest possible function f satisfying the conditions imposed on a first return map: a rotation on the circle by a constant angle α written as:

$$x \longmapsto \{x + \alpha\}$$

with the symbolic notation:

$$\{y\} = \text{fractional part of } y.$$

Here α is an arbitrary positive real, less than or equal to one. From this mapping we can then define another one, which is continuous and maps $\mathbb{R}$ into itself, by the simple translation:

$$x \longmapsto x + \alpha \quad x \in \mathbb{R}.$$

We see that for $0 < x < 1$ this second mapping yields either $\{x + \alpha\} + 1$ or $\{x + \alpha\}$. If we iterate this mapping n times, we obtain:

$$x^{(n)} = x + n\alpha$$

from which can be deduced the value of α by:

$$\alpha = \lim_{n \to \infty} \frac{x^{(n)}}{n}.$$

By an analogous construction we can define the angle of rotation of any diffeomorphism on S^1 represented by $x \longrightarrow f(x)$. The first stage consists of constructing from f a diffeomorphism $\tilde{f}$ on $\mathbb{R}$ that is to f as $x + \alpha$ is to $\{x + \alpha\}$. To do this, let x_0 be the point in $[0, 1]$ such that $f(x_0) = 1$. We simply define $\tilde{f}(x)$ by (see fig. C.2b):

* $0 < x \leqslant x_0 \quad \tilde{f}(x) = f(x)$
* $x_0 < x < 1 \quad \tilde{f}(x) = f(x) + 1$

* for all other values of $x \in \mathbb{R}$, $\tilde{f}(x)$ is completely determin by the condition of periodicity $\tilde{f}(x + n) = \tilde{f}(x) + n$, n integer.

It is easily verified that if $f(x) = \{x + \alpha\}$, we have $\tilde{f}(x) = x + \alpha$ and that, in general, $\tilde{f}(x)$ inherits the properties of continuity and invertibility of $f(x)$. The winding number is defined by:

$$\rho_{[f]} = \lim_{n \to \infty} \frac{\tilde{f}^{(n)}(x)}{n} \tag{C.1}$$

where $\tilde{f}^{(n)}(x)$ is the nth iterate of $\tilde{f}$.

The winding number is independent of the point x of departure. We can see this by the fact that, because f is monotonically increasing, $\tilde{f}(x)$ and therefore $\tilde{f}^{(n)}(x)$ vary at most by one as x varies between 0 and 1. Therefore the "residue" due to the choice of x in the definition of $\rho_{[f]}$ contributes only as $1/n$ in the right-hand side of (C.1), disappearing in the limit of large n.

The importance of the winding number comes primarily from the fact that $\rho_{[f]}$ is what is called *invariant under conjugation;* in a certain sense, it completely describes the diffeomorphism f (except if $\rho_{[f]}$ has "exceptional" arithmetic properties). Under conjugation we mean a change of parametrization (or coordinate) of the circle. Let $x \longrightarrow C(x)$ be a coordinate transformation, C being an invertible mapping of $[0, 1]$ onto itself. Under the coordinate transformation C, the diffeomorphism f becomes:

$$f_C \equiv C^{-1} \circ f \circ C$$

where $\circ$ designates composition of functions. We can construct from C a coordinate transformation $\tilde{C}$ on $\mathbb{R}$ such that:

$$\tilde{f}_C = \tilde{C}^{-1} \circ \tilde{f} \circ \tilde{C}$$

where the operation $\sim$ extending a diffeomorphism on S^1 to a diffeomorphism on $\mathbb{R}$ is defined above.

It is clear that:

$$\tilde{f}_C^{(2)} = \tilde{C}^{-1} \circ \tilde{f} \circ \tilde{C} \circ \tilde{C}^{-1} \circ \tilde{f} \circ \tilde{C}.$$

and since the composition of a function and its inverse is the identity:

$$\tilde{C} \circ \tilde{C}^{-1} = 1$$

we get:

$$\tilde{f}_C^{(2)} = \tilde{C}^{-1} \circ \tilde{f}^{(2)} \circ \tilde{C}$$

and more generally:

$$\tilde{f}_C^{(n)} = \tilde{C}^{-1} \circ \tilde{f}^{(n)} \circ \tilde{C}.$$

Applying the definition of winding number, it can now be shown that:

$$\rho_{[f_C]} = \rho_{[f]}$$

proving that ρ is an invariant under conjugation — that it is not modified by an arbitrary coordinate transformation $x \longmapsto C(x)$.

The importance of the winding number in fact results from the converse of this property: if two diffeomorphisms on S^1 have the same *irrational* winding number, then there exists a coordinate transformation $C(x)$ relating them by conjugation. The proof of this result is difficult, involving in a nontrivial way the algebraic properties of $\rho_{[f]}$; we therefore do not include it.

Hence, we see a distinction drawn between two types of situations, according to whether $\rho_{[f]}$ is rational (no conjugation in general) or irrational (conjugation). Yet this conclusion is often considered to be "non-physical" since it introduces a crucial link

IV Rational winding number. Frequency locking

Looking ahead, we can say that the situation where $\rho_{[f]}$ is rational is important because it corresponds to frequency locking, the two frequencies being those of rotation about the large and small circles, respectively, of the torus (see fig. C.1). Let $\rho_{[f]} = \{p/q\}$ (p, q integers). First of all, we can restrict consideration to the case $\rho_{[f]} = 0$. This is because, using the definition of winding number, we can show that:

$$\rho_{[f^{(k)}]} = \{k\rho_{[f]}\}$$

so that:

$$\rho_{[f^{(q)}]} = \left\{q\frac{p}{q}\right\} = \{p\} = 0.$$

Now we show that if the winding number is zero, then the equation $f(x) = x$ has at least one root in $[0, 1]$. If this were not so, since $[0, 1]$ is compact, this would imply the existence of a lower bound ξ for $|f(x) - x|$, and an upper bound ξ', such that $0 < \xi < |f(x) - x| < \xi' < 1$. The distance between the graph of $f(x)$ and the graph of the identity (or the graphs of any of the functions $x + n$) would then always be greater than min $(\xi, 1 - \xi')$. We deduce that:

$$0 < \inf(\xi, 1 - \xi') < \rho_{[f]} < 1 - \inf(\xi, 1 - \xi') < 1$$

contradicting the hypothesis that $\rho_{[f]} = 0$. Hence $f(x) = x$ has at least one root in S^1. We can show something even stronger: for a *generic* choice of f such that $\rho_{[f]} = 0$, the mapping $x \longmapsto f(x)$ has at least two fixed points, one stable, the other unstable. This property is best demonstrated graphically. Consider the case in which $f(x) - x$ has a root between 0 and the point x_0 defined by $f(x_0 - 0) = 1$, and where $f(0) = x_1 > 0$ (as in fig. C.3). Then the function $f(x) - x$ is positive between 0 and x_a, the smallest root of $f(x) = x$, and is 0 at x_a. If $(df/dx)|_{x=x_a}$ is not equal to 1, the graph of $f(x)$ is not tangent to the identity map at x_a and, for x slightly greater than x_a, the function $(f(x) - x)$ is negative. Since $(f(x) - x)$ is again positive at $x = x_0$, there certainly exists a second root x_b of $(f(x) - x)$ between x_a and x_0. This is a generic situation in the sense that we have only assumed that $(df/dx)|_{x=x_a} \neq 1$. An extension of this argument shows that the roots of $(f(x) - x)$ occur in pairs, provided that we count twice any root at which $f(x)$ is tangent to the identity map.

Let us examine what happens when the mapping $x \longmapsto f(x)$ is iterated. The roots of $(f(x) - x)$ are the fixed points of the iteration. Consider the simplest case where

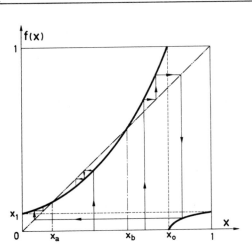

Figure C.3 Iteration of a mapping on the circle $x \longmapsto f(x)$. There are two fixed points: x_a (stable) and x_b (unstable). The iterates of all initial conditions in $[0, 1]$ converge to x_a, except for x_b.

$(f(x) - x)$ has only two roots x_a and x_b in $[0, 1]$, such that $0 < x_a < x_b < x_0$ (fig. C.3). What is the result of iterating $x \longmapsto f(x)$? Since $0 < (df/dx)|_{x=x_a} < 1$, x_a is a stable fixed point of the map and since $(df/dx)|_{x=x_b} > 1$, x_b is an unstable fixed point. Applying the graphical methods presented in Chapter IV, we see that for all initial points, the iterates of f converge to x_a, except if we start exactly from the unstable fixed point x_b.

If x_a and x_b are in fact the same tangent point of $f(x)$ with the identity map, then the iterates of f always converge towards this point. For this case, the final phase of convergence to the fixed point is much slower than if $(df/dx)|_{x=x_a} < 1$.

Recall that we have assumed that the winding number was zero, either:
— because the ratio p/q was integer, so that by definition:
$\rho_{[f]} = \{p/q\} = 0$, or:
— because we were studying the qth (rather than the first) return map $f^{(q)}$ using the modulo convention introduced in the definition of the winding number[3].

3. This case corresponds to 1:1 frequency locking (like the rotation of the moon about its axis and around the earth). The period of rotation around the small circle is a multiple of the period of rotation around the large circle. In reducing the flow on the torus to a diffeomorphism on the circle, we can no longer tell if, between two intersections with the small circle, the trajectory has made one, two, or more complete turns. Consequently, from the point of view of the diffeomorphism on S^1, the ratios between two periods of motion are defined only up to an integer.

When p and q are integers with no common divisor, we already know that $x \longmapsto f^{(q)}(x)$ has at least one pair of fixed points since $\rho_{[f^{(q)}]} = 0$. But what about the mapping f itself? Let x be one of the fixed points of $f^{(q)}$. The exponent q is the smallest number q such that $f^{(q)}(x) = x$. For suppose there were to exist an integer $q^* < q$ with $f^{(q*)}(x) = x$. Then, the winding number of $f^{(q*)}$ would be zero, and that of f would be of the form p^*/q^*. This contradicts the hypothesis that p/q is the irreducible form of the rational.

Moreover, the q successive iterates of X, that is $f(X)$, ..., $f^{(q)}(X)$, form a sequence of points $(X_1, X_2, ..., X_q)$ of S^1, called a q-cycle such that:

$$X_j = f(X_{j-1}); \quad X_j \neq X_k \quad 1 < j < k \leq q$$

where j is taken modulo q, so that $X_1 = f(X_q)$. Since $f^{(q)}$ has at least two fixed points, there exist at least two such sequences (of period q), one of which is unstable. In the marginal case where $f^{(q)}$ is tangent to the identity map at X_1 (and thus also at $X_2, X_3, ... X_q$, since $(d/dx)f^{(q)}$ has the same value at each point of a sequence), the two sequences are identical. Apart from this case, there is always one point of the unstable sequence between any two points of the stable sequence, and vice versa. Then the stable sequence attracts all initial conditions except for the points of unstable sequence. The argument can be generalized to show that if there are other stable or unstable sequences, they are also of period q.

When the Poincaré map $x \longmapsto f(x)$ possesses a stable q-cycle, the corresponding continuous flow has a limit cycle. The stable trajectory of the system intersects itself after q turns, and is therefore a simple closed curve. The Fourier spectrum consists of lines at frequencies that are multiples of some fundamental frequency. We say that there is *frequency locking* because the two fundamental frequencies, associated with motion about the large and small circles of the torus, are rationally related. Nevertheless, if q is very large, the fundamental period is very long and, since the distance between consecutive lines of the Fourier spectrum is of order $1/q$, analysis of the spectrum can prove difficult, even impossible, given the inevitable experimental noise tending to widen the lines (see Chapter III).

V Irrational winding number. Quasiperiodic behavior

The presentation here is necessarily more succinct because of the difficulty of the mathematics required to analyze the case of irrational winding numbers. We have already said that, when their winding numbers are equal *and* irrational, two diffeomorphisms f and f' of S^1 are conjugate, i.e. related by a change of variable. This allows us to consider only the simple rotation $x \longmapsto \{x + \rho_{[f]}\}$. There are many results from number theory concerning irrational rotations. One of the essential results is that if α is irrational, then the successive images $x, \{x + \alpha\}, \{x + 2\alpha\}, ..., \{x + n\alpha\}$

eventually cover the circle continuously and uniformly. The corresponding trajectory will end up covering the torus without ever intersecting itself.

In the Fourier analysis of the continuous time sequence, we will find two distinct angular frequencies ω_1 and ω_2, corresponding to the periods of rotation along the large and small circles, respectively, of the torus. Since the ratio of these two angular frequencies is irrational, the Fourier spectrum of the corresponding quasiperiodic signal exhibits — *a priori*, since the oscillators are not harmonic — a dense infinity of rays at all frequencies $|n_1\omega_1 \pm n_2\omega_2|$ with n_1, n_2 integers (see Chapter III). Things are somewhat more complicated, rigorously speaking, if ω_1/ω_2 is an irrational "very close" to a rational. This situation, while delightful to mathematicians, is probably not very important in physics.

VI Structural stability and frequency locking. Devil's staircase

Up until now, we have contented ourselves with analyzing phenomena according to the arithmetical character of the winding number. When the winding number is rational, we have seen that there is frequency locking, whereas if it is irrational, there is quasiperiodicity. To go further, we must consider the relative importance of the two situations. Is frequency locking, corresponding to a rational winding number, an exceptional situation, just as the rationals are exceptional in the set of reals? From the work by M. Herman, we know how to answer this kind of question, which involves nontrivial analysis, particularly because we place ourselves in a very general framework without specifying the model chosen. Two especially remarkable properties are worth mentioning from this point of view:

i) Frequency locking is a *structurally stable* phenomenon which explains — fundamentally — its relative universality.

ii) The variation of the winding number as a function of the control parameter is *generically* a *"devil's staircase"*.

These two assertions merit justification and more extensive explanation.

VI.1 STRUCTURAL STABILITY OF FREQUENCY LOCKING

Suppose that the winding number is rational, equal to zero for example. Then the graph of $x \longmapsto f(x)$ intersects, in the standard case, the graph of the identity map at two points. If we slightly deform the mapping f, by changing the control parameter, the double intersection remains, at least while the perturbation is not too large. This is what we mean by structural stability, not to be confused with dynamical stability in its usual sense (i.e. stability with respect to changes in the initial conditions). As long as the double intersection subsists, the winding number retains the value zero. We will arrive at a similar result whenever the winding number is rational. Therefore $\rho_{[f]}$, when rational, is a *constant* function on open intervals of the control parameter.

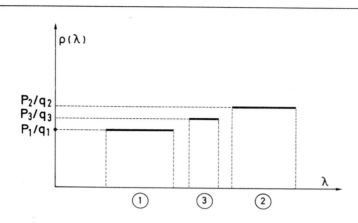

Figure C.4 Construction of a devil's staircase.
$\rho(\lambda)$ is the winding number as a function of a control parameter λ. In the open intervals ① and ②, the winding number is rational and equal to p_1/q_1 and p_2/q_2, respectively. In interval ③ between ① and ②, it is equal to p_3/q_3, a rational between p_1/q_1 and p_2/q_2. We construct in this way a function $\rho(\lambda)$ which is constant on an open interval of λ whenever its value is rational. The final result is a function containing an infinite number of steps, called for this reason a "devil's staircase".

The extremities of these intervals correspond to the marginal situation where the graph of $f(x)$ is tangent to the identity map. Here, the winding number is still zero, but an infinitesimal change in the control parameter can separate $f(x)$ from the identity map, leading to a change in winding number (cf. intermittency, Chapter IX).

VI.2 DEVIL'S STAIRCASE

The winding number $\rho_{[f]}$ is a *continuous* function of f and therefore also of any continuous parameter λ characterizing f. The variation of $\rho_{[f]}$ as a function of λ therefore has a very curious appearance: each time ρ passes through a rational value, the graph of $\rho(\lambda)$ has a horizontal step, since it is constant over an open interval of values of λ. But it is also a continuous function of λ.

Let ① and ② be two intervals of λ on which ρ has the rational values p_1/q_1 and p_2/q_2, respectively (see fig. C.4). There will certainly be another interval ③ of λ, located between ① and ②, on which $\rho = p_3/q_3$ where p_3/q_3 is any rational between p_1/q_1 and p_2/q_2, for example $(p_1 + p_2)/(q_1 + q_2)$. The same construction is repeated for a new rational between p_1/q_1 and p_3/q_3 (or p_2/q_2 and p_3/q_3), and so on for all intervals of λ remaining between the frequency-locking intervals. The λ axis is eventually densely filled with frequency-locking intervals. But this does not cover all the possibilities: since $\rho(\lambda)$ is continuous, it must *also* take on all of the irrational values between any rationals. These irrational values of ρ must be wedged in between locking intervals. The graph of $\rho(\lambda)$ then forms what is called a *devil's staircase*, because it contains an infinite number of steps between any two steps. But naturally, "most" of the steps have an infinitesimal width, since there is a (countably) infinite number of them on a bounded interval. In practice, the steps having a non-negligible width are those corresponding to frequency locking where the rational number p/q is "simple", meaning that it is a ratio between small integers, e.g. 1/1, 1/2, 2/3, etc..

A question then arises: what is the practical importance of the irrational values of ρ? In fact, the only phenomena which have physical importance are those which have non-zero measure. For, in a real physical system, the value of λ is never known exactly and it is impossible to know if we are at *the* precise value of λ corresponding to an irrational winding number, or at a very close value, located within a small frequency-locking interval. A reasonable way of taking experimental imprecision into account is to adopt a probabilistic point of view. Irrational winding numbers will have "physical" significance (as will the quasiperiodic as opposed to the frequency-locking regime) insofar as a λ chosen at random in an interval (e.g. between domains ① and ② in Figure C.4) has a significant chance of yielding an irrational $\rho_{[f]}$. We can summarize as follows the answer given by mathematicians to this question:

i) Near a *normal* (supercritical) Hopf bifurcation, that is, at the moment at which a second frequency of very weak amplitude appears in the Fourier spectrum of the signal, the two oscillators "ignore" each other, meaning that we are practically certain of landing outside of a frequency-locking interval (despite the fact that the frequency-locking intervals are dense!).

ii) As the amplitude of the second oscillator increases, that is, as we leave the Hopf bifurcation point, the probability of landing outside of a frequency-locking interval decreases, becoming zero for values of the control parameter greater than some critical finite value. As we show in Chapter VII, this value is such that the graph of the mapping $x \longmapsto f(x)$ has an inflection point of horizontal slope.

APPENDIX D

The Lorenz model

I Rayleigh-Bénard convection

Consider a layer of fluid of infinite horizontal extent, heated from below. In the Boussinesq approximation, the nondimensionalized equations describing the coupled transport of the momentum of the fluid — through its velocity $\vec{v}(\vec{r}, t)$ — and of the heat — through the temperature deviation $\theta(\vec{r}, t)$ — are following[1]:

$$\Pr^{-1}\left(\frac{\partial \vec{v}}{\partial t} + \vec{v} \cdot \nabla \vec{v}\right) = -\nabla p + \theta \vec{\lambda} + \nabla^2 \vec{v} \tag{D.1 a}$$

$$\nabla \cdot \vec{v} = 0 \tag{D.1 b}$$

$$\frac{\partial \theta}{\partial t} + \vec{v} \cdot \nabla \theta = \operatorname{Ra} \vec{\lambda} \cdot \vec{v} + \nabla^2 \theta \tag{D.1 c}$$

with the Prandtl number: $\Pr = v/D_T$,
and the Rayleigh number:

$$\operatorname{Ra} = (\rho_0 g \alpha d^3 / \eta D_T) \delta T.$$

The presence of the spatial vector differential operator ∇ means that these are partial differential equations, not very practical to manipulate. It is therefore desirable to transform them into a finite set of differential equations, i.e. a flow (in a mathematical, not fluid, sense). To this end, we use the Galerkin method, which might be said to be an extrapolation of a Landau-type expansion. The essential problem here is to determine the unstable modes that the Galerkin truncation must retain, for a Landau-type approach cannot satisfactorily include all of the nonlinearities when fluctuations of finite amplitude enter in. Several rules help to correctly carry out the selection of modes.

a) Divergent singularities should not be introduced. This is not necessarily an easy task in nonlinear dynamics. For example, the simplest nonlinear differential equation:

$$\frac{dx}{dt} = x^2$$

---

1. Here we again employ the notation of Chapter V, to which the reader should refer for definitions.

has the solution:

$$x(t) = \frac{x(0)}{1 - t \cdot x(0)}$$

therefore diverging at $t = 1/x(0)$, a positive finite time if $x(0) > 0$. Of course one should be able to use the physics of a given problem to eliminate all divergences from the initial equations. But recourse to approximations is almost always necessary in order to complete the calculations. And here it is very difficult to insure that the introduction of approximations is not accompanied by that of singularities.

 b) Quantitative properties of the equations should be preserved, in particular, the conservation laws of extensive variables such as the mass, momentum, etc.

 c) One should derive truncated equations that are as close as possible to the initial equations, If we seek to model partial differential equations, as we do here for the R.B. problem, it is therefore desirable to take the maximum number of gridpoints (or of Fourier modes) compatible with effective of the calculation.

Derivation of the Lorenz model from (D.1) emphasizes points (*a*) and (*b*). As for point (*c*), it is sacrificed in favor of maximum simplification of the dynamics. This procedure does not seem unreasonable *a posteriori*, at least judging by the abundant harvest of results that have been obtained[2].

II Derivation of the Lorenz model

The most immediate simplification consists of assuming that the convection rolls appearing above the R.B. instability threshold are all parallel (fig. D.1). Then the velocity vector of any fluid element is always perpendicular to the axis of the rolls. There is translational invariance along the roll axis, so that the variables of (D.1) depend only on two spatial coordinates: the height z, and the horizontal coordinate x perpendicular to the roll axis (see fig. D.1).

Let $u(x, z, t)$ and $w(x, z, t)$ be the x and z components, respectively, of the velocity field. The equation of incompressibility (D.1 *b*) is then[3]:

$$\frac{\partial u}{\partial x} + \frac{\partial w}{\partial z} = u_x + w_z = 0.$$

---

2. Recall that originally in 1963, Lorenz's goal was to set up a model, even a crude one, of the terrestrial atmospher, allowing him to perform calculations on the relatively small computer (LPG 30) available to him at the time.

3. We use the usual notation:

$$y_x = \left(\frac{\partial y}{\partial x}\right)_{z,t}.$$

II DERIVATION OF THE LORENZ MODEL

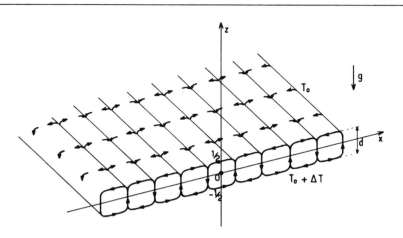

Figure D.1 Schematic representation of the convection rolls in R.B. convection. All of the rolls are rectilinear and their axes are parallel. The structure is periodic in the x direction. The height d along the z axis is taken as the unit of length so that the upper and lower surfaces are defined by:

$$z = 1/2 \quad \text{and} \quad z = -1/2.$$

The temperature gradient is in the direction opposite to the gravitational field g.

By introducing the Lagrange stream function $\psi(x, z, t)$, we see that this equation is satisfied if:

$$u = -\frac{\partial \psi}{\partial z} = -\psi_z; \quad w = +\frac{\partial \psi}{\partial x} = +\psi_x.$$

The velocity field must obey the conditions imposed at the upper and lower surfaces of the fluid defined by $z = +1/2$ and $z = -1/2$ in units of length d. The first requirement is that there be no fluid flux across the surfaces, expressed by:

$$w|_{z=\pm 1/2} = 0.$$

Additionally, if both surfaces are considered as being "free" — which amounts to neglecting the forces of surface tension — the shear component of the pressure tensor vanishes:

$$-\eta \frac{\partial u}{\partial z}\bigg|_{z=\pm 1/2} = 0$$

implying:

$$\frac{\partial u}{\partial z}\bigg|_{z=\pm 1/2} = 0.$$

It is fairly easy to see that these conditions are met when ψ satisfies:

$$\psi|_{z=\pm 1/2} = 0$$
$$\psi_{zz}|_{z=\pm 1/2} = 0$$

These two relations are fulfilled by a term like $\cos(\pi z)$. Since the level curves of ψ are the streamlines of the velocity field, they must adequately reproduce the system of rolls in the x direction. This can be done with a term like $\sin(qx)$, of period $2\pi/q$ in the x direction.

Using the Galerkin method, we arbitrarily limit the description of ψ to a single term:

$$\psi(x, z, t) = \psi_1(t) \cos(\pi z) \sin(qx)$$

which leads to the following expressions for the velocity field components:

$$u = \pi \psi_1(t) \sin(\pi z) \sin(qx) \qquad (D.2\,a)$$
$$w = q \psi_1(t) \cos(\pi z) \cos(qx) \qquad (D.2\,b)$$

Let us now write Equation (D.1 a) for the x and z, components[4]:

$$\Pr^{-1}(u_t + uu_x + wu_z) = -p_x + \Delta u \qquad (D.3\,a)$$
$$\Pr^{-1}(w_t + uw_x + ww_z) = -p_z + \Delta w + \theta \qquad (D.3\,b)$$

Taking the curl of (D.1 a) by calculating $\partial/\partial z$ (D.3 a) $- \partial/\partial x$ (D.3 b), we obtain:

$$\Pr^{-1}\left[-(\Delta\psi)_t + \frac{\partial}{\partial z}(uu_x + wu_z) - \frac{\partial}{\partial x}(uw_x + ww_z)\right] = -\Delta^2\psi - \theta_x \qquad (D.3\,c)$$

using the relation:

$$\Delta\psi = -\left(\frac{\partial}{\partial z}u - \frac{\partial}{\partial x}w\right).$$

Still following the Galerkin method, it remains for us to choose an arbitrary but appropriate form for θ. Given that θ is coupled to the velocity field by Equations (D.1 a) and (D.3 c), it is natural to impose on it the same periodicity in x as has the field itself; this is why we adopt for θ an x dependence like that of w, the z component of the velocity. In addition, we assume that the temperature is perfectly fixed on the two surfaces, which means that the temperature deviation θ vanishes there:

$$\theta|_{z=\pm 1/2}.$$

---

4. ∇ denotes the Laplacian operator, expressed in Cartesian coordinates (x, z) as:

$$\Delta = \nabla^2 = \frac{\partial^2}{\partial x^2} + \frac{\partial^2}{\partial z^2}.$$

II DERIVATION OF THE LORENZ MODEL

We are thus led to take the following form for θ:

$$\theta(x, z, t) = \theta_1(t) \cos(\pi z) \cos(qx) + \theta_2(t) \sin(2\pi z) \tag{D.4}$$

which satisfies the boundary conditions. The second term, independent of x, turns out to be necessary for including at least some part of the nonlinearities of the system (D.1), as we will see later.

By substituting into (D.3 c) the expressions given for u, w, ψ and θ we arrive at:

$$\Pr^{-1}\dot{\psi}_1 = \frac{q\theta_1}{\pi^2 + q^2} - (\pi^2 + q^2)\psi_1 \tag{D.5}$$

a first-order linear differential equation[5] for ψ_1.

From the heat transport Equation (D.1 c), we get:

$$\dot{\theta} + \psi_x \theta_z - \psi_z \theta_x = \text{Ra}\,\psi_x + \Delta\theta \tag{D.6}$$

The quantity $\psi_x \theta_z - \psi_z \theta_x$ comes from the nonlinear term $\vec{v} \cdot \nabla\theta$ in (D.1 c) and is equal to:

$$\psi_x \theta_z - \psi_z \theta_x = \pi q \psi_1 \left[-\frac{\theta_1}{2} \sin(2\pi z) + \theta_2 \cos(qx)(\cos(\pi z) + \cos(3\pi z)) \right].$$

Substituting into (D.6) the expressions for the velocity field (D.2) and for the temperature (D.4) we obtain an equation whose only solutions are non-convective ($\psi_1 = 0$). Therefore, rather than require exact equality as we would normally do, we will impose equality of the $\cos(qx)\cos(\pi z)$ and $\sin(2\pi z)$ terms, neglecting the term in $\cos(qx)\cos(3\pi z)$. We then obtain two equations:

$$\dot{\theta}_1 = -\pi q \psi_1 \theta_2 + q\,\text{Ra}\,\psi_1 - (\pi^2 + q^2)\theta_1 \tag{D.7 a}$$

$$\dot{\theta}_2 = \frac{1}{2}\pi q \psi_1 \theta_1 - 4\pi^2 \theta_2 \tag{D.7 b}$$

concerning the terms in $\cos(qx)\cos(\pi z)$ and $\sin(2\pi z)$, respectively. The set (D.5), (D.7 a), and (D.7 b) form a system of three ordinary nonlinear differential equations for the amplitudes ψ_1, θ_1, and θ_2. It is convenient to use the change of variables:

$$t' = (\pi^2 + q^2)t\,; \qquad X = \frac{\pi q}{\sqrt{2(\pi^2 + q^2)}}\,\psi_1$$

$$Y = \frac{\pi q^2}{\sqrt{2(\pi^2 + q^2)^3}}\,\theta_1\,; \quad Z = \frac{\pi q^2}{(\pi^2 + q^2)^3}\,\theta_2$$

---

5. The truncation applied by choosing the functional forms for ψ and θ has made the nonlinear contribution $\vec{v} \cdot \nabla\vec{v}$ of Equation (D.1 a) disappear.

and to introduce the two parameters:

$$r = \frac{q^2}{(\pi^2 + q^2)^3} \text{Ra}; \quad b = \frac{4\pi^2}{\pi^2 + q^2}$$

We have arrived at what is called the Lorenz model:

$$\dot{X} = \text{Pr}(Y - X) \qquad \text{(D.8 a)}$$
$$\dot{Y} = -XZ + rX - Y \qquad \text{(D.8 b)}$$
$$\dot{Z} = XY - bZ \qquad \text{(D.8 c)}$$

III Coherence of the model

Let us first verify that the truncation has not surreptitiously introduced an undesirable singularity in the model. It suffices to show that no solution of (D.8) goes to infinity; in other words, that the velocity field in $\mathbb{R}^3$ is everywhere directed towards the origin on a surface surrounding, and at a large distance from the origin. Let $f(X, Y, Z) = 0$ be the equation of such a surface. It is necessary and sufficient that the scalar product of the velocity vector and of the normal pointing out of the surface be everywhere negative:

$$Df = \dot{X}f_x + \dot{Y}f_y + \dot{Z}f_z < 0.$$

By substitution from (D.8) we obtain:

$$Df = \text{Pr}(Y - X)f_x + (-XZ + rX - Y)f_y + (XY - bZ)f_z.$$

From the large range of possible surfaces f, we can choose the ellipsoid whose equation is:

$$f(X, Y, Z) = \frac{X^2}{2\,\text{Pr}} + \frac{Y^2}{2} + \frac{Z^2}{2} - (r+1)Z - \mu = 0$$

with $\mu \geq 0$ arbitrarily large. Substituting this formula for f, we have:

$$Df = -X^2 - Y^2 - bZ^2 + (r+1)bZ.$$

From the definition of the surface, the quadratic terms dominate the linear term in Z, for μ sufficiently large. The quantity Df is then always negative: no trajectory originating from a point a finite distance away from the origin can go off to infinity[6].

---

6. The proof above used a method specific to the model. Establishing the regularity for all time of the solutions to a nonlinear flow is not always such an easy task.

III COHERENCE OF THE MODEL

Another point deserves to be emphasized: volumes in phase space contract, meaning that the flow defined by (D.8) is dissipative, like the original equations. Volume contraction follows from the fact that the velocity field has a constant negative divergence:

$$\frac{\partial \dot{X}}{\partial X} + \frac{\partial \dot{Y}}{\partial Y} + \frac{\partial \dot{Z}}{\partial Z} = -(\text{Pr} + 1 + b).$$

If, at a given moment $t = 0$, we consider a set of initial conditions occupying a volume $\Omega(0)$, the endpoints of the ensuring trajectories will at a time t fill a volume $\Omega(t)$ equal to:

$$\Omega(t) = \Omega(0) \exp\left[-(\text{Pr} + 1 + b)t\right].$$

Thus the volume decreases monotonically in time. In passing, it is worth noting that since the Lorenz model is a flow in $\mathbb{R}^3$, its exponential volume contraction forbids its having as an attractor a torus T^2, since otherwise the volume inside the torus would have to be conserved by the flow. Intrinsically, then, the Lorenz model cannot have a quasiperiodic solution [7].

It remains for us to specify the wavenumber q of the rolls along the x axis. Since one of the objectives is to model the convection threshold, it seems natural to choose — as did Lorenz himself — the wavenumber of the fluctuations which destabilise the conductive state. By definition, the conductive state is:

$$\dot{X} = \dot{Y} = \dot{Z} = 0.$$

Its linear stability is determined by the evolution of an infinitesimal fluctuation δX, δY, δZ, which obeys:

$$\delta \dot{X} = \text{Pr}(\delta Y - \delta X)$$
$$\delta \dot{Y} = r\, \delta X - \delta Y$$
$$\delta \dot{Z} = -b\, \delta Z$$

obtained by linearizing the flow (D.8) about the conductive state. The δZ component is always damped since b is positive. As for the δX and δY components, they are of the form:

$$\delta X = \delta X_0 \exp(\sigma t); \quad \delta Y = \delta Y_0 \exp(\sigma t)$$

where σ is a root of the characteristic equation:

$$(\sigma + \text{Pr})(\sigma + 1) - r\,\text{Pr} = 0$$

or:

$$\sigma^2 + (\text{Pr} + 1)\sigma + \text{Pr}(1 - r) = 0$$

---

7. This property is common to all flows in $\mathbb{R}^3$ whose divergence is everywhere negative.

Since Pr is positive, we see that one of the real roots of this quadratic equation becomes positive as soon as r is greater than 1. The conductive state is then linearly unstable for $r \geqslant 1$, that is for:

$$\mathrm{Ra} \geqslant \frac{(\pi^2 + q^2)^3}{q^2}.$$

The corresponding fluctuations at the threshold of instability are those leading to minimum Ra, i.e. those for which the right hand side is least:

$$q^2 = \frac{1}{2}\pi^2.$$

Hence: $\mathrm{Ra} = 27\pi^4/4$.

These are the well-known values first calculated by Lord Rayleigh. The parameter b is then $8/3$, a value often used in numerical simulations of the Lorenz model.

IV Bifurcation diagram (Pr = 10; b = 8/3)

Once established, the flow (D.8) can of course be the subject of study in it own right independent of any reference to its hydrodynamic origin. As it is not integrable in general, its solutions are determined numerically by computer, once the three parameters Pr, b, and r are fixed. Many strudies have been carried out using the values initially adopted by Lorenz himself[8]:

$$\mathrm{Pr} = 10; \quad b = 8/3; \quad r \text{ positive}.$$

The parameter r — directly related to Ra in the R.B. problem, that is, to the applied temperature difference — serves as the bifurcation, or control, parameter. Let us quickly summarize the bifurcation diagram.

The stationary solutions are by definition such that:

$$\dot{X} = \dot{Y} = \dot{Z} = 0$$

$$Y = X; \quad Z = \frac{1}{2}X^2; \quad X = \pm[b(r-1)]^{1/2}.$$

For $0 \leqslant r \leqslant 1$ there exists only one solution of this kind, the pure conducting state:

$$X = Y = Z = 0$$

When r crosses 1, this solution becomes unstable. Two stationary solutions succeed it:

$$X = Y = \pm[b(r-1)]^{1/2}; \quad Z = r - 1.$$

---

8. A number of other combinations, notably Pr = 16, $b = 4$, have also been explored. The results obtained with Lorenz's values largely suffice for our purposes.

IV BIFURCATION DIAGRAM (Pr = 10; b = 8/3)

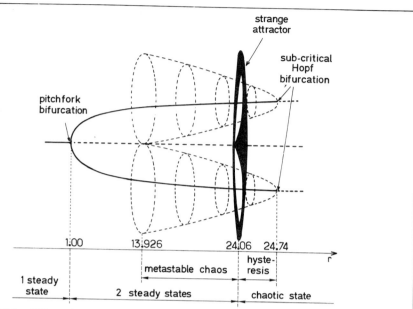

Figure D.2 Bifurcation diagram for the Lorenz model (Pr = 10; $b = 8/3$). The first bifurcation ($r = 1$) is a supercritical pitchfork bifurcation, due to the symmetry $(X, Y, Z) \longrightarrow (-X, -Y, Z)$ of the model. The second bifurcation ($r = 24.74$) is a subcritical Hopf bifurcation. The quantity plotted along the ordinate varies like X or Y. Following the usual convention, the dashed lines represent unstable states and the solid lines stable states.

From C. Tresser, *Modèles simples de transitions vers la turbulence*, Ph. D. Thesis, University of Nice (1981).

We have here a pitchfork bifurcation, where a stable fixed point gives rise to two other unstable fixed points. This results from the invariance of the flow (D.8) under the symmetry:

$$(X, Y, Z) \longrightarrow (-X, -Y, Z).$$

A fairly simple calculation verifies that the two solutions for $r > 1$ are linearly stable. Physically, they correspond to the onset of convection, each being associated with one of the two possible directions of rotation. They lose linear stability at $r = 24.74$, where each gives rise to a subcritical Hopf bifurcation; above, only an aperiodic solution exists. Even though a mathematically rigorous proof has yet to be produced, this is probably a strange attractor. Hysteresis is exhibited at the Hopf bifurcation and three attractors coexist for values of r between 24.06 and 24.74: the two stationary solutions, and the strange attractor, sometimes called "non-standard". We mention also that metastable chaos discussed in Chapter VI is obtained for smaller

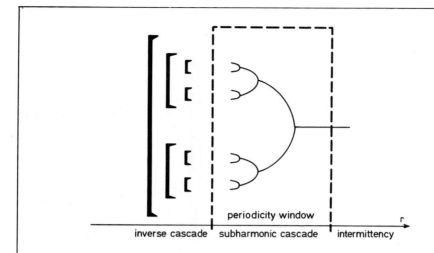

Figure D.3 Typical window of periodicity and its modes of appearance and disappearance.
Numerical study, necessarily limited to the windows of largest width in r, shows that an inverse cascade always precedes their appearance. A direct subharmonic cascade terminates with the basic limit cycle, which is later destroyed by Type I intermittency.

values of r, as low as $r = 13.926$. Figure D.2 recapitulates schematically the bifurcation diagram. Note that here, chaos does not appear through loss of stability of a periodic trajectory. This mode of transition therefore does not fit into one of the three routes described in the second part of the book.

For $r \in [24.74, 30.1]$, the strange attractor constitutes the only stable solution of the flow (D.8). We can easily understand why many numerical studies have been carried out for $r = 28$, located right in the center of this domain. Above $r = 30.1$, and until $r \approx 214$, the diagram of solutions becomes extremely complex, with alternation of chaotic and periodic regimes whose general appearance is reminiscent of those obtained from quadratic mapping of the interval onto itself for $\mu \in [0.892..., 1]$ (see fig. VIII.7). There are good reasons to believe that the "windows of periodicity" here too are infinite in number. This is why we mention only the three largest[9]:

$$99.524... < r < 100.795...$$
$$145.... < r < 166....$$
$$214.364... < r.$$

9. The varying number of decimal digits indicated depends on the results available and is without significance.

IV BIFURCATION DIAGRAM (Pr = 10; b = 8/3)

These windows are the domains of different periodic attractors, as explained in Section VIII.3.5. And it is altogether remarkable that their appearance and disappearance seem to follow the same sequence of events, which we will now describe (see fig. D.3). Slightly before the actual beginning of the window of periodicity, an inverse cascade of bifurcations is observed, of the kind described in Section VIII.3.3. In the Fourier spectrum, there is a progressive decrease of noise and therefore, also of chaos. The inverse cascade is succeeded by a direct subharmonic cascade, which ends with the basic limit cycle, the whole constituting the window of periodicity[10]. In the same way that a window of periodicity is always preceded by an inverse cascade (at least in all cases sufficiently well studied up until now), Type I intermittency each time signals its end[11]. When the basic limit cycle loses stability, it is due to the passage through (+ 1) of one of the eigenvalues of its Floquet matrix. The transition to chaos takes place from a periodic solution and does indeed fit into one of the scenarios of the theory.

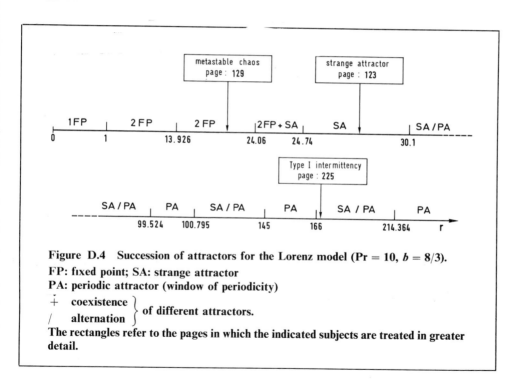

Figure D.4 Succession of attractors for the Lorenz model (Pr = 10, b = 8/3).
FP: fixed point; SA: strange attractor
PA: periodic attractor (window of periodicity)
+ coexistence ⎫
/ alternation ⎬ of different attractors.
The rectangles refer to the pages in which the indicated subjects are treated in greater detail.

---

10. This is the situation explained in detail in Chapter VIII, where we discuss the subharmonic cascade encountered in the quadratic mapping of the interval into itself. To re-establish the same sequence, a direct cascade followed by an inverse cascade, it would suffice to choose as the bifurcation parameter μ a decreasing function of r, such as $1/r$.
11. It is, in fact, numerical simulations for r greater than 166 which led to the discovery and to the study of Type I intermittency.

Analytic study of (D.8) in the limit of large values of r shows that the ultimate attractor is necessarily a limit cycle, a result that is strongly confirmed by numerical calculation. The last window of periodicity begins at $r = 214.364$, terminating an inverse cascade which begins at $r = 197.4$. The direct cascade ends at $r = 313$, above which there remains only the corresponding basic limit cycle as stable attractor. Figure D.4 recapitulates the succession of attractors along the r axis, while specifying the phenomena described in greater detail in other parts of this book.

APPENDIX E

Mathematical complements

I Matrices

Here we will describe several properties of square matrices illustrating the principles of Floquet matrices as they are encountered in various parts of the book.

I.1 GENERALITIES

Let S_0 be the origin of a Poincaré section of a stable limit cycle and let $S_1, S_2, ..., S_n$ be the successive points of intersection of a trajectory initially close to the limit cycle.

Let us see how $S_2, ..., S_n$ are deduced from S_1 according to the type of mapping applied. Let $(S_0 x, S_0 y)$ be a Cartesian frame and let the coordinates of the point S_1 be x_1 and y_1 (see fig. E.1). Consider a linear transformation defining the coordinates of the image S_2 of S_1:

$$S_2 \begin{cases} x_2 = ax_1 + by_1 \\ y_2 = cx_1 + dy_1 \end{cases}$$

and, more generally, those of the nth iterate S_n:

$$S_n \begin{cases} x_n = ax_{n-1} + by_{n-1} \\ y_n = cx_{n-1} + dy_{n-1} \end{cases}$$

The matrix M of this linear transformation is:

$$M = \begin{bmatrix} a & b \\ c & d \end{bmatrix}$$

and we may write:

$$\begin{pmatrix} x_n \\ y_n \end{pmatrix} = M^{n-1} \cdot \begin{pmatrix} x_1 \\ y_1 \end{pmatrix} \quad \text{or} \quad \overrightarrow{S_0 S_n} = M^{n-1} \cdot \overrightarrow{S_0 S_1}.$$

It is usually cumbersome and difficult to calculate M^{n-1}. We can, however, in general find another coordinate frame (x', y') and two numbers λ_1 and λ_2 (which can be

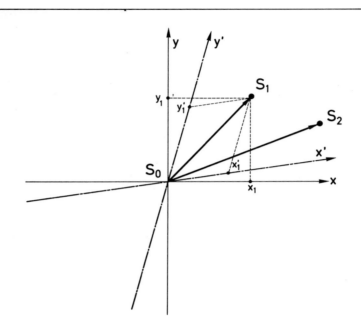

Figure E.1 Two coordinate frames. $S_0 x'$ and $S_0 y'$ are the two eigendirections of the matrix (if it has real eigenvalues).

complex) such that the coordinates of S_2 are related simply to the coordinates x'_1, y'_1, of S_1:

$$S_2 \begin{cases} x'_2 = \lambda_1 x'_1 \\ y'_2 = \lambda_2 y'_1 \end{cases}$$

Similarly:

$$S_n \begin{cases} x'_n = \lambda_1^{n-1} x'_1 \\ y'_n = \lambda_2^{n-1} y'_1 \end{cases}.$$

The matrix M' in this new frame is *diagonal* : $M = \begin{bmatrix} \lambda_1 & 0 \\ 0 & \lambda_2 \end{bmatrix}$ and it is now elementary to calculate $M'^{n-1} = \begin{bmatrix} \lambda_1^{n-1} & 0 \\ 0 & \lambda_2^{n-1} \end{bmatrix}$.

I MATRICES

We call λ_1 and λ_2 the *eigenvalues* of the matrix M; they are the roots of the equation:

$$\begin{vmatrix} a-\lambda & b \\ c & d-\lambda \end{vmatrix} = 0$$

that is: $\lambda^2 - (a+d)\lambda + (ad - bc) = 0$.

We now understand how the eigenvalues can be complex; moreover, for this matrix they are necessarily complex conjugates:

$$\lambda_1, \lambda_2 = \alpha \pm i\beta.$$

The eigenvectors $\vec{V}$ of the matrix M are such that $M \cdot \vec{V} = \lambda \cdot \vec{V}$. One is parallel to the $S_0 x'$ axis, the other to the $S_0 y'$ axis, of the new coordinate frame, so that these two axes are called the eigendirections of M.

The axes $S_0 x'$ and $S_0 y'$ are defined (in the x, y frame) by the relations:

$$\begin{cases} ac + by = \lambda x \\ cd + dy = \lambda y \end{cases} \text{with} \quad \lambda = \lambda_1, \lambda_2.$$

Since:

$$S_n \begin{cases} x'_n = \lambda_1^{n-1} x'_1 \\ y'_n = \lambda_2^{n-1} y'_1 \end{cases}$$

we can see that it is the value of λ of largest modulus $|\lambda|$, called $|\lambda|^+$, which will determine the asymptotic evolution of S_n:

if $|\lambda|^+ < 1$, $\lambda^n \longrightarrow 0$ and S_n converges towards the origin S_0
if $|\lambda|^+ > 1$, $\lambda^n \longrightarrow \infty$ and S_n goes to infinity.

I.2 ROTATION MATRIX

A rotation matrix is a matrix M_r such that:

$$M_r = \begin{bmatrix} a & -b \\ b & a \end{bmatrix} \quad \text{with:} \quad a^2 + b^2 = 1.$$

We can verify that the modulus of λ (the two λ's are complex conjugates) is 1, and that the effect of this matrix on a vector is to rotate it by an angle γ such that $\cos \gamma = a$, $\sin \gamma = b$. The rotation matrix can be written in the form $M_r = \begin{bmatrix} \cos \gamma & -\sin \gamma \\ \sin \gamma & \cos \gamma \end{bmatrix}$. We will consider several concrete examples to illustrate the evolution of successive images of an initial point in the (x, y) plane — assumed to be the plane of a Poincaré section.

I.3 MATRIX WHOSE EIGENVALUE OF LARGEST MODULUS IS REAL AND GREATER THAN ONE

This case occurs when there is loss of stability of a limit cycle by traversal of the unit circle through $+1$. Suppose that $|\lambda_1| < 1 < \lambda_2$.

At each iteration, the abscissa x' will be multiplied by a number λ_1 whose absolute value is less than 1, while the ordinate y' is multiplied by a number λ_2 greater than 1. The iterates of any initial point S_1 will thus converge rapidly (exponentially) to the y' axis along which they diverge, also exponentially, to infinity. Note the close similarity with the description in Section VI.3.2 of hyperbolicity in Poincaré maps. Here too there is a contracting direction x' and a dilating direction y' and two (local) Lyapunov coefficients λ_1 and λ_2. We also note that the contracting direction is not necessarily perpendicular to the dilating direction.

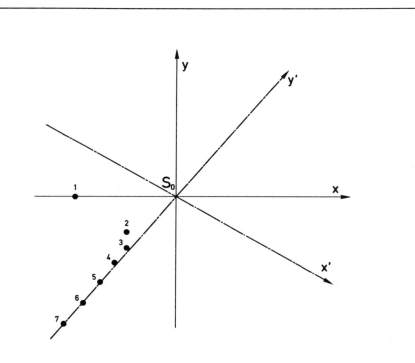

Figure E.2 The first few iterates for a matrix $M = \begin{bmatrix} 0.5 & 0.5 \\ 0.5 & 0.9 \end{bmatrix}$ with eigenvalues $\lambda_1 = 0.16$ and $\lambda_2 = 1.23$. In this and the following figures (up until fig. 4 b), the initial point is numbered 1 and the iterates are numbered in increasing order.

I MATRICES

We have represented in Figure E.2 the position of the first few iterates of an initial point under a linear transformation whose matrix: $M = \begin{bmatrix} 0.5 & 0.5 \\ 0.5 & 0.9 \end{bmatrix}$ has the eigenvalues:

$\lambda_1 = 0.16$
$\lambda_2 = 1.23$.

I.4. MATRIX WHOSE EIGENVALUE OF LARGEST MODULUS IS REAL AND LESS THAN ONE

$\lambda_2 < -1 \qquad -1 < \lambda_1 < 1$.

This case occurs when there is loss of stability by traversal of the unit circle through -1. At each iteration, the abscissa x' is multiplied by a number λ_1 whose absolute value is less than 1, while the ordinate y' is multiplied by a *negative* number λ_2

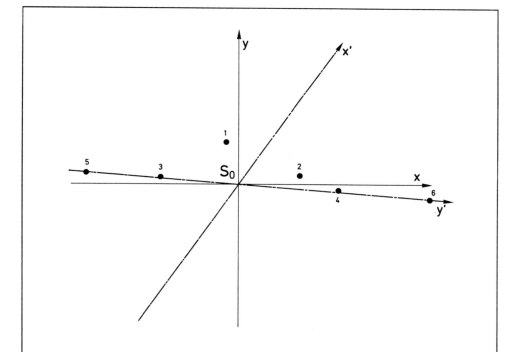

Figure E.3 a Representation of the first iterates for a matrix $M = \begin{bmatrix} -1.3 & 1 \\ 0.1 & 0.3 \end{bmatrix}$ with eigenvalues $\lambda_1 = 0.36$ and $\lambda_2 = -1.36$.

of absolute value greater than 1. As in the previous case there is exponential convergence towards the y' axis, along which exponential divergence takes place, but this time in an alternating fashion due to multiplication by a negative λ_2 which changes the sign of the ordinate y' with each iteration.

We can see in Figure E.3 a the action of the linear transformation whose matrix $M = \begin{bmatrix} -1.3 & 1 \\ 0.1 & 0.3 \end{bmatrix}$ has the eigenvalues:

$$\lambda_1 = 0.36$$
$$\lambda_2 = -1.36.$$

I.5 MATRIX WHOSE EIGENVALUE OF LARGEST ABSOLUTE VALUE IS REAL AND EQUAL TO −1

We might think that this case would be of little practical interest. It can in fact be considered as the asymptotic situation obtained when the linear growth of an instability has been saturated by the effect of nonlinear terms.

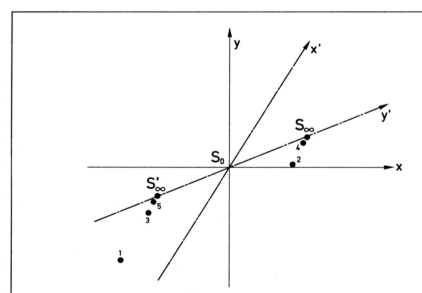

Figure E.3 b Convergence of successive iterates towards the two limits S_∞ and S'_∞ visited on alternate iterations. The linear transformation has matrix $M = \begin{bmatrix} -1.5 & 1 \\ 1 & 1 \end{bmatrix}$ with eigenvalues $\lambda_1 = 0.5$ and $\lambda_2 = -1$.

I MATRICES

We can see in Figure E.3*b* the evolution of the successive iterates under multiplication by the matrix $M = \begin{bmatrix} -1.5 & 1 \\ -1 & 1 \end{bmatrix}$ whose eigenvalues are:

$\lambda_1 = 0.5$
$\lambda_2 = -1.$

After having reached the line $S_0 y'$ (the effect of λ_1) the points alternate between S_∞ and S'_∞. This graph illustrates a Poincaré section after a subharmonic bifurcation while Figure E.3*a* illustrates the phase of exponential growth (linear theory).

1.6 MATRIX WITH COMPLEX CONJUGATE EIGENVALUES OF MODULUS GREATER THAN ONE

This case occurs when a limit cycle is destabilized by traversal of the unit circle through $\alpha \pm i\beta$. We can no longer represent the eigendirections x' and y' and we must

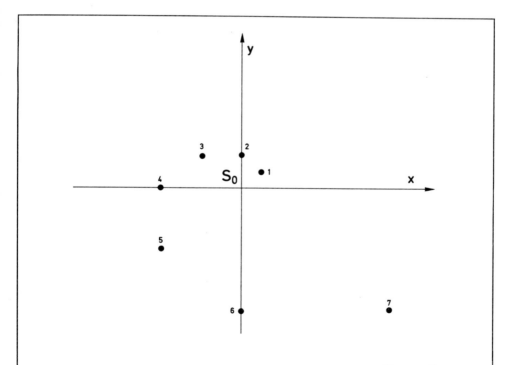

Figure E.4 a Representation of the first iterates for a matrix $M = \begin{bmatrix} 1 & -1 \\ 1 & 1 \end{bmatrix}$ with eigenvalues $\lambda_1, \lambda_2 = 1 \pm i$.

instead calculate the rotation matrix and the angle γ from the given matrix. Consider, for example, the transformation whose matrix $M = \begin{bmatrix} 1 & -1 \\ 1 & 1 \end{bmatrix}$ has the eigenvalues:

$$\lambda_1, \lambda_2 = 1 \pm i$$

of modulus $\sqrt{2}$.

We then write:

$$M = \sqrt{2} \begin{bmatrix} 1/\sqrt{2} & -1/\sqrt{2} \\ 1/\sqrt{2} & 1/\sqrt{2} \end{bmatrix}$$

and we recognize the rotation matrix:

$$M_r = \begin{bmatrix} \cos \gamma & -\sin \gamma \\ \sin \gamma & \cos \gamma \end{bmatrix} \quad \text{with} \quad \gamma = \pi/4.$$

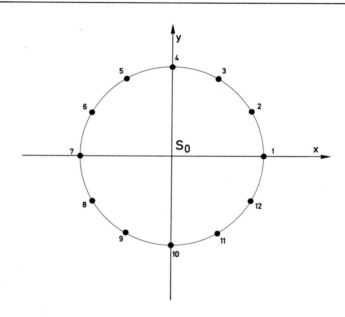

Figure E.4b Representation of the iterates for a matrix with complex eigenvalues of modulus 1.

$$M = \begin{bmatrix} \dfrac{\sqrt{3}}{2} & -\dfrac{1}{2} \\ \dfrac{1}{2} & \dfrac{\sqrt{3}}{2} \end{bmatrix}$$

II THE JACOBIAN

Multiplication by M consists of a dilation by $\sqrt{2}$, followed by a rotation of angle $\pi/4$, illustrated by Figure E.4 a.

The marginal case in which the modulus of the eigenvalues is exactly one corresponds to the asymptotic state of the Poincaré section of a regime after the nonlinear effects have cancelled the exponential growth due to linear instability (e.g. fig. E.4 a). We can see in Figure E.4 b the action of the transformation whose matrix is:

$$M = \begin{bmatrix} \dfrac{\sqrt{3}}{2} & -\dfrac{1}{2} \\ \dfrac{1}{2} & \dfrac{\sqrt{3}}{2} \end{bmatrix}$$

The modulus of the eigenvalues is equal to one and M is a simple rotation matrix ($\gamma = \pi/6$). Figure E.4 b shows that the S_i visit twelve equidistant points on the circumference of the unit circle. They can be seen as the Poincaré section of a flow which, after a Hopf bifurcation, exhibits frequency locking of order twelve.

II The Jacobian

II.1 LINEAR TRANSFORMATION

Consider two coordinate frames (Ox, Oy) and (Oz, Ot), and a linear transformation that maps a point M with coordinates x, y in the first frame into a point P with coordinates z, t in the second frame (see fig. E.5), defined by:

$$\begin{cases} z = \alpha x + \beta y \\ t = \gamma x + \delta y \end{cases}$$

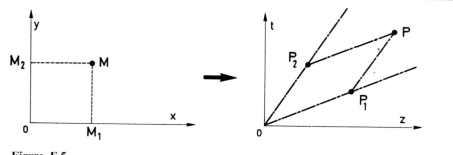

Figure E.5
Schematic representation of a linear transformation.

where $\alpha, \beta, \gamma, \delta$ are real numbers. Let M_1 and M_2 be the two points $M_1 \begin{vmatrix} 1 \\ 0 \end{vmatrix}$ and $M_2 \begin{vmatrix} 0 \\ 1 \end{vmatrix}$ and let their images be $P_1 \begin{vmatrix} \alpha \\ \gamma \end{vmatrix}$ and $P_2 \begin{vmatrix} \beta \\ \delta \end{vmatrix}$, respectively.

The unit square OM_1MM_2 is thus transformed into a parallelogram OP_1PP_2 whose area (magnitude of the cross product) is the absolute value of the determinant:

$$\begin{vmatrix} \alpha & \beta \\ \gamma & \delta \end{vmatrix} = \alpha\delta - \beta\gamma$$

(the sign of the determinant indicates the orientation of the transformation, a + sign conserving the sense of angles). If $|\alpha\delta - \beta\gamma| > 1$, there is dilation of areas, whereas if $|\alpha\delta - \beta\gamma| < 1$, there is contraction.

Note that the singular case $\alpha\delta - \beta\gamma = 0$ corresponds to the transformation of the unit square into a line segment. This transformation is no longer bijective, that is x and y cannot be uniquely determined from z and t; the transformation is no longer invertible[1].

II.2 LOCAL STUDY OF AN ARBITRARY TRANSFORMATION

Let $z = f(x, y)$ and $t = g(x, y)$ be a nonlinear transformation. In the *neighborhood* of a point $M_0 \begin{vmatrix} x_0 \\ y_0 \end{vmatrix}$, that is, for *small* increments $\Delta x, \Delta y$, we can neglect the terms of second order (see fig. E.6). Given that:

$$df = \frac{\partial f}{\partial x} dx + \frac{\partial f}{\partial y} dy = f'_x\, dx + f'_y\, dy$$

we have:

$$\Delta z \approx f'_x\, \Delta x + f'_y\, \Delta y$$

and similarly for Δt and $g(x, y)$. If $\overrightarrow{\delta P}$ represents the image of a vector $\overrightarrow{\delta M}$ in the neighborhood of M_0, then:

$$\overrightarrow{\delta P} \approx \begin{bmatrix} f'_x(x_0, y_0) & f'_y(x_0, y_0) \\ g'_x(x_0, y_0) & g'_y(x_0, y_0) \end{bmatrix} \overrightarrow{\delta M}$$

---

1. We say that a mapping from a set A to a set B is invertible or bijective if each element of B has a *unique* antecedent. For example, the mapping of the circle onto itself mapping a point M of angle θ to a point M' of angle 2θ is not invertible (M' rotates twice as fast as M) whereas the rotation mapping:

$$\theta \longmapsto \theta + \alpha$$

is invertible. Note that if the mapping is a numerical function of a single variable, it must be monotonic in order to be invertible (otherwise some images have at least two antecedents: see Chapter VIII). However it needs not be continuous to be invertible (see Section VII.4 and Appendix C).

II THE JACOBIAN

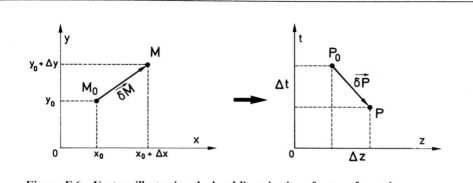

Figure E.6 Vectors illustrating the local linearization of a transformation.

The matrix is called the Jacobian matrix and defines a linear mapping[2], bringing us back to the case of linear transformations examined in the beginning of this section. If the determinant:

$$J = \begin{vmatrix} f'_x(x_0, y_0) & f'_y(x_0, y_0) \\ g'_x(x_0, y_0) & g'_y(x_0, y_0) \end{vmatrix}$$

called the *Jacobian* at (x_0, y_0) is nonzero, the mapping is locally invertible. If $|J| > 1$, there is dilation of areas around M_0; if $|J| < 1$, areas are contracted.

Example : Hénon mapping:

$$\begin{cases} x_{k+1} = y_k + 1 - \alpha x_k^2 \\ y_{k+1} = \beta x_k. \end{cases}$$

In the notation above:

$$\begin{cases} f(x, y) = -\alpha x^2 + y + 1 \\ g(x, y) = \beta y. \end{cases}$$

the Jacobian is:

$$J = \begin{vmatrix} -2\alpha x & 1 \\ \beta & 0 \end{vmatrix} = -\beta.$$

For the Hénon mapping, the Jacobian is constant. When iterating the mapping, the area is multiplied each time by β, and after k iterations is therefore equal to:

$$a = a_0 |\beta^k|.$$

For the case $\beta = 0.3$, areas are contracted.

2. The mapping is called the linear tangent mapping.

The limiting case $J = 0$, i.e. $\beta = 0$, must correspond to loss of invertibility: indeed, the y coordinate disappears and the mapping becomes a nonlinear one-dimensional map which we have already indicated (see Chapter VIII) is non invertible.

III Homeomorphisms, diffeomorphisms

Let $f(x, y, z, ...)$ be a function of n real-valued variables, *a mapping* from a part of n-dimensional space, denoted $\mathbb{R}^n$, to the set of reals $\mathbb{R}$:

$$f : \mathbb{R}^n \longmapsto \mathbb{R}$$

(if there were p functions of n variables, the mapping would be from $\mathbb{R}^n$ into $\mathbb{R}^p$).

III.1 HOMEOMORPHISMS

A mapping like f from a subset A of $\mathbb{R}^n$ to a subset B of $\mathbb{R}^p$ is a homeomorphism of A onto B if it is invertible and bicontinuous (i.e. continuous, along with its inverse f^{-1}: $B \longrightarrow A$). Continuity here is defined relative to the natural topology of the n-dimensional space (defined from the Euclidean distance) and assumes that the subsets A and B both contain a neighborhood of each of their points.

Consider, for example, a circle inside an equilateral triangle whose center of mass G is the same as the center of the circle. Any line beginning at G intersects the circle at M and the triangle at M'. The mapping:

$$f : M \longmapsto M'$$

assigns to each point of the circle a point of the triangle and vice versa. In addition, two *neighboring* points of one shape are mapped to two *neighboring* points of the other. The mapping f is a homeomorphism. We can also say that the circle and the triangle are homeomorphic (cf. Section VII.3).

III.2 DIFFEOMORPHISMS

A diffeomorphism must satisfy a more restrictive condition than a homeomorphism: f is a diffeomorphism from A to B if it is invertible and has continuous partial derivatives of order one on A. (It can then be shown that its inverse also has continuous partial derivatives of order one on B.) Any diffeomorphism is a homeomorphism, but not vice versa. A real valued function f of a real variable, defined on an interval I on which it has a continuous derivative of constant sign, is a diffeomorphism of I onto the image interval $f(I)$.

The mapping circle $\longmapsto$ triangle of the previous section is not a diffeomorphism, due to the discontinuities in its derivative at the vertices of the triangle. However an analogous mapping from a circle to an ellipse is a diffeomorphism. Another example of a diffeomorphism which occurs in Section VII.4 and in Appendix C, is the bijective mapping of the circle to the interval $[0, 1[$.

Some General References

I   Books

R. H. Abraham, C. D. Shaw, *Dynamics: the Geometry of Behavior*, Aerial Press, Santa Cruz (1983).
V. Arnol'd, *Equations différentielles ordinaires*, Mir. Moscou (1974).
V. Arnol'd, *Chapitres supplémentaires sur la théorie des équations différentielles ordinaires*, Mir, Moscou (1980).
V. Arnol'd, *Méthodes mathématiques de la mécanique classique*, Mir, Moscou (1972).
V. Arnol'd, A. Avez, *Ergodic Problems of Classical Mechanics*, Benjamin, New York (1968).
P. Collet, J.-P. Eckmann, *Iterated Maps on the Interval as Dynamical Systems*, Birkhaüser, Boston (1980).
I. Ekeland, *Le calcul, l'imprévu*, Le Seuil, Paris (1983).
J. Guckenheimer, P. Holmes, *Non-linear Oscillations, Dynamical Systems and Bifurcations of Vector Fields*, Springer-Verlag, New York (1983).
H. Haken, *Advanced Synergetics*, Springer-Verlag, Heidelberg (1983).
M. H. Hirsch, S. Smale, *Differential Equations, Dynamical Systems and Linear Algebra*, Academic Press, New York (1974).
D. Joseph, *Stability of Fluid Motion*, Springer-Verlag (1976).
L. Landau, E. Lifchitz, *Mécanique des fluides*, Mir, Moscou (1971).
C. C. Lin, *The Theory of Hydrodynamic Stability*, Cambridge University Press, Cambridge (1955).
B. B. Mandelbrot, *Les objets fractals: forme, hasard et dimension*, Flammarion, Paris (1975).
J. E. Mardsen, M. McCracken, *The Hopf Bifurcation and its Applications*, Springer-Verlag, New York (1976).
N. Minorsky, *Nonlinear Oscillations*, Van Nostrand, New York (1962).
A. M. Nayfeh. D. T. Mook, *Nonlinear Oscillations*, Wiley, New York (1979).
I. Prigogine, *Physique, temps et devenir*, Masson, Paris (1980).
Y. Rocard, *Dynamique générale des vibrations*, Masson, Paris (1949).
S. Smale, *The Mathematics of Time : Essays on Dynamical Systems, Economic Processes and Related Topics*, Springer-Verlag, New York (1980).
C. Sparrow, *The Lorenz Equations*, Springer-Verlag, New York (1982).
R. Thom, *Stabilité structurelle et morphogénèse*, Benjamin, New York (1972).

II   Reviews

N. B. Abraham, J. P. Gollub, H. L. Swinney, "Testing non linear dynamics", *Physica* **11 D,** p. 252 (1984).
J. Bass, "Contribution à l'étude de certaines fonctions susceptibles de présenter la vitesse d'un fluide turbulent", *Journal de Mathématiques*, **38,** (1958).
P. Bergé, Y. Pomeau, "La turbulence", *La Recherche*, **110,** p. 422 (1980).
V. Croquette, "Déterminisme et chaos", *Pour la Science*, **62,** p. 62 (1982).
M. Dubois, "Attracteurs étranges et dimension fractale", *Images de la Physique*, p. 92 (1984).
J. P. Eckmann, "Roads to turbulence in dissipative dynamical systems", *Review of Modern Physics*, **53,** p. 643 (1981).
E. Ott, "Strange attractors and chaotic motions of dynamical systems", *Review of Modern Physics*, **53,** p. 655 (1981).

M. I. Rabinovich, "Stochastic self-oscillations and turbulence", *Soviet Physics Uspekhi*, **21**, p. 443 (1978).
D. Ruelle, "Déterminisme et prédictibilité", *Pour la Science*, **82**, p. 58 (1984).
D. Ruelle, "Les attracteurs étranges", *La Recherche*, **108**, p. 132 (1980).
R. S. Shaw, "Strange attractors, chaotic behavior and information flow", *Zeitschrift für Naturforschung*, **A36**, p. 80 (1981).
H. L. Swinney, J. P. Gollub, "The transition to turbulence", *Physics Today*, **31**, p. 41 (1978).
C. Vidal, J.-C. Roux, "Comment naît la turbulence", *Pour la Science*, **39**, p. 50 (1981).

Index

A

Aeolian noise 6
Adiabatic expansion, or approximation 260
Aperiodic 58
Arnol'd mapping 183
Attractor 22
Aspect ratio 87
Autocorrelation function 46
Autonomous flow 63

B

Baker's transformation 234
Basin of attraction 142
Bénard-von Karman vortex street 5
Belousov-Zhabotinsky reaction 91
Bifurcation 38
Bifurcation diagram 38

C

Cantor set 147
Chaos, chaotic 103
Characteristic exponent 32
Circle, crossing of unit 107
Codimension 272
Colored noise 60
Compass 79
Conservation of area 16
Conservative system 16
Contraction of area 22
Convective structures 87
Correlation function 117
Coupled oscillators 31
Curry-Yorke model 168

D

Decibel 50
Degree of freedom 13
Devil's staircase 298
Diffeomorphism 324
Discretization of signals 44
Dissipative system 22
Divergence of trajectories 105 & 126

E

Ergodic theorem 280

F

Fast Fourier Transform (F.F.T.) 62
First return map 70
Fixed point 71
Floquet matrix 68
Flow 63
Folding 119
Fourier transform 44
Fractal dimension 146
Frequency locking 58
Friction 18

G

Glycolysis 7

H

Hamiltonian system 16
Harmonic 52

Hausdorff-Besicovitch dimension   146
Hénon attractor   135
Hill equation   31
Homeomorphism   324
Hopf bifurcation   38
Horseshoe attractor   155
Hyperbolicity   120
Hysteresis   41

I

Incommensurability   69
Incompressibility (equation of)   89
Information (creation or loss of)   118
Integrability   30
Intermittence, intermittent   223
Invariant of conjugation   294
Inverse bifurcation   40
Inverse cascade   205
Invertible mapping   65
Iteration of a mapping   71

J

Jacobian   321

K

Koch's curve   149

L

Landau theory   103
Lie derivative   114
Limit cycle   24
Lorenz attractor   123
Lorenz model   301
Lyapunov exponent   279

M

Manifold (stable or unstable)   122 & 234
Marangoni effect   85
Marginal stability   22
Mathieu equation   31
Mathieu functions   32
Matrix   313

N

Navier-Stokes equation   89
Non-autonomous (flow)   63
Non integrability   30
Non invertible (mapping)   205
Non periodic   58
Normal bifurcation   40
Normal forms   275
Nyquist noise   60

O

Oscillator (damped)   18
Oscillator (forced)   25
Oscillator (free)   11
Oscillator (parametric)   30

P

Parseval-Plancherel relation   45
Period doubling   195
Periodic (function, signal, ...)   50
Phase portrait   13
Phase space   13
Phase trajectory   13
Pitchfork bifurcation   275
Poincaré section   63
Power spectrum   48
Prandtl number   89

Q

Quasi periodic (function, signal...)   54

R

Random   58
Rayleigh number   85
Rayleigh-Bénard instability   83
Relaminarization   225 & 231
Relaxation oscillation   27
Resonance (exact)   33
Reversible dynamics   18
Reynolds number   4
Rössler model   74
Route toward chaos   159
Ruelle-Takens-Newhouse theory   161

INDEX

S

Saddle-node bifurcation 275
Scaling law 207
Sensitivity to initial conditions (S.I.C.) 104 & 118
Separatrix 15
Shift map 234
Smale attractor 237
Stability of trajectories 18
Strange attractor 123
Stretching 119
Structural stability 102
Subcritical bifurcation 40
Subharmonic cascade 191
Subharmonic instability 33
Supercritical bifurcation 40
Synchronization 289

T

Tangent bifurcation 208
Time delay 76

Torus 55
Transcritical bifurcation 274
Turbulence, turbulent 103

U

Universal sequence 208

V

Van der Pol equation 25

W

White noise 58
Wiener-Khintchine theorem 46
Winding number 293
Window of periodicity 207

Imprimé en France, Jouve, 18, rue Saint-Denis, 75001 Paris
Dépôt légal : Août 1986
Numéro d'édition : 5980
Numéro d'impression : 15468
Hermann, éditeurs des sciences et des arts